World Scientific Series in Current Energy Issues Volume 5

Critical Materials
Underlying Causes and Sustainable Mitigation Strategies

World Scientific Series in Current Energy Issues

Series Editor: Gerard M Crawley *(University of South Carolina & Marcus Enterprise LLC, USA)*

World Scientific Series in Current Energy Issues   Volume 5

Critical Materials

Underlying Causes and Sustainable Mitigation Strategies

Editor

S. Erik Offerman

Delft University of Technology, The Netherlands

 **World Scientific**

NEW JERSEY · LONDON · SINGAPORE · BEIJING · SHANGHAI · HONG KONG · TAIPEI · CHENNAI · TOKYO

Published by

World Scientific Publishing Co. Pte. Ltd.

5 Toh Tuck Link, Singapore 596224

USA office: 27 Warren Street, Suite 401-402, Hackensack, NJ 07601

UK office: 57 Shelton Street, Covent Garden, London WC2H 9HE

Library of Congress Cataloging-in-Publication Data
Names: Offerman, S. Erik, editor.
Title: Critical materials : underlying causes and sustainable mitigation strategies /
 S. Erik Offerman, Delft University of Technology, The Netherlands.
Description: New Jersey : World Scientific, [2018] | Series: World Scientific series in
 current energy issues ; volume 5 | Includes bibliographical references and index.
Identifiers: LCCN 2018028408 | ISBN 9789813271043 (hardcover)
Subjects: LCSH: Raw materials--Research. | Strategic materials--Research. |
 Mineral industries--Environmental aspects. | Sustainable engineering.
Classification: LCC TA404.2 .O43 2018 | DDC 333.8--dc23
LC record available at https://lccn.loc.gov/2018028408

British Library Cataloguing-in-Publication Data
A catalogue record for this book is available from the British Library.

For any available supplementary material, please visit
https://www.worldscientific.com/worldscibooks/10.1142/11007#t=suppl

Desk Editor: Tay Yu Shan

Typeset by Stallion Press
Email: enquiries@stallionpress.com

Printed in Singapore

Foreword to the World Scientific Series on Current Energy Issues

Sometime between four hundred thousand and a million years ago, an early humanoid species developed the mastery of fire and changed the course of our planet. Still, as recently as a few hundred years ago, the energy sources available to the human race remained surprisingly limited. In fact, until the early nineteenth century, the main energy sources for humanity were biomass (from crops and trees), their domesticated animals and their own efforts.

Even after many millennia, the average per capita energy use in 1830 only reached about 20 Gigajoules (GJ) per year. By 2010, however, this number had increased dramatically to 80 GJ per year.[1] One reason for this notable shift in energy use is that the number of possible energy sources increased substantially during this period, starting with coal in about the 1850s and then successively adding oil and natural gas. By the middle of the twentieth century, hydropower and nuclear fission were added to the mix. As we move into the twenty-first century, there has been a steady increase in other forms of energy such as wind and solar, although presently they represent a relatively small fraction of world energy use.

Despite the rise of a variety of energy sources, per capita energy use is not uniform around the world. There are enormous differences from country to country, pointing to a large disparity in wealth and opportunity. (See Table 1) For example, in the United States the per capita energy use per year in 2011 was 312.8 million Btu[a] (MMBtu) and in Germany, 165.4 MMBtu. In China, however, per capita energy use was

---

[a]Note 1 GJ = 0.947 MMBtu.

Table 1: Primary Energy Use Per Capita in Million
Btu (MMBtu)[2].

Country	2007 (MMBtu)	2011 (MMBtu)	Percentage Change
Canada	416.1	393.7	−5.4
United States	336.9	312.8	−7.2
Brazil	52.7	60.2	14.2
France	175.7	165.9	−5.6
Germany	167.8	165.4	−1.4
Russia	204.0	213.4	4.6
Nigeria	6.1	5.0	−18.0
Egypt	36.4	41.6	14.3
China	57.1	77.5	35.7
India	17.0	19.7	15.9
World	**72.2**	**74.9**	**3.7**

only 77.5 MMBtu, despite its impressive economic and technological gains. India, weighs in even lower at 19.7 MMBTU per person.[2] The general trends over the last decade suggest that countries with developed economies generally show modest increases or even small decreases in energy use, but that developing economies, particularly China and India, are experiencing rapidly increasing energy consumption per capita.

These changes, both in the kind of resource used and the growth of energy use in countries with developing economies, will have enormous effects in the near future, both economically and politically, as greater numbers of people compete for limited energy resources at a viable price. A growing demand for energy will have an impact on the distribution of other limited resources such as food and fresh water as well. All this leads to the conclusion that energy will be a pressing issue for the future of humanity.

Another important consideration is that all energy sources have disadvantages as well as advantages, risks as well as opportunities, both in the production of the resource and in its distribution and ultimate use. Coal, the oldest of the "new" energy sources, is still used extensively to produce electricity, despite its potential environmental and safety concerns in mining both underground and open cut mining. Burning coal releases sulphur and nitrogen oxides which in turn can leads to acid rain and a cascade of detrimental consequences. Coal production requires careful regulation and oversight to allow it to be used safely and without damaging the environment. Even a resource like wind energy using large wind turbines has its critics because of the potential for bird kill and noise pollution. Some critics also find large wind turbines an unsightly addition to the landscape, particularly when the wind farms are erected in pristine environments. Energy from nuclear fission, originally believed to be "too cheap to meter"[3] has not had the growth predicted because of the problem with long term storage of the waste from nuclear reactors and because of the public perception regarding the danger of catastrophic accidents such as happened at Chernobyl in 1986 and at Fukushima in 2011.

Even more recently, the measured amount of carbon dioxide, a greenhouse gas, in the global atmosphere has steadily increased and is now greater than 400 parts per million (ppm).[4] This has raised concern in the scientific community and has led the majority of climate scientists to conclude[5] that this increase in CO_2 will produce an increase in global temperatures. We will see a rise in ocean temperature, acidity and sea level, all of which will have a profound impact on human life and ecosystems around the world. Relying primarily on fossil fuels far into the future may therefore prove precarious, since burning coal, oil and natural gas will necessarily increase CO_2 levels. Certainly for the long term future, adopting a variety of alternative energy sources which do not produce CO_2 seems to be our best strategy.

The volumes in the World Scientific Series on Current Energy Issues explore different energy resources and issues related to the use of energy. The volumes are intended to be comprehensive, accurate, current, and include an international perspective. The authors of the various chapters are experts in their respective fields and provide reliable information that can be useful to scientists and engineers, but also to policy makers and the general public interested in learning about the essential concepts related to energy. The volumes will deal with the technical aspects of energy questions but will also include relevant discussion about economic and policy matters. The goal of the series is not polemical but rather is intended to provide information that will allow the reader to reach conclusions based on sound, scientific data.

The role of energy in our future is critical and will become increasingly urgent as world population increases and the global demand for energy turns ever upwards. Questions such as which energy sources to develop, how to store energy and how to manage the environmental impact of energy use will take center stage in our future. The distribution and cost of energy will have powerful political and economic consequences and must also be addressed. How the world deals with these questions will make a crucial difference to the future of the earth and its inhabitants. Careful consideration of our energy use today will have lasting effects for tomorrow. We intend that the World Scientific Series on Current Energy Issues will make a valuable contribution to this discussion.

References

1. Our Finite World: World energy consumption since 1820 in charts. March 2012. Accessed in February 2015 at http://ourfiniteworld.com/2012/03/12/world-energy-consumption-since-1820-in-charts/
2. U.S. Energy Information Administration, Independent Statistics & Analysis. Accessed in March 2015 at http://www.eia.gov/cfapps/ipdbproject/iedindex3.cfm?tid=44&pid=45&aid=2&cid=regions&syid=2005&eyid=2011&unit=MBTUPP.
3. The quote is from a speech by Lewis Strauss, then Chairman of the United States Atomic Energy Commission, in 1954. There is some debate as to whether Strauss actually meant energy from nuclear fission or not.
4. NOAA Earth System Research Laboratory, Trends in Atmospheric Carbon Dioxide. Accessed in March 2015 at http://www.esrl.noaa.gov/gmd/ccgg/trends/
5. IPCC, Intergovernmental Panel on Climate Change, Fifth Assessment report 2014. Accessed in March 2015 at http://www.ipcc.ch/

Acknowledgement

I am very proud of and thankful for the high-quality contributions that I received from the authors of the chapters. I would like to sincerely thank professor Gerard (Gary) Crawley for critically reviewing the entire book in great detail. I am happy to collaborate with the people at World Scientific and thank them for their patience. I am grateful for the loving support of Liesbeth and for the joy that Casper and Daniël are giving me by just being here with us.

S. Erik Offerman, Delft, 2018

Contents

Part IV: Recycling as a Critical Material Mitigation Strategy 265

Chapter 1

General Introduction to Critical Materials

S. Erik Offerman

Department of Materials Science & Engineering,
Delft University of Technology,
Mekelweg 2, 2628 CD Delft, The Netherlands

A growing world population and rising levels of prosperity are driving up the global demand for energy and materials and are increasing the negative impact on the environment.[1] Challenges related to energy use, materials consumption, and climate change are closely intertwined. On the one hand, producing materials consumes about 21% of global energy use and is responsible for about the same percentage of carbon emitted to the atmosphere.[2] On the other hand, the transition from a fossil to a non-fossil electricity mix — to mitigate climate change — would result in a much higher usage of metals. The increase in the usage of metals would range from a few percent to a factor of a thousand for certain metals.[3] Concerns over the future security of the supply of raw materials has led to the identification of critical raw materials for the USA, Japan, and the EU.[4-8] As part of the World Scientific Series on Current Energy Issues, this book is focused on '*Critical Materials*'.

A united and worldwide effort to build and share knowledge about the consumption and production of materials appears to be a more recent development than equivalent efforts for energy-related issues and for climate change when the founding dates of the relevant intergovernmental organizations are considered. The International Resource Panel (IRP) of the United Nations Environmental Program (UNEP) was founded fairly recently in 2007, whereas the International Energy Agency (IEA) was founded in 1974 and the Intergovernmental Panel on Climate Change (IPCC) was founded in 1988.

Since its establishment, the IRP has published a number of reports that provide insight into the grand societal challenge of materials. In 2011, the International Resource Panel stated that "the annual resource extraction would need to triple by 2050, compared to extraction in 2000, in case the levels of resource use per head for all global citizens reached the levels of current resource use of the average European".[1] Further work of the IRP shows that the material footprint per capita is not uniform around the world.[9] For example, North America required about 30 metric tons of material per capita in 2017. In contrast, in Africa the material footprint was just below 3 metric tons per person in 2017. These large differences in the material footprint between North America and Africa point to a large disparity in wealth and opportunity.

Furthermore, the IRP provides information about the changes in the material footprint per capita per region in the world over the last 27 years, which gives insight into the development of material demand per region.[9] The average per capita material footprint of Asia and the Pacific has grown from 4.8 metric tons per capita in 1990 to 11.4 metric tons per capita in 2017, a 3.2% average yearly growth. This can be related to rapid economic growth underpinned by the region's unprecedented industrial and urban transitions (in scale and speed).[9] The average growth of the per capita material footprint of Latin America and the Caribbean and West Asia was half that of Asia and the Pacific, at around 1.4% average growth per year in the period from 1990 to 2017. Africa, on the contrary, has seen no growth in the per capita material supply for final demand over the past three decades, which coincides with a stagnating material standard of living of large parts of the population. The material footprint of Europe has remained approximately constant at values that are just above 20 metric tons per capita per year between 2010 and 2017. North America has even seen a decrease in material footprint per capita over the last 17 years from about 35 metric tons to about 30 metric tons. This points to a saturation level of the material footprint in developed economies.

The general trends over the last decades suggest that the material footprints per capita in countries in Europe and North America — with developed economies — generally remain constant or even decrease, but that the average per capita material footprints of developing economies are rapidly increasing. The growth in material footprint per capita in countries with developing economies will have enormous effects in the near future, both economically and politically, as greater numbers of people compete for limited material resources at a viable price.

The first signs of geopolitical tensions related to resources have already appeared in the recent past. In September 2010, following a diplomatic clash with Japan, China briefly suspended exports of rare-earth minerals (REM).[10] In January 2011, China reduced its export quota by 35% for REM. Following a World Trade Organization ruling, China officially raised production for the rest of the year but began closing dozens of rare-earth producers in August that year, while forcing private companies to close or to merge with Bao Gang, a state-controlled monopoly. This resulted in a sharp increase in the price of the rare-earth metals. The price of the rare-earth metals also became more volatile. The price for certain rare-earth metals (e.g. dysprosium) temporarily increased by 10–50 times. This was the result of the near monopoly (95%) of the supply of REM by China at the time and the limited availability of substitutes for some of the REM, given their unique properties. The Obama administration filed a complaint to the World Trade Organization at that time, which eased the export restrictions for the time being. However, the underlying causes have not diminished.

At present, we live in a largely linear materials economy of 'take-make-use-dispose': raw materials are extracted from the environment, converted into (high-tech) materials, used in products and disposed of at the end of the useful life of the product. This is illustrated by the recycling rates of materials, which may be less than 1% for certain elements in the periodic table (e.g. the rare-earth metals) and which generally decrease for higher-grade materials.[11] The linear economy is not sustainable in the long term, since the world has a finite capacity to provide resources and to absorb waste. A circular economy, in which material loops are closed, promises to be a more sustainable way of using materials.[12,13] Several chapters in this book describe the different aspects of the circular materials economy.

The aim of this book is to give the reader a deeper understanding of the underlying causes of what is nowadays termed '*Critical Materials*' and to give the reader insight into possible sustainable mitigation strategies. The topic of critical materials requires both a '*systems view*', which considers the geopolitical, economic, energy and environmental aspects of materials, and an '*in-depth materials view*', which considers the mechanical, chemical, and physical properties, the processing, and the microscopic structure of materials. Parts I and II of the book are mainly related to the systems perspective of critical materials, whereas Parts III and IV mainly focus on the in-depth materials perspective. However, '*zooming in*' and '*zooming out*' is inherent to the complexity of the topic of critical materials and therefore present throughout this book.

The following sections describe the coherency between the different chapters and the structure of this book.

Part I: Geopolitics and the Energy–Materials Nexus

Raw and high-tech materials are an important commodity for most economies in the world and are therefore of geopolitical importance. The combination of population growth and economic development can be a driver for resource nationalism, which is centered around the availability and control of raw materials, as presented in Chapter 1 by Rademaker. The rare-earth-metals-crisis in 2011, which was the result of export restrictions of rare-earth-metals imposed by China, is an illustrative example of this.

The geopolitical role of raw and high-tech materials cannot be fully understood without considering the changing geopolitics of energy, which is presented by De Jong in Chapter 2. Access to cheap, reliable sources of energy has always been a key requirement for economic development. Throughout history episodes of economic growth have been underpinned by a reliance on particular types of fuel. History shows several disruptive changes in global energy production. A major disruptive change that occurred recently in the global energy landscape has been the rapid increase in renewable sources of energy, which are considered to be our future because they are sustainable.

This has important implications, since 'energy' and 'materials' are two different sides of the same coin. On one hand, materials are needed to convert the different primary forms of energy into electricity and other usable forms of energy. On the other hand, about 21% of global energy production is needed to produce and process materials from ore and waste into products.[2] The intimate relationship between energy and materials becomes stronger with the transition from fossil fuels to renewable energy, since renewable energy technologies are more material intensive due to the more diffuse nature of renewable energy sources compared to the high energy density of fossil fuels.[3] Certain materials which are used in renewable energy technologies are critical in terms of scarcity, geopolitics, supply risk, competition with the food industry, carbon footprint, and/or conflict minerals. This is illustrated in Chapter 3, in which Kelder shows that the global quest for intermittent renewable energy sources (wind, solar) requires a strong increase in the use of rechargeable energy storage devices, such as batteries, and the associated materials.

Part II: Defining Critical Materials

Geopolitical developments around materials and energy stimulated scientific efforts to address the lack of understanding of and the lack of data on nonfuel minerals that are important to the economy. This has led to the identification of critical materials for the USA and Japan in 2008 and for Europe in 2010. The work of Peck, which is presented in Chapter 4, shows a historical perspective to critical materials thinking, which led to the defining of critical materials from 2006 onwards. Critical materials thinking has been present through the Second World War and the Cold War and includes concerns over energy availability and environmental impacts. Chapter 4 shows how the historical military–energy framework for assessing strategic materials has evolved into critical materials approaches to help address the challenges of energy, materials, and the environment in the 21st century.

Criticality can be defined as "the quality, state, or degree of being of the highest importance". But how can we understand what is meant by "highest importance"? In Chapter 5, Graedel and Reck define and describe a multi-parameter approach to the criticality issue that involves (as do the efforts of other researchers and governments) a variety of geological, economic, technological, environmental, and social concerns. Their results suggest that the highest level of concern should be for metals whose processing and use involves extensive separation from parent ores, high levels of embodied energy, little opportunity for substitution, and low levels of recyclability. Improved approaches to material use should thus involve the preferential utilization of non-critical materials, attention to the potential for material reuse at the design stage, and a focus on increasing the efficiency and the total amount of recycling.

Lists of critical materials may change from country to country, from business to business and from time to time. In Chapter 6, Goddin identifies supply chain risks for critical materials from a business perspective. For companies, understanding the environmental impacts of their products and operations is steadily rising in their business agenda. Common business drivers include:

1. Legislation on energy consumption, hazardous substances and conflict minerals.
2. Volatile material and energy prices.

3. Product marketing, brand value and Corporate Social Responsibility (CSR)
4. Stimulus for product innovation.

The approach presented by Goddin aims to integrate product sustainability into the strong culture for business risk management that already exists within most advanced manufacturing organizations.

In Chapter 7, Rietveld and Bastein present the search for an appropriate criticality assessment of raw materials related to the Dutch economy. Past events and predictions suggest the need for a methodology to assess the criticality of raw materials to national economies. Existing criticality methodologies were combined to develop a raw materials criticality methodology for the Dutch economy, including materials embedded in intermediate or finished goods as well.

Part III: Critical Material Mitigation Strategies

The over-arching vision of critical material mitigation strategies may be summarized as a transition from a linear 'take-make-use-dispose' economy to a circular economy.[12,13] Critical material mitigation strategies are often technical in nature. However, the implementation of these strategies into the economy may depend strongly on the development of novel business models (e.g. leasing products instead of selling products), existing and future legislation, and public acceptance. These non-technical aspects of the circular economy are considered to be essential, but they are beyond the scope of this book. Instead, Parts III and IV present critical material mitigation strategies that are of a technical nature.

In general, critical material mitigation strategies that are of a technical nature may include:

1. Circular product design
2. Substitution of critical materials by

   a. non-critical materials,
   b. alternative technologies that do not rely on critical materials
   c. replacing a product that contains critical materials by a service that does not rely on critical materials.

3. Improve the resource efficiency of materials
4. Maximize the properties (functionality) per unit of material to minimize material and/or energy use for a particular function.

5. Sustainable mining
6. Materials design for recycling
7. Minimize the embodied energy of the material
8. Valorization of by-products/waste of materials
9. Improve the recycling and the recyclability of materials

In Part III, three different critical material mitigation strategies are discussed:

1. Circular product design
2. Substitution of critical materials
3. Sustainable mining

Part IV is specifically focused on the different aspects of recycling, which is considered to be an essential critical material mitigation strategy.

In Chapter 8, Bakker, Den Hollander, Peck and Balkenende explore how embedding circular economic principles into product design practice and education could help product designers to take critical material problems into account. They introduce four product design strategies that address materials criticality: (1) avoiding and (2) minimizing the use of critical materials, (3) designing products for prolonged use and reuse, and (4) designing products for recycling.

In Chapter 9, Arechabaleta Guenechea and Offerman present a case study related to the substitution of the critical alloying element Niobium that is used in certain nano-steels. Nano-steels are a novel grade of advanced high-strength steels that are suitable for use in the chassis and suspension of cars. The high strength and ductility per unit mass make the nano-steels resource-efficient and reduce vehicle weight while maintaining crash worthiness. The excellent mechanical properties of certain nano-steels rely on the addition of small amounts (up to 0.1 wt.%) of Niobium as an alloying element to the steel. Niobium is considered to be a critical raw material by the European Union due to its high economic importance as an alloying element in advanced, high-strength steel grades and due to the high supply risk related to the high degree of monopolistic production within the supply chain. This chapter describes the fundamental materials science that is needed for the substitution of the critical alloying element Niobium by Vanadium as an alloying element in nano-steels.

In Chapter 10, Kasry and Maarouf, present another case study that is related to the substitution of the critical element Indium, which is used in transparent conducting layers for solar cells and smart phones. Both applications require the use of transparent conducting electrodes with very low electrical sheet resistance and very high transparency. In conventional thin film solar cells, the transparent conducting electrodes consist of Indium Tin Oxide, which includes the critical element Indium. Hence, the development of alternative TCE materials is desirable to achieve the performance metrics of low cost and compatibility with flexible substrates, while maintaining acceptable engineering performance characteristic of the sheet resistance and optical transparency. In this chapter, Kasry and Maarouf describe their efforts to use carbon nanomaterials, specifically graphene, as transparent conducting electrodes.

A growing world population and rising levels of prosperity will lead to an increasing demand for raw materials in the future in a business as usual scenario.[1] As long as the increased demand for raw materials cannot be mitigated with increased material efficiency and recycling alone, it requires — in turn — a continued supply of raw materials from mining. In Chapter 11, Voncken and Buxton investigate sustainability in mining: meeting the resource and service needs of current and future generations without compromising the health of the ecosystems that provide them. A number of aspects of sustainability in mining are addressed in this chapter: the use of energy, the use of water, land disruption, reducing waste (involving solid waste, liquid waste and gaseous waste), acid rock drainage when dealing with sulfide minerals, and restoring environmental functions at mine sites after mining has been completed. To do everything in an environmentally sound way is costly, but in the end necessary. Regarding this, it is concluded that governmental regulations concerning the emission of waste, the storage of waste, and the re-use of the land after mining are essential to provide a sustainable form of mining and mineral processing.

Part IV: Recycling as a Critical Material Mitigation Strategy

Part IV of this book is dedicated to recycling as a critical material mitigation strategy. This part of the book follows the main steps involved in the process of recycling. The recycling process starts with the collection of waste. In the second step, the mixed solid waste is separated into different streams to enhance the concentration of the different target (to be recycled) materials, which are subjected to further processing. In the third step (in this case the focus is on the recycling of metals), the extraction and refining of metals from scrap and residues takes place. The subsequent processing of the refined metals to high-value alloys can follow the same metallurgical principles that relate to the case study about the substitution of niobium in steel which is described in Chapter 10. Part IV ends with an example in which a waste stream is turned into a resource, i.e. the recovery of rare-earths elements from Bauxite residue (red mud).

The first step that is needed in order to recycle (or re-use) products that contain critical materials is the collection of waste. In Chapter 12, Welink presents how the collection of waste from electrical and electronic equipment (WEEE) from consumers and professional organizations is organized and stimulated. Lessons can be learned from the collection of WEEE, such as how to influence and encourage consumers to collect WEEE separate from other waste, and how to stimulate companies in separating waste. These lessons could also be applied to other products containing critical materials.

The second step in the recycling process is the separation of solid waste into different streams to enhance the concentration of the different target (to be recycled) materials, which is presented in Chapter 13 by Bakker. Efficient mechanical and sensor-based separation of mixed solid waste into valuable secondary materials is a critical step in recycling and in the preservation of primary resources. The chapter gives examples of the physical principles that are behind many contemporary separation technologies.

The third step in the recycling process (in this case the focus is on the recycling of metals) is the extraction and refining of metals from scrap and residues. In Chapter 14, Yang presents the main technologies that are available for extraction and refining of metals: pyrometallurgy, hydrometallurgy, and electrolysis (electrowinning and electro-refining).

The last chapter of Part IV of this book describes an example in which a waste stream is turned into a resource. In Chapter 15, Borra, Blanpain, Pontikes, Binnemans, and Van Gervend show how the recovery of rare earth elements can be realized from bauxite residue, which is a by-product of aluminium production.

References

1. Fischer-Kowalski, M., Swilling, M., von Weizsäcker, E.U., Ren, Y., Moriguchi, Y., Crane, W., Krausmann, F., Eisenmenger, N., Giljum, S., Hennicke, P., Romero Lankao, P., Siriban Manalang, A., and Sewerin, S. (2011). *Decoupling natural resource use and environmental impacts from economic growth*. Nairobi, Kenya: UNEP.
2. IEA. (2007). *Tracking industrial energy efficiency and CO2 emission*. Paris, France: IEA.

3. Kleijn, R., van der Voet, E., Kramer, G.J., van Oers, L., and van der Giesen, C. (2011). Metal requirements of low-carbon power generation. *Energy*, 36(9), 5640–5648.
4. European Commission. (2010). *Critical raw materials for the EU*. Brussels, Belgium: European Commission.
5. European Commission. (2014). *Report on critical raw materials for the EU.* Brussels, Belgium: European Commission.
6. European Commission. (2017). *Study on the review of the list of critical raw materials*. Brussels, Belgium: European Commission.
7. US DOE. (2011). *Critical materials strategy*. Washington, D.C., USA: US DOE.
8. Kawamoto, H. (2008). Japan's policies to be adopted on rare metal resources. *Science and Technology Trends Quarterly Review*, 27, 57–76.
9. Bringezu, S., Ramaswami, A., Schandl, H., O'Brien, M., Pelton, R., Acquatella, J., Ayuk, E., Chiu, A., Flanegin, R., Fry, J., Giljum, S., Hashimoto, S., Hellweg, S., Hosking, K., Hu, Y., Lenzen, M., Lieber, M., Lutter, S., Miatto, A., Singh Nagpure, A., Obersteiner, M., van Oers, L., Pfister, S., Pichler, P., Russell, A., Spini, L., Tanikawa, H., van der Voet, E., Weisz, H., West, J., Wiijkman, A., Zhu, B., and Zivy, R. (2017). Assessing global resource use: A systems approach to resource efficiency and pollution reduction. Nairobi, Kenya: IRP
10. Van den Berg, D.A. and Offerman, S.E. (2011, September 29). 'Rare earth' policy omission threatens European prosperity. *European Voice*. Retrieved from https://www.politico.eu.
11. Reuter, M.A., Hudson, C., van Schaik, A., Heiskanen, K., Meskers, C., and Hagelüken, C. (2013). *Metal recycling: opportunities, limits, infrastructure*. Nairobi, Kenya: UNEP.
12. Dutch government. (2016). *A circular economy in the Netherlands by 2050* (in Dutch). The Hague, the Netherlands: Dutch government.
13. Yuan, Z., Bi, J., and Moriguichi, Y. (2006). The circular economy: A new development strategy in China. *Journal of Industrial Ecology*, 10(1–2), 4–8.

Part I

Geopolitics and the Energy — Materials Nexus

Part I

Chapter 2

The Geopolitics of Materials: How Population Growth, Economic Development and Changing Consumption Patterns Fuel Geopolitics[1]

Michel Rademaker MTL
Deputy Director
The Hague Centre for Strategic Studies
Lange Voorhout 1
2514 AE Den Haag

Raw Materials are an important commodity for most economies in the world. Due to a combination of population growth, economic developments, and quality of life expectations, resource nationalism centred around the availability and control of raw materials has the attention of the highest political levels worldwide, fuelling the geopolitical usage of materials for other purposes.

2.1 Introduction

Population growth, economic development and, as a consequence, changing consumption patterns, are important drivers for the demand for natural resources. Whereas demand because of population growth is growing rapidly, especially in Asia, supply is growing more slowly due to a complex mix of factors, such as technological challenges, financial barriers or hindering legislation. The imbalance since the beginning of the twenty first century between booming demand and limited supply resulted in high prices and increased competition between countries over access to natural resources. At the moment, slowing economic growth in China enforces the opposite trends. At the same time, the international system is in transition. The relative power of emerging economies is growing and the influence or shaping power of the international system of the Western countries is in decline. The world is moving from a Western-dominated order to a multipolar world, faster than expected. In this context, state capitalist autocratic tendencies are becoming more prominent, making the economic environment harsher as well as the geopolitical climate.

Concerns about mitigating climate change, the artic route, depletion of fossil fuels, worries regarding the security of supply of resources, and a non-level playing field regarding economic competitiveness and innovation have pushed governments around the world to formulate more nationalistic strategies, among them natural resource strategies. At the same time other countries, especially China, have developed a focussed, strong and comprehensive strategy taking care of the country's economic growth in combination with political influence, in which logistics, resources, and infrastructure in combination with a higher assertiveness in international relations play a prominent role. Because of this China secures its resources as a basis for industrialisation. Strengthening its internal market has become a priority for Chinese policy makers and companies.

2.2 Lack of Urgency

The European Union (EU) and most of its Member States' responses partially lack urgency. Its strategies are inward looking and its policy measures insignificant although for many critical materials countries are sometimes up to 100% import dependent, especially from China.[1-4] The policy measures countries take vary. Whereas import dependent countries, including some of the emerging economies, aim to secure the necessary resources for economic growth, producing countries aim to reap the benefits from their natural resource endowment. Certain policy measures have negative effects on international trade in resources.[5] Increasing protectionism and other trade barriers like export quotas, a measure taken by China on rare earth elements. Although reversed by the WTO, export quotas pose a real challenge for the European Union (EU) and for European companies which are to a large degree dependent on imports.[6]

This article describes some international trends that are shaping the geopolitics of natural resources and looks at the implications for Europe and the Netherlands. First, it looks at the position of Europe and the Netherlands in international trade flows of natural resources and the vulnerability associated with import dependence for certain resources, including energy, minerals, and food commodities. Second, the article looks at the changes in the international system that are shaping the economic and political world order in which trade in natural resources takes place.

Third, the article identifies challenges and opportunities for the EU and the Netherlands.

2.3 Geographically uneven Distribution

Natural resources are geographically unevenly distributed over the globe. Whereas some countries enjoy rich resource deposits, others have limited or no domestic supplies. Trade has helped alleviate some of these disparities.[7] Trade in natural resources flows from sourcing or mining countries via production centres, often in different countries, to consuming countries. Importing countries need resources as input to production processes and to maintain economic growth and well-being. For exporting countries, especially those with less diversified economies, the revenues from resource exports are an important source of income.

For the EU it is important to have free access to materials, because for many materials the EU is 100% dependent on imports from outside the EU.[8] Figure 2.1 shows the list of critical raw materials for the EU. The criticality of materials is discussed in more detail in part II.

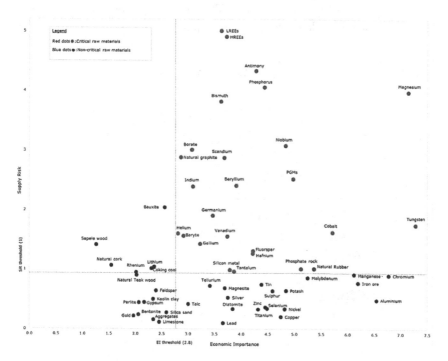

Fig. 2.1.   EU critical materials list, 2017.[9]

Overall, the EU is a net importer of raw materials, including energy. In 2013, imports of raw materials, including energy, made up approximately one third of total EU imports.[10]

There is a broad geographical spread of raw materials between exporting and importing countries although it is possible to identify a clear group of three to five leading exporting countries that account for most of the EU raw materials imports. For many commodities, the top three exporting countries represent at least 50% of the total of the EU imports. For minerals, metals, and fossil fuels, this concentration is even bigger.

The countries featuring most regularly in the top three sources of EU imports of agricultural commodities (both food and non-food), minerals and fossil fuels are Russia, US, Brazil, Norway, Canada, Australia, and Ukraine.

2.4 Agricultural and Non-food Agricultural Commodities

The EU is a major player on the world's agricultural markets, being the second biggest exporter of agricultural products and also the largest importer. The Netherlands is even the second largest exporter of fresh fruit and vegetables worldwide, see Fig. 2.2.[11] The high import dependence for soy and vegetable oils makes the EU a net importer.

Top ten exporting countries agricultural products, 2014

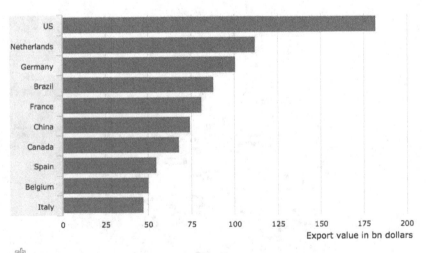

Source: WTO

Fig. 2.2.   The Netherlands is the second biggest exporter of agricultural goods.[12]

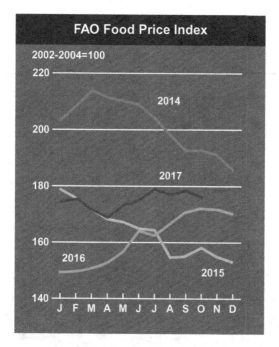

Fig. 2.3. FAO food price index.

Source: http://www.fao.org/worldfoodsituation/foodpricesindex/en/, 2017.

2.5 Price Developments

The growing demand for and struggling supplies of resources resulted in an unprecedented commodities price boom in 2008. According to the World Bank (2009), the commodity price spike between 2005 and 2008 was the highest and longest since 1900. The financial crisis brought down prices only temporarily as growth in the emerging economies picked up sooner than expected. Between 2003 and 2008, nominal prices of some abiotic resources increased by as much as 230%. In some years prices dropped again, but in 2016 a slight increase in prices took place.

2.6 Security of Supply

Although economic growth slowed down because of the economic crisis from 2008 on, shortly after, due to population growth, rising income levels and enhanced quality of life and healthcare are fuelling an unprecedented demand for natural resources. Because of the interconnectivity (resource nexus) between energy, minerals, food, water and land, changes in one part of the system have impacts on the others, see Fig. 2.4.[13] Resource

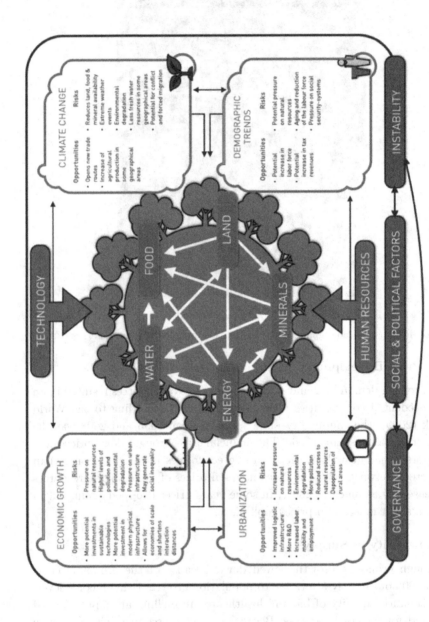

Fig. 2.4.   HCSS Conceptual framework for understanding and analyzing the global resource nexus.[14]

extraction, human-induced pollution, and climate change are weakening ecosystems and contributing to land degradation and water scarcity. Environmental degradation can limit economic productivity and agricultural output, as the food system relies on ecosystem services and natural resources such as soils, water and biodiversity. These developments have raised fears of disruption of supply of resources, low availability, price spikes and economic losses for both non-renewable and renewable resources. At the same time, aggressive technology development policies are put in place to try to mitigate these circumstances. As a consequence of these risks, governments are no longer solely relying on market forces but are interfering strategically in resource markets while responding to changes in the international system.

2.7 The Emerging International System

The scarcity of resources is not unique in history. However, the strain seen in the last decade on the natural resource markets is still taking place at a time of global political and economic transition; i.e. the emergence of a multipolar world. This new global system has been particularly challenging for importing countries, as state capitalist tendencies have become more prominent in the natural resource sector spurring resource nationalism.

2.8 Multipolarity

The international economic and political order was until recently dominated by the triad of powers, the US, Europe, and Japan, of which the US was considerably the strongest.[15] Their power is in decline as economic and political power is shifting towards multiple emerging power centres. The most prominent new centres of power are the emerging economies, most notably Brazil, Russia, India, and China. China and India in particular have shown impressive economic growth rates in the last decade, leading some to refer to the 21st century as the Asian or Pacific century.

The economic and financial crisis has accelerated this power shift from West to East and the transition to a multipolar world. The crisis hit the developed countries hardest, resulting in the depreciation of the dollar and the euro, a decline in GDP growth, mounting government debt, and depleted national reserves. The federal debt crisis in the US has undermined its geopolitical muscle and has made the country dependent on foreign, mainly Chinese, creditors. Europe's position has been undermined by the euro crisis and the failure of its political leaders to resolve it. The crisis has highlighted the economic and political divisions that exist among EU member states, and has threatened the continuation of the monetary union itself. Recent migration flows to the EU as well as conflicts at its borders do not help either.[16]

Alongside these financial and economic changes, a paradigm shift has occurred at the international political level. Economic growth has in the last seven or eight years encouraged the emerging economies to also bolster their political influence.[17] The BRICS countries and other emerging economies such as Turkey are increasingly challenging the Western-dominated international order, including models of economic growth that were developed in the previous century.

2.9 State Capitalist and Autocratic Tendencies

China's economic success has brought state capitalism to the fore as a competing model for economic development. The Chinese model is based on state-led economic growth without political liberalization. As an authoritarian regime, China can do business with resource-rich developing countries much more quickly than liberal democracies, which often lack a long-term strategic vision and decisiveness, or international organizations.[18] This point was illustrated by Senegalese President Abdoulaye Wade: "If I wanted to do five kilometres of road with the World Bank, or one of the international financial institutions, it takes five years. One year of discussions. One year of back and forth. One year of I don't know what. With the Chinese it is a few days and I say yes or no, they send a team and we sign".[19]

China's quick way of doing business and its beneficial terms of trade, aid, and investment make it an attractive economic partner for many resource-rich developing countries. In addition, its pragmatic foreign policy, which is characterized by adherence to the principle of non-interference and its 'no (political) strings attached' development assistance policy, yields China a lot of political support from developing countries and less democratic regimes. As a result, China's state-owned enterprises have gained access to many natural resources reserves in the developing world, particularly in Africa.

The economic crisis also has broken the taboo against economic government interference in liberal market democracies (Qasem *et al.*, 2011c). Whereas government intervention in the resource sector has been widespread in non-democratic states for decades, leaders in the developed world are now also increasing state control over their economies. To deal with the economic crisis, measures of deregulation, privatization and trade, and financial liberalization have been reduced or reversed and nationalisations temporarily enforced.

2.10 Implications

What are the implications of the transition to a multipolar world and the rise of state capitalist tendencies for the geopolitics of natural resources? First, in a multipolar world, countries tend to turn inwards, prioritizing their own national interest over cooperative policies on public good issues. This means that multilateral approaches and international institutions, such as the United Nations or the World Trade Organization (WTO), are weakened. In a multipolar world interests are diffuse rather than shared, which makes it more difficult to advance policy agendas that aim to benefit the international system as whole.[20] Multipolarity also increases uncertainty and instability in international relations. In such a context the likelihood of international friction grows. Competition for resources makes the world more conflict-prone and increases the likelihood of war between major powers, as securing access to resources becomes a national security interest that may even justify the use of military means.[21]

The relationship between violence and the supply of vital resources has always been a persistent reality of international relations (Dalby, 2003). Second, the rise of state capitalist tendencies in a multipolar world leads to more government interference in the natural resource sector and to more resource nationalism. Resource nationalism refers to a situation in which control over natural resources shifts from foreign to domestic state-owned companies. It also means that governments align their natural resource policies more explicitly with the national interest.[22] In resource-producing and exporting countries, governments are putting increasing emphasis on maximizing revenues from the resource sector, which may result in access restrictions, trade barriers, export quotas and other manifestations of protectionism.

In countries that rely on imports of natural resources, governments are primarily focused on the security of supplies, which may result in the proactive acquisition of resources abroad, trade restrictions, and the creation of stockpiles.[23]

Both the rise of state capitalist tendencies and the transition to a multipolar world are putting pressure on the rules and values that are in place to safeguard free trade in natural resources. The growing competition for natural resources and high prices are encouraging governments to assess their resource policies through a filter. The policy measures, however, are based on national interests and are at times worsening the situation by contributing to market distortions that further drive up prices and increase price volatility.[24]

2.11 Policy Trends

Increased government interference is a response to growing competition for natural resources, but may in turn have an effect on natural resource markets and global trade.

Trade barriers and other distortions of the free market may heighten resource scarcity, encourage price gouging, and possibly lead to increased international instability.[25]

2.12 Securing Resources

Countries with a limited or no natural resource endowment are most vulnerable to the effects of high prices and supply disruptions. High prices and supply disruptions are a threat to economic security, as they may reduce a country's economic and innovative competitiveness and negatively affect national employment and prosperity. In some countries, economic security is also important for social and political security. In China, for example, securing resources for economic growth is key to avoiding instability as a result of domestic opposition to the authoritarian regime. Some metals are also of strategic importance for military security, as they are used in high-tech defence technologies. As a consequence, import-dependent countries show a tendency to securitize their policy, meaning that they tend to prioritize these policies as important to national security. The securitization trend is reflected in three categories of policy measures. First, import-dependent countries employ policy instruments aimed at securing stable and affordable supplies from abroad. These include, for example, concluding strategic bilateral agreements with producing countries, establishing joint ventures in resource-rich countries, pursuing upstream integration in value chains or the strategic use of development assistance. In 2010,

Japan gained access to Bolivia's lithium reserves in exchange for financial support and the construction of solar panels, energy plants and hospitals (Chambers, 2010). Another example is that China, and also other Asian and Middle East states, are actively purchasing and leasing land in Africa for food production (Smith, 2009).

Second, import-dependent countries adopt policy measures to reduce their import dependency by reducing domestic consumption, promoting reuse and recycling of materials, exploiting alternative domestic sources and developing substitutes. Import dependence can also be reduced by stockpiling strategic materials, like in South Korea and the US. Third, in the most extreme case, import-dependent countries may use their military capabilities to secure resources, as is happening in the Arctic region[26] and in the South China Sea.[27]

2.13 Maximizing the Benefits of High Prices

As a consequence of the relatively slower growth of production compared to consumption, resource markets transformed from buyers' markets into sellers' markets, in which the producing countries can determine market prices. It also implies that they can use their resource endowment as an asset to achieve their economic and political objectives, both domestically and internationally. The de facto power position of producing countries depends on their level of economic development, the size of the export flows and the extent to which a particular resource can be substituted.

The common characteristic of producing countries is their aim to maximize the economic and political benefits of their resources. Resources are generally considered state property and therefore some countries apply strict resource ownership rights. As many countries struggle with the consequences of the economic downturn, the commodity price boom has meant that the resource sector has become an important source of government revenue. This has led to the politicization of resource policy and increased resource nationalism. Resource nationalism has been a well-known policy in the oil and gas sector and has also in the last few years spilled over to the mineral sector.

Policy measures that governments enact in order to maximize profits from high prices include: increased taxation on extraction revenues and export quotas (such as the Chinese export restrictions on rare earth elements).[28] Governments are also increasingly limiting access to the domestic resource sector to foreign companies through licence fees, tariffs for mining, and agricultural permits, making it more difficult for them to invest and gain access to resources in producing countries. In Canada, for example, the government ruled against the takeover of the Canadian potash-producing company Potash of Saskatchewan by mining multinational BHP Billiton on the grounds that the takeover was not in Canada's interest (BBC, 2010). Resource nationalism can also result in the creation of state-owned enterprises or the nationalization of an entire industry.

To sum up, there is a trend towards more government interference in both import-dependent and producing countries. It should be noted that the degree of government control over the resources sector varies. In state capitalist systems government control is stronger than in countries that have a market capitalist tradition. Nonetheless, the general trend towards more government interference has important implications for trade and the availability of resources on markets.

2.14 Impact on Trade and Investment

Increased government interference and protectionism have negative effects on the free trade of resources, which may result in reductions in global welfare. It is widely accepted that liberalized trade helps to lower prices and broadens the range and quantity of goods available on the market. Free trade facilitates competition and investment, and increases productivity.

The free trade in commodities has most prominently been distorted by export restrictions. The governance regime on export restrictions is weak. Export restrictions are often poorly motivated and lack transparency, and are generally prohibited under the WTO regulations. Several complaints have already been filed at the WTO against Chinese practices that are allegedly protectionist and do not comply with WTO rules. In several rulings, the WTO stated that China's export duties, quotas, and minimum export price of coke, fluorspar, manganese, zinc and other commodities are in conflict with WTO trade practices.[29]

According to Article XI on the General Elimination of Quantitative Restrictions of the General Agreement on Tariffs and Trade (GATT), however, exceptions are allowed if export restrictions are applied to 'prevent or relieve critical shortages of foodstuffs or other products essential to the exporting contracting party.' The use of export restrictions during the food crisis of 2007–2009, however, exacerbated the crisis (FAO *et al.*, 2011).

After over 40 countries imposed various forms of export restrictions, food prices rose sharply (Korinek and Kim, 2010). Export restrictions contribute to higher prices for foreign consumer countries when producing countries divert material from export to the domestic market (OECD, 2010). Prices may rise further because export restrictions create uncertainty about future prices and consequently discourage investments in extracting and producing raw materials. Investments in the mining industry, for example, are long term and require large amounts of capital and expertise. The possibility of sharp fluctuations in world prices due to the imposition or sudden removal of export restrictions, or an outbreak of political instability, represents a significant risk for investors (Korinek and Kim, 2010). A lack of investment will reduce the overall supply of resources in the long term. The negative effects of export restrictions on the availability and price of resources are especially problematic since export restrictions by one country may lead to a spiral of restrictions by others that will further increase competition and fears of resource scarcity.

The disruption and fragmentation of the trade in resources will have a negative impact on global prosperity, which in turn will increase the likelihood of conflict. This is amplified by the tendency among import-dependent states to align their natural resource policy with other policy areas, such as development assistance and foreign policy, which means that countries risk having conflicting interests in a growing number of policy domains.

2.15   The EU: Challenges and Policy Responses

The international dynamics of natural resources also affects the EU, although the challenges vary from country to country, depending on their natural resource endowment and their industrial and economic profile. The situation is also very different depending on whether one looks at biotic or abiotic resources.

2.15.1   *Biotic resources*

Overall, the EU is nearly self-sufficient when it comes to biotic resources. During the 1960s and 1970s the EU introduced agricultural policies that were aimed at ensuring food security in Europe by maintaining an agricultural sector large enough for self-sufficiency. The EU is a major player on the world's agriculture markets, being a producer of large quantities of a wide range of products. Globally, the EU is the second largest exporter of agricultural products but also the largest importer. The high dependence on soy and vegetable oils, which are essential to the European food industry, make the EU a net importer. The European Commission has identified climate change as a potential future threat to European food security in the long run, in addition to increasing water scarcities in Southern Europe. According to the Commission, the rising food prices in recent years have had a limited effect on the daily lives of European consumers. The share of household expenditure on food has been gradually declining and is currently estimated at around 14% of total expenditures. For Europe, the risks related to food insecurity are therefore limited.

2.15.2 Mineral supply disruption risks

The major issue of concern for the EU is the supply of abiotic resources. The EU is self-sufficient when it comes to several base metals and construction and industrial minerals. Minerals such as barytes, kaolin, potash, salt, silica, and talc, which are used in a wide range of industries, are extracted within the EU. From 2000 to 2007 the import share in direct material input (i.e. domestic extraction and imports) of metal ores has increased regularly, reaching two thirds. It then dropped suddenly and significantly in the year 2009, and it has remained at a lower average level of 53.6 % in the period 2009–2014. Import dependency for fossil energy materials increased steadily from 2000 (47.4%) to 2014 (58.6 %).[30] For some strategic minerals, the import dependence is as high as 100%.[31] The EU has recently assessed the strategic importance of 41 abiotic resources for the European economy. Of this group, 14 materials or material groups were identified as critical due to their importance to European industries and the high risk of supply disruption, see Fig. 2.1.

The group of critical materials includes, for example, rare earth elements. Europe is especially concerned with its dependence on imports of Chinese rare earth elements.

Due to China's export restrictions on rare earth elements, world prices are now typically 20–40% higher than Chinese domestic prices, a price difference that is negatively affecting Europe's competitiveness.

Another issue of concern is the EU's dependence on phosphate imports. Phosphate is a crucial element in agricultural fertilizers, there is no substitute, and so far little is retrieved from waste flows. Phosphate reserves are located in a limited number of countries.

Morocco, China, the US, and Russia are among the most important producers. Political instability in the producing countries is a major threat to supplies of phosphate to the EU. For example, the EU used to import a significant share of its phosphate from Syria, but the supply has been disrupted due to the civil war and international sanctions against Syria's repressive regime. In Morocco, which currently produces about 70% of global phosphate supply, the resistance movement Polisario is claiming sovereignty over Western Sahara, a conflict that could potentially cause supply disruptions to the EU. In addition, China, the US and Russia have implemented trade restrictions on phosphate, such as high export tariffs, to protect their domestic reserves in light of the increased pressure on the phosphate market caused by the growing demand for food. This has resulted in price increases that negatively affect the EU's agricultural sector.

The EU's nuclear power industry is also highly dependent (90%) on imported uranium. In 2016, 25.6% came from Russia followed by Kazakhstan. Russia and Kazakhstan together provide 44.1% of the EU uranium.

Europe imports many of its minerals from politically or economically unstable regimes such as China, Russia, and African states. The demand for minerals is projected to rise in the future, which means that the EU's dependence on these external suppliers will increase.

Another obstacle to the security of supply is that some minerals are not traded on major exchanges, such as the London Metal Exchange, but through non-transparent long-term bilateral contracts based on prices negotiated between parties. Bilateral mineral trade fragments the market and may result in increased inflation, higher price volatility, lower investment levels, and fuel fears about supply constraints.[32]

2.15.3 *Security risks related to political instability*

The international resource dynamics do not only pose direct economic risks for European countries and industries, such as increased supply disruptions and higher prices. They also bring about indirect security risks for Europe in the form of political instability elsewhere.

High prices or physical scarcity of resources may trigger social segmentation, migration, conflict and insurgencies, especially in countries where there is popular resentment about the political or institutional status quo. High food prices, for example, have major consequences for living standards and quality of life, especially in the developing world, where most households spend a high proportion of their income on food (Motaal, 2011). Poor urban populations are disproportionately affected since they cannot grow their own food, which means that high prices have a direct impact on their consumption patterns. Rising food inflation is therefore not only an economic concern, forcing governments and central banks to tighten their monetary policies, but also a political one, as high food prices can trigger social unrest (Blas, 2011). The ongoing events in the Middle East, popularly known as the Arab Spring, are an example of a trigger event which shows how rising food prices can contribute to social unrest and regime change. The spread of revolutions in North Africa and the Middle East following the outbreak of protests in Tunisia in December 2010 has shown how food scarcity can have consequences for broader regional stability (de Ridder, 2011b).

In the context of high resource prices, corruption may also lead to instability. In producing countries, elites are often heavily involved in the resource sector and may be tempted to use resource revenues for personal purposes, which may increase inequality in society and sharpen social unrest. Especially when the elites are affiliated with specific population groups, this can lead to social and political instability in countries with strong ethnic divisions or with a history of ethnic conflict.

These shocks are not, however, confined to states where poor governance is the norm. Even states with strong institutions and rich traditions of good governance can slip into domestic disarray if policy measures are implemented by the national government that fail to consider the interests of regional governments. For example, tax increases on resource extraction may lead to intra-governmental friction, as the governments from resource-rich regions are less inclined to favour the redistribution of resource wealth to poorer regions.

In summary, high prices, price volatility, and physical scarcity of natural resources may lead to economic stagnation, increase poverty and hunger, intensify domestic tensions, and trigger migration. Local instability has the potential to spill over to neighbouring countries and amplify the security threat to the European Union.

2.16 Conclusions

Current consumption and production patterns and trade in international resources have major impacts on our prosperity and wellbeing, and also affect international relations, peace, and security. In the coming decades, the world will be characterized by tight markets and competition for natural resources, due to the underlying drivers of demand and supply of natural resources, such as demographic shifts, changing consumption patterns, economic growth, and climate change. Scarcity of natural resources is largely a dynamic, mostly economic concept.

The availability of natural resources is a function of current market conditions and technological means. Nonetheless, the risks of physical scarcity are not negligible due to geopolitical developments and the risks of supply disruptions, export restrictions, and price developments, all of which pose real challenges to economic growth and competitiveness.

One such development is the increasing prevalence of resource nationalism, both in resource-rich countries and producer states where the state has traditionally played an important role, as well as in parts of the world that take a more liberal view of government interference with the market. The emergence of state capitalist tendencies and resource nationalism means that the EU needs to respond strategically and may have to adjust its view of the role of government with regard to natural resources.

References

1. HCSS and TNO. (2013). *Resources for our Future*. Amsterdam University Press, Amsterdam, the Netherlands: HCSS and TNO. This is a shortened version and a partial update of a chapter of the book *Resources for our Future*.
2. European Commission. (2014). On the review of the list of critical raw materials for the EU and the implementation of the Raw Materials Initiative. Retrieved from http://eur-lex.europa.eu/legal-content/EN/TXT/?uri= COM:2014:0297:FIN
3. Dutch government. (2015). *Materials in the Dutch economy* (in Dutch). The Hague, the Netherlands: Dutch government. Retrieved from https://www.rijksoverheid.nl/documenten/rapporten/2015/12/11/materialen-in-de-nederlandse-economie
4. European Commission. (n.d.). Critical raw materials. Retrieved from http://ec.europa.eu/growth/sectors/raw-materials/specific-interest/critical_nl
5. OECD. (2014). *Export restrictions in raw materials trade: Facts, fallacies and better practices*. Paris, France: OECD. Retrieved from http://www.oecd.org/trade/benefitlib/export-restrictions-raw-materials-2014.pdf
6. European Commission. (2014). On the review of the list of critical raw materials for the EU and the implementation of the Raw Materials Initiative. Retrieved from http://eur-lex.europa.eu/legal-content/EN/TXT/?uri=CEL EX:52014DC0297
7. OECD. (2016). *OECD Quarterly International Trade Statistics, Volume 2015 Issue 4*. OECD Publishing, Paris, France: OECD.
8. European Commission. (n.d.). Critical raw materials. Retrieved from http://ec.europa.eu/growth/sectors/raw-materials/specific-interest/critical/index_en.htm
9. Ibid.
10. European Commission. (n.d.) Raw materials. Retrieved from http://ec.europa.eu/trade/policy/accessing-markets/goods-and-services/raw-materials/

11. Value-wise, the Netherlands is the second largest exporter of fresh fruit and vegetables, after Spain. Quantity-wise, the Netherlands takes third place, following Spain and Mexico.
    The Netherlands is the biggest exporter of fresh vegetables though, both looking at quantity and value. With fresh fruit, the Netherlands is the third (re)exporter value-wise, but quantity-wise the Netherlands takes 11th place, even though the Netherlands, on an international scale, is a very modest producer not only of fruit but also of vegetables.
    Netherlands world's biggest (re-)exporter for 11 fruit & veg products. (2015, November 2). *Fresh Plaza.* Retrieved from http://www.freshplaza. com/article/135083/Netherlands-worlds-biggest-(re-)exporter-for-11-fruit-a nd-veg-products

12. Based on data sources provided by the World Trade Organization. (2014).

13. HCSS and TNO. (2014). *The global resource nexus.* The Hague, the Netherlands: HCSS and TNO. Retrieved from http://www.hcss.nl/reports/the-glo bal-resource-nexus/157/

14. Ibid.

15. de Wijk, R. (2016). *Power Politics: How Russia and China Reshape the World,* Amsterdam: Amsterdam University Press

16. ECFR. (2016). *Return to instability: How migration and great power politics threaten the Western Balkans.* London, UK: ECFR. Retrieved from http:// www.ecfr.eu/publications/summary/return_to_instability_6045

17. de Wijk, R. (2016), *Power Politics: How Russia and China Reshape the World,* Amsterdam: Amsterdam University Press

18. Kaplan, R. D. (2010), *Monsoon: The Indian Ocean and the Future of American Power,* New York: Random House.

19. Comment by President Abdoulaye Wade at the EU-Africa summit in Lisbon. (2007, December 8–9). Retrieved from https://www.ft.com/content/cf2cf16 4-a683-11dc-b1f5-0000779fd2ac.

20. de Wijk, R. (2016), *Power Politics: How Russia and China Reshape the World,* Amsterdam: Amsterdam University Press

21. Ibid.

22. HCSS and TNO. (2013). *Resources for our Future.* Amsterdam University Press, Amsterdam, the Netherlands: HCSS and TNO.

23. HCSS. (2011). *Op weg naar een Grondstoffenstrategie , quick scan ten behoeve van de Grondstoffennotitie.* The Hague, the Netherlands: HCSS.

24. Ibid.

25. This section is adapted from the conceptual framework for understanding natural resource policy in a multipolar world, originally developed by Weterings et al. (2013), in *Resources for our future: Key issues and best practices in resource efficiency,* Amsterdam University Press, Amsterdam, the Netherlands.

26. Kuo, M. A. and Tang, A. O. (2015, December 16). China's Arctic Strategy: The Geopolitics of Energy Security. *The Diplomat.* Retrieved from http:// thediplomat.com/2015/12/chinas-arctic-strategy-the-geopolitics-of-energy-s ecurity/

27. Playing Chicken in the South China Sea. (2016, May 20). *The New York Times*. Retrieved from http://www.nytimes.com/2016/05/21/opinion/playing-chicken-in-the-south-china-sea.html?_r=0

28. Retrieved from https://hcss.nl/report/rare_earth_elements_and_strategic_mineral_policy_1.

29. Dispute settlement DS431. (2015, May 20). China — Measures Related to the Exportation of Rare Earths, Tungsten and Molybdenum, WTO.

30. Eurostat. (2016). *Physical imports and exports*. Luxembourg City, Luxembourg: Eurostat.

31. European Commission. (2014). *Critical Materials Profiles*. Brussels, Belgium: European Commission. Retrieved from http://ec.europa.eu/growth/sectors/raw-materials/specific-interest/critical/index_en.htm

32. De Ridder, M. (2013). *The Geopolitics of Mineral Resources for Renewable Energy Technologies*. The Hague, the Netherlands: HCSS.

Chapter 3

The Changing Geopolitics of Energy

Sijbren de Jong*
The Hague Centre for Strategic Studies (HCSS)
Lange Voorhout 1
2514 EA
The Hague
The Netherlands
shdejong@gmail.com

Geopolitics and energy markets have a close relationship, as the outbreak of geopolitical tension has traditionally caused fluctuations in the oil price. Similarly, politically motivated disruptions in the supply of natural gas have caused security concerns and shortages. Much has changed however since the mid-20th century. The development of unconventional fuel types such as shale gas and oil in the United States (U.S.) has meant that the U.S. has morphed into a major energy producer. Particularly the onset of shale oil has had a major impact on the international oil market. Similarly, the onset of Liquefied Natural Gas (LNG) has caused gas markets to be more globally connected. A third major disruptive change in global energy has been the rapid increase in renewable sources of energy since half-way through the past decade. This chapter presents an analysis of how the role of geopolitics in energy markets has changed over time and why political upheaval has a markedly different impact today than it has had throughout history.

3.1 Introduction

Access to cheap, reliable sources of energy has always been a key requirement for economic development. Throughout history episodes of economic growth have been underpinned by a reliance on particular types of fuel. During the 1700s and early 1800s, for example, the primary source of energy was wood. This changed dramatically due to the onset of coal during the

---

*At the time of writing Dr. Sijbren de Jong was a Strategic Analyst with the Hague Centre for Strategic Studies in the Netherlands.

industrial revolution of the 19th century. The mid-20th century ushered in the age of oil and gas. Increasingly, the rapid rise of renewable power sources such as wind and solar and others are acting as the next disruptive force to existing energy paradigms.

Geopolitics and energy markets have a close relationship, as the outbreak of geopolitical tension in oil-rich regions of the world has been one of the prime reasons for oil price fluctuations. Given the interconnectedness of the international oil market, events in one part of the world can have ramifications around the globe. Examples in history include the 1973 Arab oil embargo and the 1979 Iranian revolution. Similarly, access *to* and control *over* energy resources has often been a reason for the outbreak of conflicts and wars. Disagreements on oil production levels within the Organization of Petroleum Exporting Countries (OPEC) and the 1990 Iraqi invasion of Kuwait serves as a case in point. Other notable episodes in history which had an impact on the price of oil were the outbreak of the second Gulf War in 2003 and the Arab uprisings that began in North Africa in 2010.

Gas markets function differently in the sense that, despite the tremendous growth in LNG in many parts of the world — with the exception of Asia — natural gas is shipped via pipeline infrastructure. This means that disruptions in natural gas flows often tend to have regional, rather than global implications. A particular example of a type of geopolitical risk that manifests itself in natural gas markets is that differences can exist between legal and regulatory regimes to which a pipeline is subjected when it crosses the territory of several states. By the same token, the interests of transit-states, whose territory a pipeline transects, do not always overlap with those of producer and consumer states. Illustrative of these types of tensions were the interruptions in the supply of Russian natural gas through Ukraine that took place in 2006, 2009 and 2014.

Energy markets have undergone some fundamental changes in the past couple of years. The development of unconventional fuel types such as shale gas and oil in the U.S. has meant that the U.S. has morphed into a major energy producer and has seen its imports significantly go down. Particularly the onset of shale oil has had a major impact on the international oil market. Whereas the oil price was as high as $115 in June 2014, the massive oversupply in the market in part caused by the production of shale oil set in motion an oil price decline that is still felt today. By late January 2018 the international oil price hovered around $70, buoyed by geopolitical tensions in Libya and elsewhere, but reached below $30 in the beginning of 2016. The impact on oil-producing countries worldwide has been profound, with many producers in the Middle East experiencing budgetary constraints. The ability of the OPEC cartel to intervene in the market in order to bring about a price increase has proven difficult, prompting some commentators to proclaim the end of OPEC.

Gas markets have similarly undergone major changes. The growth of LNG means that gas markets become more and more interconnected, allowing for more flexibility compared to the inert nature of fixed pipeline infrastructure. What is more, the construction of numerous pipeline inter-connections in the European market has meant that the flexibility in times of a supply disruption akin to the ones experienced in 2006 and 2009 is greatly enhanced.[1] Also, the enforcement of market liberalisation rules and competition law has meant that the EU internal gas market has attained much greater resilience compared to a decade ago. The threat that therefore emanates from the Russian 'energy weapon' has greatly subsided over the past years.

A third major disruptive change in global energy markets has been the rapid increase in renewable sources of energy since half-way through the past decade. 2015 in particular was a boom year for renewable energy where for the first-time renewables accounted for more than half of the total net annual additions to globally installed power capacity.[2] Investments in renewable energy capacity also set a new record, reaching a total of $286 billion.[3] What is remarkable about this large increase is that this took place against the backdrop of a prolonged period of low fossil fuel prices. The main reason why the use of renewables grew so much therefore lies in notable cost reductions of particular renewable energy technologies such as solar PV, and favourable government policies stimulating the use of renewables over fossil fuel resources.[4]

This chapter presents an analysis of how the role of geopolitics in energy markets has changed over time and why political upheaval today has a markedly different impact today than it has had throughout history. The chapter consists of four parts. Part one delves into the geopolitics of oil markets and how OPEC's role today is different from its heydays in the 1970s. The second part analyses the various changes that took place in natural gas markets and how geopolitical tensions today have a differ-ent impact on energy security in Europe compared to roughly a decade ago. Part three looks ahead at the new geopolitical dynamics that can be expected as a result of the energy transition and the displacement of fossil fuels by renewable energy. The fourth and final part presents a number of conclusions.

3.2   Geopolitics and the International Oil Market: Then and Now

The 1973 Arab oil embargo has gone down in history as the defining moment when the OPEC cartel flexed its muscles. Disgruntlement over Western support to Israel during the Yom Kippur War was the reason for the escalatory move. The decision introduced limits in oil production and banned the export of petroleum products to the U.S., the Netherlands, Great Britain, Canada, and Japan. The act sparked a major increase in the price of oil.[5] Following the imposition of the embargo, the OPEC cartel started to increase prices. The move was illustrative of the long-term dissatisfaction among cartel members over the low prices they received for oil sales up to that point.[6] The embargo caused consuming nations to form a counterweight to OPEC in the form of the International Energy Agency (IEA) in 1974.[7]

Fast forward almost three decades, the rise of China and India fuelled a global commodity boom that saw oil prices grow steadily and reach close to $150 per barrel in July 2008. The onset of the financial crisis afterwards caused a massive slide in the price of oil, seeing it plummet to as low as $35 per barrel. The period of 1998–2008, with its consistently rising prices, of course made it appealing for oil-producing countries to pump more oil thus contributing to a greater supply on the market. The ferocious demand coming from emerging economies did little to dampen prices up to that point.

3.2.1   *What goes up, must come down*

At the same time however, this period of consistently high prices caused hitherto more expensive fuel resources to become economical, as well as inspired ways to conserve energy and invest in alternative technologies. The biggest upset to OPEC in this regard came from the development of the shale industry in North America, where innovations in existing techniques allowed the extraction of oil and gas trapped in shale rock formations at profitable rates. The onset of a process known as hydraulic fracturing, or 'fracking' for short, injects a mix of water, sand and chemicals into the ground under pressure which leads the rock to crack, thus releasing the oil or gas trapped inside. Horizontal drilling, the second technique used, drills a well downwards just like a traditional oil or gas well, only to move sideways afterwards, thus allowing a greater area of rock holding valuable resources to be exposed.

From 2009 onwards the production of shale oil in America started to increase rapidly. At the same time, this caused a reduction in the amount of imported oil. Imports began to drop so much in fact that in September 2013 China overtook America as the world's largest net importer of petroleum and other liquids in the world. The combination of an increasingly self-sufficient US, weak global economic activity, increased efficiency, a growing switch from oil to other fuels, *and* the relentless production on the part of OPEC countries meant something had to give way. The large supply excess was simply more than the oil market could bear. The price of oil began to slide as a result, dropping even below $30 a barrel in early 2016.

3.2.2 Cutting production with reluctant members

Another major factor why oil today is still priced at less than half of its June 2014 high of $115 is that OPEC hitherto has been unable to foment an adequate response to tackle the oversupply in the market. Back in 2014, many OPEC nations expected Saudi Arabia — OPEC's de facto leader — to step up to the plate and instigate a cut in production. Member countries such as Venezuela, Nigeria, and Angola were hard hit by the oil price decline. However, given how much they need the revenue of every additional barrel sold, they themselves were unwilling to make the necessary cuts, looking to Saudi Arabia and its Gulf allies instead. Moreover, Iran, which looked to emerge from decades of western sanctions and keen to regain its position on the international oil market, was not in a mood to make any cuts either. The Saudis read the writing on the wall and knew that by intervening they would bear the brunt of the costs, incentivizing other OPEC and non-OPEC members to step into the market share they left behind. As a result, the OPEC meeting of 27 November 2014 ended with Saudi-Arabia resisting a call from OPEC's poorer members to cut production, sending the oil price in a tailspin.

What ensued was a prolonged battle about market share between OPEC and non-OPEC supply, U.S. shale in particular. The Saudis were banking on the idea that given shale's higher cost of production; they could simply 'sweat the Americans' out by continuing to flood the market. That resulted in oil prices dipping below the break-even point for U.S. shale producers. In doing so, the Saudis gravely underestimated shale's resilience, as ferocious cost-cutting and increased productivity meant that a large portion of the shale plays were in fact economical at oil prices below $50 per barrel.[8] Furthermore, this strategy created a lot of resentment within

OPEC from the poorer members who were struggling to cope with the lower prices.

After a bruising battle, it was ultimately Saudi Arabia that blinked first. In November 2016, OPEC agreed to reduce output by around 1.2 million barrels. Under the agreement, which exempts Nigeria and Libya due to ongoing domestic unrest, Iran is allowed to increase its production to 3.8 million barrels per day as it recovers from sanctions.[9] Following the accord, major non-OPEC oil producers such as Russia, Mexico, Oman, Azerbaijan and others agreed to reduce their production by 558,000 barrels per day.[10] Several months later, both OPEC and non-OPEC producers agreed to extend the deal to March 2018, after the agreement had hitherto failed to deliver a decisive blow to the global oil supply glut.[11]

One of the reasons why the production cut has failed to bring U.S. shale to its knees is because not every country has been fully implementing the cuts. In June 2017 total OPEC compliance with the targets slipped below 100%, back to levels last seen in February of that year. Ecuador, Algeria, Iraq, Gabon, Kuwait, and the UAE all failed to reach their respective targets in June 2017.[12] In February of 2017, non-OPEC nations recorded a 31% compliance level, also well below target. By June 2017 this had improved to 85%.[13] With crude prices hovering around $50 a barrel, Saudi Arabia and Russia have not been thoroughly pleased with the impact that the deal has had, leading both nations to utter threats to 'free-riders', demanding full participation of all countries involved.[14] By late November 2017 OPEC and Russia agreed to roll over the deal until the end of 2018.

The battle between OPEC and U.S. shale shows the limitations of the oil cartel to successfully intervene in today's market. Non-compliance of its own members and greater resilience and efficiency on the part of shale producers — a production technique spurred on by a decade of high prices — all serve to undermine the cartel's efficacy. The likelihood that the world will witness a rerun of the 1973 oil embargo thus appears a very distant threat.

3.3 The 'Gas Weapon' Revisited

In January 2006 and January 2009 disputes between Ukraine and Russia over supplies, transit fees and outstanding debt prompted Russian state-owned gas producer Gazprom to cut off the gas supplied to Ukraine. Whereas the interruption in 2006 lasted only a few days, the cut-off in 2009 was much more severe, lasting almost three weeks and causing disruptions in numerous EU countries in south-eastern Europe.[15] In 2014, Russia again cut off the gas to its neighbour, an escalatory move amidst a rapidly growing conflict in which Ukraine's Crimean peninsula was annexed by Russia and Moscow fomented a proxy war in the eastern Donbas region of Ukraine.

The willingness on the part of Russia to use its energy resources as a tool to apply (political) pressure on countries has not been limited to Ukraine alone. As much as forty politically motivated energy cut-offs were recorded between 1991 and 2004, well before the much publicised interruptions of 2006, 2009 and 2014.[16] When Ukraine attempted to lessen its dependence on Russian natural gas by contracting so-called 'reverse flow' gas via Hungary, Poland and Slovakia, Russia threatened to cut off gas deliveries to these countries if they continued to sell to Ukraine.[17] Hungary buckled under the pressure and announced it would freeze its gas deliveries to Ukraine.[18] Although these examples make clear that Russia is not hesitant to resort to using its energy deliveries as a political tool when it considers this expedient, Europe has in fact come a long way in mitigating the risks that were laid bare in 2009. One such way by which the implications of a gas cut-off have been lessened is by building interconnector pipelines.

3.3.1 Connecting Europe

Gazprom has traditionally been able to use its power as a monopolistic supplier to make use of a lack of alternative delivery systems and thus maximise the prices paid by numerous European consumers. By way of example, in 2013 Macedonia paid $564 per thousand cubic metres of gas, whereas Germany paid only $379.[19] Recognising the multitude of alternative suppliers that Germany has compared to the tiny Balkan nation, the European Commission has been a staunch advocate of building interconnectors. To that effect the Commission has created a number of initiatives. First, it established a list of so-called 'Projects of Common Interest' (PCIs); a list of 195 key energy infrastructure projects deemed essential for completing the EU internal energy market and for meeting the Union's objectives of affordable, secure and sustainable energy.[20] Second, it created the €5.35 billion large 'Connecting Europe Facility', aimed to ensure the development of 'non-competitive' infrastructure which has strategic significance despite not immediately considered commercially attractive.[21] Third, it launched the Trans-European Networks for Energy (TEN-E) process, focused on linking the energy infrastructure of EU countries.[22] As part of these programs the construction of pipeline interconnections and LNG terminals in areas that are poorly connected have been prioritised.

The Baltic region saw the arrival of a floating regasification and storage unit (FRSG) to Lithuania in December 2014. The effect of the unit's arrival was immediate; Gazprom agreed to lower its gas price by 20%.[23] In 2014 Finland and Estonia agreed to the construction of two LNG terminals, set to be connected to a sub-sea gas pipeline between the two countries.[24] In September 2016, Finland's first LNG terminal commenced commercial operations and received the first cargo.[25] Poland, similarly, saw the first arrival of American LNG at its terminal in Swinoujscie in June 2017.[26]

Arguably the most important work is being done in south-eastern Europe, as part of the Southern Gas Corridor.[27] The signature project in the region is the Trans-Adriatic-Pipeline (TAP) which connects Turkey with Greece and onwards to Italy. The project, set to be completed by early 2020, delivers 10 bcm of natural gas from Azerbaijan to south-eastern Europe, thus breaking Gazprom's hold on the region.[28] In the future the addition of two extra compressor stations can enable the pipeline to double its capacity to 20 bcm, to allow for other sources from the Caspian region.[29] Other notable improvements in the region are the planned interconnectors between Greece and Bulgaria (IGB) and between Moldova and Romania.[30] With respect to LNG, Croatia is slated to finish construction on a floating terminal in the northern Adriatic in 2019.[31]

With the construction of pipeline interconnectors and LNG terminals, Europe has come a long way to increase the resilience of the European gas market in the case of an interruption in supplies coming from Russia. The increased options for hitherto captive markets in parts of Europe have also meant that Gazprom was forced to change its ways. These changes arguably helped pave the way for the settlement of the anti-trust case, as Gazprom's old business model whereby it provided long-term contracts with so-called 'take or pay' clauses[32] simply could no longer hold.[33]

3.3.2 *The battle for market share*

The risk posed by a cut-off in the supply of Russian gas to Europe has therefore somewhat subsided compared to 2009, at least insofar as when there is an interruption, gas from other markets in Europe can now more easily reach the areas affected. The increase in competition from LNG and the greater resilience of the EU internal market made possible through inter-connectors has meant that Gazprom has been forced to change its tactics. That however does not mean that the threat has disappeared. Instead, it has morphed. Gazprom's latest tactic, perhaps in anticipation of a greater share of renewable energy that one day will make up Europe's energy mix, is to ensure that it creates a 'lock-in' of gas demand in Europe in the coming decades by building new pipelines to key-markets. However, the problem with this strategy is that these pipelines are — strictly speaking — diver-sionary in nature, aimed at circumventing transit countries that Russia deems problematic. These pipelines are in and of themselves not necessary, as existing pipeline capacity is more than sufficient.

Take the expansion of the Nord Stream II pipeline between Russia and Germany for example. The $10 billion pipeline, when constructed, would double Nord Stream's existing 55 bcm capacity to 110 bcm. An often heard argument by proponents of the pipeline is that the project is needed as gas production in north-western Europe is in decline. It is true that the production in the Netherlands and the UK is going down, yet the Nord Stream II pipeline does little to add new capacity. In fact, all that the pipeline will do is replace gas that is currently being transited through Ukraine, thereby adding no new supplies.[34] Moreover, the project does little to promote market liberalisation as it would concentrate as much as 80% of all Russian gas supplied to Europe in a single route.[35]

Similarly, the proposed 31.5 bcm large Turkish Stream pipeline which would run from Russia and underneath the Black Sea to Turkey, does little to promote diversification. The pipeline, which consists of two strings, would first of all replace the trans-Balkan pipeline currently traversing Ukraine, Moldova, Romania and Bulgaria.[36] Turkish Stream therefore does in the South what Nord Stream II hopes to do in the North: to get rid of the Ukrainian gas transit route by making European customers directly reliant on Russian supplies. Any additional flows will have to come from Turkish Stream's second string. This is however where another strategy comes into play. The second string of 15.75 bcm that goes to Turkey delivers more gas than Turkey needs for its own consumption, which means that some of this gas will have to find its way to other nearby markets in south-eastern Europe. The only available pipeline infrastructure in the area that does not demand third-party access is part of the Italy-Turkey-Greece interconnector (ITGI) which has a 25-year exemption for this rule. The pipeline could provide 8 bcm of capacity if connected to the planned offshore section of ITGI that runs to Italy known as ITGI-Poseidon. If one compares the capacity of the TAP pipeline that forms part of the Southern Corridor, which can be expanded to hold 20 bcm, then it becomes clear that Gazprom is attempting to pre-empt future deliveries from other parties that border the Caspian Sea.[37]

These two stories make clear that with Gazprom switching tactics, the quest for diversification of natural gas supplies is far from over. Looking towards the future therefore it is imperative that the work on building interconnectors is finalised and the European Commission holds firm in what EU regulations demand from energy companies operating within the EU internal market.

3.4 The Next Big Disruptor: Energy Transition

The transition to a low carbon economy due to concerns over climate change is likely to affect countries in an uneven manner. Countries rich in oil and gas reserves stand to lose part of their revenues if energy transition means that countries will replace fossil fuel imports with renewable sources. Problematic in this regard is the fact that many nations in the Middle East, North Africa and the region belonging to the former Soviet Union earn much of their GDP through the sale of hydrocarbons. Countries such as these are often qualified as 'rentier states', owing to the high share of resource rents in their public finances. If successful climate mitigation policies cause the demand for fossil fuels to decline, the financial and socio-political stability of these countries stands to suffer, particularly if they do not succeed at reforming their domestic economies.[38]

3.4.1 *Lessons from the shale revolution*

The oil price decline, partly brought about by the U.S. shale revolution, is illustrative of the kinds of disruptive effects that an energy transition might have on the global system. As mentioned in section 1, the oil price dropped to below \$30 per barrel in the beginning of 2016. Many large hydrocarbon exporting countries experienced a large reduction in their government revenues as a result. Only Iran and Egypt were able to escape the dance. For Iran this was due to the removal of sanctions and the expansion of its oil production following the accord struck between Iran and the P5+1 powers in April 2015. For Egypt, it had to do with the fact that the country had already become a net oil importer of oil and gas in 2012.[39] In the case of Saudi Arabia, the decline in revenue was as much as 16% of GDP. Algeria,

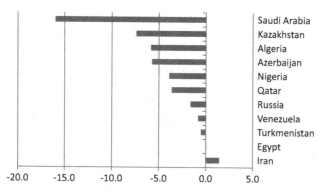

Fig. 3.1. Change in general government revenue between 2013 and 2015, in percentage points of GFP.

Source: IMF.

Azerbaijan and Kazakhstan also saw a major decline, with more than 5% of their GDP evaporating (Fig. 3.1).

In dealing with the budget shortfall, many countries rolled back government spending, causing the cancellation of major long-term investment projects. Azerbaijan, Kazakhstan, Russia, and Saudi Arabia chose this way of economizing on their expenditures. Egypt, Qatar, Nigeria, and Saudi Arabia, cut energy and water subsidies. Russia, Algeria, and Saudi Arabia also slashed public sector salaries. Finally, Algeria, Azerbaijan, Iran, Russia, and Turkmenistan all chose to raise taxes.[40] Some of the abovementioned countries were able to accrue significant financial reserves in the form of foreign exchange and sovereign wealth funds and use these to plug any budgetary shortfalls. Countries such as Qatar resorted to these reserves to make sure the spending for the FIFA 2022 World Cup was not put in jeopardy.[41] Countries such as Venezuela or Egypt however do not have such deep coffers as Qatar.

A measure frequently employed in the face of worsening economic circumstances is currency depreciation. Although, Azerbaijan, Egypt, Kazakhstan, and Nigeria initially all resisted doing so, they were eventually forced to devalue under the pressure of international financial markets. Only the countries with large financial reserves, such as Saudi Arabia and Qatar, were able to steer through the oil price rout without depreciating their currencies. This came at the cost of a significant depletion of their financial reserves however. In addition to devaluing the currency, resource rich countries at times raise interest rates during a period of low commodity prices. The downside of employing these measures is that the combination of currency devaluation and raising interest rates create a significant risk for the stability of the banking system. To address this challenge, Azerbaijan, Algeria, and Russia purchased non-performing loans from banks and increased the supply of liquidity.[42]

Looking towards the future, Saudi Arabia launched a bold reform plan dubbed 'Vision 2030'.[43] The document lays out plans for long-term economic diversification, whereby the Saudi leadership invests more in high-tech and raises the share of employment in the private sector. The Vision aims to propel the Saudi economy into the global top 15.[44] Crucial to accomplishing this feat is the partial privatization of Saudi Aramco, the national oil company, and its transformation into a global industrial conglomerate. The plan is to transfer the ownership of Saudi Aramco and other state-owned companies to the Public Investment Fund, which then will become the largest sovereign wealth fund in the world.[45] In addition, the plans aim to diversify the Saudi economy away from its excessive reliance on oil, planning to raise the share of non-oil exports in non-oil GDP from 16% to 50% by 2030. To achieve this ambitious goal, the government plans to generate 9.5 gigawatts of renewable energy. Energy pricing is also set to be reformed, bringing prices up to the market level.[46] In order to fund this all, the Saudi leadership will issue tenders for major solar and wind programs worth between $US 30 and $US 50 billion.[47]

Although highly ambitious, Vision 2030 lacks specifics on how to achieve its goals. A year after its launch, Riyadh has already redrafted the plans, stripping out some areas earmarked for change and extending the timeline of other targets in some cases by as much as a decade. The move is an indication that some of the tabled ideas were overly ambitious and lacked realism. Amid the delays there is concern that the reform efforts have focused too heavily on revenue-increasing measures such as subsidy cuts and tax raises, rather than initiatives to boost growth.[48] Disgruntlement about the partial flotation of Saudi Aramco is increasing. The Saudi crown prince Mohammed bin Salman suggested that the 5% initial public offering would comfortably raise US$ 2 trillion. However, there is widespread scepticism in industry circles that such an amount of money will actually be raised. Within Saudi Arabia itself, members of the royal family oppose the sell-off as they view it as a needless handover of family silver and institutional power.[49] All in all, there are ample indications that the implementation of Vision 2030 is not going as smoothly as initially hoped.

3.4.2 *Expectations for the future*

If the stated ambitions of the Paris Climate Accord are anything to go by, then one thing is certain, the share of renewable energy worldwide needs to greatly increase in the future. According to a report by consultancy DNV GL global energy demand is set to plateau by 2030. The greater electrification of our energy supply and greater energy efficiency contribute to the rapid decarbonisation of the world's energy system. By 2050 this would imply that renewables make up almost half of the global energy mix.[50] Should this indeed materialise, it constitutes a truly watershed moment. In this context, one should not read the announcement by the Chinese government to end the sale of all fossil-fuel-powered vehicles as an overly ambitious dream. Although they will take several decades to implement, such measures are increasingly becoming a realistically attainable goal. The UK announced its plan to ban the sale of diesel- and gasoline-fuelled cars by 2040, two weeks after France had announced a similar plan. Countries such as Norway and the Netherlands too are contemplating more aggressive ways to end fossil fuel cars years ahead of other European nations.[51]

The implications of such plans are far-reaching as they would displace a major source of fossil fuel demand from the transport sector. That said, passenger cars and light vehicles are not the only gas guzzlers in town. Freight transport, air traffic and the petrochemical sector are major sources of oil demand for which few substitutes currently exist. Since demand in these sectors is actually increasing, rather than decreasing, the likely implication is that energy demand will face a long plateau.[52]

What is worrying however is that beyond this plateau decline will inevitably set in, and major hydrocarbon exporters have hitherto shown little appetite for adjustment. A number of factors are important in determining whether countries are vulnerable to the ensuing loss of revenue. Indicators that mark a heightened vulnerability include a high share of resource rents in a country's GDP in combination with limited financial reserves, a high national debt as a percentage of GDP or a rapid increase in a state's national debt, a young and/or relatively fast-growing population, and a relatively high share of youth unemployment (particularly prevalent in the Middle East and North Africa). Other factors that render a country more vulnerable are uncertainty about the continuation of the existing political leadership, a combination of high subsidies and a high domestic energy demand, and a low credit rating and difficulties with borrowing in international markets.[53] The experience from the shale revolution has already laid bare some of the effects that such vulnerabilities generate and the energy transition is unlikely to be more merciful in its impact. In the absence of far-reaching reforms, therefore, major hydrocarbon exporters in Europe's vicinity run a higher risk of witnessing socio-political upheaval.[54]

3.5　Conclusion

Since the early 1970s numerous structural shifts in global energy have led to profound changes in the way geopolitics affect the world's energy markets. Whereas the threat emanating from 'the oil weapon' was certainly a profound one, as witnessed by the 1973 Arab oil embargo, today's international oil market is marked by oversupply and the ability for US shale producers to react to any production cut instigated by OPEC. Put differently, years of high prices and the subsequent rise of shale has taken the sting out from OPEC.

Similarly, the advent of LNG and the construction of interconnector pipelines within the European gas market have eroded the ability for the Russian state to utilise gas exports as a geopolitical tool vis-à-vis its neighbours. That said, Gazprom's doubling down on capital-intensive pipeline infrastructure means that the EU's quest for diversification of natural gas supplies is far from over. With construction set to be finished by the end of this decade the next few years will prove crucial in determining whether the European Commission's attempts at increasing the resilience of Europe's internal gas market will succeed.

Finally, the energy transition to renewables is likely to usher in a new period of geopolitics; one whereby the geopolitical effects are the result of the transition itself. When seen from that perspective, it is crucial therefore to keep a close eye on the various factors that cause hydrocarbon exporters to be vulnerable to the destabilising effects of the energy transition and seek to reduce these vulnerabilities in tandem with wanting to increase the share of renewable energy in the global energy mix.

References

1. Roberts, J. (2016). *Completing Europe: Gas interconnections in Central and Southeastern Europe — an update*. Washington D.C., USA: Atlantic Council. Retrieved from http://www.atlanticcouncil.org/publications/reports/comple ting-europe-update.

2. IEA. (2016). *Medium- Term Renewable Energy Market Report 2016*. Paris, France: IEA. Retrieved from https://www.iea.org/newsroom/news/2016/october/medium-term-renewable-energy-market-report-2016.html..

3. Frankfurt School-UNEP Centre. (2016). *Global Trends in Renewable Energy Investment 2016*. Frankfurt, Germany: Frankfurt School-UNEP Centre. Retrieved from https://www.actu-environnement.com/media/pdf/news-264 77-rapport-pnue-enr.pdf.

4. IEA. (2016). *World Energy Outlook 2016*. Paris, France: IEA. Retrieved from http://www.worldenergyoutlook.org/publications/weo-2016/.

5. OPEC. (n.d.). OPEC: Brief History. Retrieved from http://www.opec.org/opec_web/en/about_us/24.htm; Goldthau, A. and Witte, J.M. (2011). Assessing OPEC's performance in global energy. *Global Policy*, 2(s1), 31–39.

6. Florini, A. (2011). The international energy agency in global energy governance. *Global Policy*, 2(s1), 40–50.

7. Decision of the Council Establishing an International Energy Agency of the Organisation, OECD. (1974).

8. Leaner and meaner: US shale greater threat to OPEC after oil price war. (2016, November 30). *CNBC*. Retrieved from https://www.cnbc.com/2016/11/30/leaner-and-meaner-us-shale-greater-threat-to-opec-after-oil-price-war.html; Meyer, G. (2017, January 12). US shale oil output remains resilient despite rig count fall. *Financial Times*. Retrieved from https://www.ft.com/content/73c5297e-d813-11e6-944b-e7eb37a6aa8e.

9. Razzouk, N., Rascouet, A. and Motevalli, G. (2016, November 30). OPEC confounds skeptics, agrees to first oil cuts in 8 years. *Bloomberg*. Retrieved from https://www.bloomberg.com/news/articles/2016-11-30/opec-said-to-agree-oil-production-cuts-as-saudis-soften-on-iran.

10. Raval, A. and Sheppard, D. (2016, December 10). Non-OPEC producers agree to cut oil output. *Financial Times*. Retrieved from https://www.ft.com/content/4cd8dce2-beec-11e6-9bca-2b93a6856354.

11. Meredith, S. (2017, May 25). OPEC and non-OPEC members agree to extend production cuts for nine months. *CNBC*. Retrieved from https://www.cnbc.com/2017/05/25/opec-agrees-to-extend-oil-production-cuts-for-nine-months-delegate-tells-reuters.html.

12. Voss, S. and Dodge, S. (2017, July 17). OPEC didn't manage to keep its promises last month. *Bloomberg*. Retrieved from https://www.bloomberg.com/graphics/2017-opec-production-targets/.

13. Ibid.

14. Sheppard, D., and Raval, A. (2017, July 24). Saudis and Russians threaten to step up oil deal battle. *Financial Times*. Retrieved from https://www.ft.com/content/e322f302-7064-11e7-aca6-c6bd07df1a3c.

15. For an in-depth study of the 2009 dispute, see Pirani, S., Stern, J. and Yafimava, K. (2009). *The Russo-Ukrainian gas dispute of January 2009: A comprehensive assessment*. Oxford, UK: Oxford Institute for Energy Studies.

Retrieved from https://www.oxfordenergy.org/wpcms/wp-content/uploads /2010/11/NG27-TheRussoUkrainianGasDisputeofJanuary2009Acomprehen siveAssessment-JonathanSternSimonPiraniKatjaYafimava-2009.pdf

16. Rendahl, J. (2006). Russia's energy policy: security dimensions and Russia's reliability as an energy supplier," *ResearchGate*. Retrieved from https://w ww.researchgate.net/publication/242714626_Russia's_Energy_Policy_Securit y_Dimensions_and_Russia's_Realibility_as_an_Energy_Supplier.

17. LeVine, S. (2014, June 27). Russia is threatening to cut off European countries' gas if they don't do its bidding. *Quartz*. Retrieved from https://qz.com/227484/russia-is-threatening-to-cut-off-european-countries-gas-if-they-dont-do-its-bidding/.

18. Hungary suspends gas supplies to Ukraine under pressure from Moscow. (2014, September 26). *The Guardian*. Retrieved from http://www.thegua rdian.com/world/2014/sep/26/hungary-suspends-gas-supplies-ukraine-press ure-moscow.

19. Roberts, J. (2016). *Completing Europe: Gas interconnections in Central and Southeastern Europe — an update*. Washington D.C., USA: Atlantic Council; Kates, G. and Li L. (2014, July 1). Russian gas: How much is that? *RadioFreeEurope/ RadioLiberty*. Retrieved from https://www.rferl.org/a/ru ssian-gas-how-much-gazprom/25442003.html.

20. European Commission (n.d.) Projects of common interest — Energy — European Commission. Retrieved from https://ec.europa.eu/energy/en/topics/in frastructure/projects-common-interest.

21. Roberts, J. (2016). *Completing Europe: Gas interconnections in Central and Southeastern Europe — an update*. Washington D.C., USA: Atlantic Council; European Commission. (n.d.) Connecting Europe Facility — Innovation and Networks Executive Agency — European Commission. Retrieved from https://ec.europa.eu/inea/en/connecting-europe-facility.

22. European Commission. (n.d.). Trans-European Networks for Energy — Energy — European Commission. Retrieved from https://ec.europa.eu/e nergy/en/topics/infrastructure/trans-european-networks-energy; European Commission. (n.d.). Energy Priority Corridors. Retrieved from https://ec. europa.eu/inea/en/connecting-europe-facility/cef-energy/cef-energy-project s-and-actions.

23. Grigas, A. (2014, September 22). Klaipeda's LNG terminal: A game changer. *EURACTIV*. Retrieved from http://www.euractiv.com/section/energy/opin ion/klaipeda-s-lng-terminal-a-game-changer/.

24. European Commission. (n.d.). Construction of the Gas Interconnection Poland-Lithuania [GIPL] including supporting infrastructure. Retrieved from https://ec.europa.eu/inea/sites/inea/files/8.5-0046-pllt-p-m-2014_-_m ay_2016_-_final_jms.pdf; Finland and Estonia to build LNG terminals. (2014, November 18). *EURACTIV*. Retrieved from http://www.euractiv.com/sectio n/energy/news/finland-and-estonia-to-build-lng-terminals/; Pagni, J. (2017, May 24). The Finland-Estonia gas pipeline is a 'high-level project for the EU' — breaking the dependency on Russia. *Business Insider Nordic*. Retrieved from http://nordic.businessinsider.com/the-finland-estonia-gas-pipeline-is-a

-high-level-project-for-the-eu-breaking-the-dependency-on-russia-and-conn
ecting-finland-with-the-continental-market-2017-5/.

25. Gasum. (2016). Finland's first LNG import terminal opened. Retrieved
from https://www.gasum.fi/en/About-gasum/for-the-media/News/2016/Fi
nlands-first-LNG-import-terminal-opened-provides-access-to-LNG-deliverie
s-also-outside-the-gas-pipeline-network/.

26. US LNG cargoes arrive in Poland, Netherlands, first into Northern Europe.
(2017, June 8). *Platts*. Retrieved from https://www.spglobal.com/platts/e
n/market-insights/latest-news/shipping/060817-us-lng-cargoes-arrive-in-po
land-netherlands-first-into-northern-europe.

27. European Commission. (n.d.). Energy Priority Corridors. Retrieved from
https://ec.europa.eu/inea/en/connecting-europe-facility/cef-energy/cef-ene
rgy-projects-and-actions.

28. Roberts, J. (2016). *Completing Europe: Gas interconnections in Central and
Southeastern Europe — an update*. Washington D.C., USA: Atlantic Council.

29. Trans Adriatic Pipeline. (n.d.). TAP at a glance. Retrieved from https://ww
w.tap-ag.com/the-pipeline.

30. Hope, K. (2017, October 17). Bulgarian and Greek pipeline aims to cut gas
dependency on Russia. *Financial Times*. Retrieved from https://www.ft.
com/content/f6752fde-44f4-11e6-9b66-0712b3873ae1?mhq5j=e4; Rosca, O.,
Ondrejička. D. and Onofreiciuc, V. (2016, December 19). EU bank
and EBRD support gas interconnection between Moldova and Romania. European Bank for Reconstruction and Development. Retrieved
from http://www.ebrd.com/news/2016/eu-bank-and-ebrd-support-gas-inter
connection-between-moldova-and-romania-.html.

31. Ilic, I. (2017, January 25). Croatia floating LNG terminal taking a year longer
to finish. *Reuters*. Retrieved from https://www.reuters.com/article/us-croat
ia-lng-idUSKBN15912O.

32. A take-or-pay clause is a rule in a contract that stipulates that the recipient
country is obliged to take X-amount of natural gas irrespective of market
demand. If it fails to do so, it pays a penalty.

33. Toplensky, R. and Foy, H. (2017, March 13). Gazprom Reaches Draft
Antitrust Deal with EU. *Financial Times*. Retrieved from https://www.ft.c
om/content/575f8d2e-07f2-11e7-ac5a-903b21361b43; Tagliapietra, S. (2017,
March 15). The EU Antitrust Case: No Big Deal for Gazprom. *Bruegel*.
Retrieved from http://bruegel.org/2017/03/the-eu-antitrust-case-no-big-de
al-for-gazprom/.

34. Riley, A. (2017, March 29). Smoke and mirrors: Russian disinformation meets
pipeline politics. *CEPA*. Retrieved from https://www.ceep.be/russian-disinf
ormation-pipeline-politics/.

35. Denková, A. and Gotev, G. (2015, December 18). Tusk joins 'Visegrad Four'
in attack on Nord Stream 2. *EURACTIV*. Retrieved from https://www.eur
activ.com/section/energy/news/tusk-joins-visegrad-four-in-attack-on-nord-
stream-2/; European Parliament. (2015, December 10). Written Question —
Nordstream Pipeline Project in the Baltic — E-015658/2015. Retrieved
from http://www.europarl.europa.eu/sides/getDoc.do?type=WQ&reference
=E-2015-015658&language=EN.

36. Cutler, R.M. (2016, November 16). The Turkish stream agreement and what it means. *Intersection.* Retrieved from http://intersectionproject.eu/article/ economy/turkish-stream-agreement-and-what-it-means.
37. De Jong, S. (2016, October 24). Turkish stream natural gas pipeline deal signed — but obstacles remain. *RUSI.* Retrieved from https://rusi.org/c ommentary/turkish-stream-natural-gas-pipeline-deal-signed-%E2%80%93-o bstacles-remain.
38. De Jong, S., et al.. (2017). *The Geopolitical Impact of Climate Mitigation Policies. How Hydrocarbon Exporting Rentier States and Developing Nations Can Prepare for a More Sustainable Future.* The Hague, the Netherlands: HCSS.
39. Adly, A. (2016, August 2). Egypt's oil dependency and political discontent. *Carnegie Middle East Center.* Retrieved from http://carnegie-mec.org/2016 /08/02/egypt-s-oil-dependency-and-political-discontent-pub-64224.
40. De Jong, S., et al.. (2017). *The geopolitical impact of climate mitigation policies. How hydrocarbon exporting rentier states and developing nations can prepare for a more sustainable future.* The Hague, the Netherlands: HCSS.
41. Sommer, M. et al. (2016). *Learning to live with cheaper oil: policy adjustment in oil-exporting countries of the Middle East and Central Asia.* Washington D.C., USA: IMF. Retreived from http://www.imf.org/external/pubs/ft/dp/ 2016/mcd1603.pdf.
42. De Jong, S., et al.. (2017). *The geopolitical impact of climate mitigation policies. How hydrocarbon exporting rentier states and developing nations can prepare for a more sustainable future.* The Hague, the Netherlands: HCSS.
43. Kingdom of Saudi Arabia. (2016). *Vision 2030.* Saudi Arabia: Kingdom of Saudi Arabia.
44. Ibid.
45. De Jong, S., et al.. (2017). *The geopolitical impact of climate mitigation policies. How hydrocarbon exporting rentier states and developing nations can prepare for a more sustainable future.* The Hague, the Netherlands: HCSS.
46. Kingdom of Saudi Arabia. (2016). *Vision 2030.* Saudi Arabia: Kingdom of Saudi Arabia.
47. Kerr, S. (2017, January 16). Saudi Arabia seeks $30bn-$50bn solar and wind energy investment. *Financial Times.* Retrieved from https://www.ft.com/co ntent/d370829e-dbfe-11e6-86ac-f253db7791c6.
48. Kerr, S. (2017, September 7). Saudi Arabia redrafts crown prince's transformation plan. *Financial Times.* Retrieved from https://www.ft.com/content /2cd73084-92e4-11e7-a9e6-11d2f0ebb7f0.
49. Gardner, D. (2017, September 19). Saudi Arabia's crown prince scales back his ambitions. *Financial Times.* Retrieved from https://www.ft.com/conten t/068b21a0-9d2b-11e7-9a86-4d5a475ba4c5.
50. Haugen, U. (2017, September 4). World energy demand to plateau from 2030, says DNV GL's inaugural energy transition outlook. *DNV GL.* Retrieved from https://www.dnvgl.com/news/world-energy-demand-to-plateau-from-2030-says-dnv-gl-s-inaugural-energy-transition-outlook--99848.

51. China fossil fuel deadline shifts focus to electric car race. (2017, September 10). *Bloomberg News*. Retrieved from https://www.bloomberg. com/news/articles/2017-09-10/china-s-fossil-fuel-deadline-shifts-focus-to-el ectric-car-race-j7fktx9z.
52. Butler, N. (2017, September 18). Beyond peak oil. *Financial Times*. Retrieved from https://www.ft.com/content/4b7f788f-87af-3295-afb3-a2e6ad1eddf5.
53. De Jong, S., *et al.*. (2017). *The geopolitical impact of climate mitigation policies. How hydrocarbon exporting rentier states and developing nations can prepare for a more sustainable future.* The Hague, the Netherlands: HCSS.
54. Ibid.

Chapter 4

Materials for Electrochemical Energy Storage Devices

Erik M. Kelder

Department of Radiation, Science and Technology,
Faculty of Applied Sciences, Delft University of Technology,
Mekelweg 15, 2629JB, Delft, The Netherlands

The global quest for intermittent renewable energy sources (wind, solar), consumer goods (mobile phones, notebooks), and electrification of the transport sector (electric vehicles), requires a strong increase in the use of rechargeable energy storages devices, such as batteries. Hence, novel types of rechargeable batteries are to be found that are cost-effective, consist of environmentally friendly materials and non-critical materials, and deliver a high performance in terms of energy density and power, during a long cycle and service life, taking current legislation laws into account. Here we concentrate on various types of batteries and their materials, their typical sectors of application, and their forecast, including potential market share. It turns out that metals will still remain the predominant type of battery material, irrespective of the choice of battery system. From the required materials that are important for future rechargeable batteries, a number are critical in terms of scarcity, geopolitics, supply risk, competition with the food industry, carbon footprint, and/or ethical mining. With respect to the critical elements, particularly once those concern scarcity and supply risks, recycling is of utmost relevance, and needs to be regarded as an industrial sector of paramount importance.

4.1 Introduction

Cheap, reliable, safe and high energy density electricity storage systems have always been of paramount importance for large-scale introduction of portable consumer electronics and of (hybrid) electric vehicles (EVs) today. From an economical point of view, companies in the battery and battery materials market are investing time and money to develop systems to fulfil the above goals. In addition, because of current legislation which emphasizes environmental protection and recycling, companies are coming up with less polluting materials, which may be easier to obtain and worthwhile to recycle. Obviously this often increases the cost of the materials. Hence, it is necessary to find adequate trade-offs. According to a report by Transparency Market Research,[1] the battery materials market is expected to rise by 13.6% Compound Annual Growth Rate (CAGR) from 2017 to 2025.

Today, consumer goods such as mobile phones and notebooks dominate the end-use market for battery materials. However, the transportation sector, e.g. cars, trains, boats and aircrafts, is rapidly catching up.

Storage systems can be divided into primary (non-rechargeable) and secondary (rechargeable) batteries. Primary systems are found mainly in consumer goods, whereas secondary systems find use in both the transportation and consumer goods sector. A third sector, medical appliances, is becoming more and more important because of the increasing amount of advanced implants, such as pacemakers, defibrillators, and neurostimulators. A last sector is the industrial one, which is increasingly knocking on the door. An example of activity in the industrial sector is when utility companies search for systems required for large-scale electricity storage to permit peak shaving and load levelling. These four sectors of consumer goods, medical, transportation and industry require different types of batteries all with their own particular specifications and priorities. These are briefly presented in Table 4.1.

In this chapter, we concentrate on various types of batteries and their materials requirements. Other electrochemical devices, such as (super)capacitors and/or fuel cells are not considered here. The chapter

Table 4.1:   Relevance of a Rechargeable Battery Issue for the Specific Sector.

	Issue							
Sector	Costs	Energy density	Power density	Life time	Cycle life	Safety	Reliability	Recycle-ability
Consumer Goods	☺[1]	☹	😐	😐	☹	😐	☹	☹
Medical	☺	😐[1]	😐	☹	☹	☹	☹	☺
Transportation	☹	☹	☹	☹	☹	☹	☹	☹
Industry	☹	☺	☺	☹		☹	☹	☺

☹: very critical   😐: critical   ☺: less critical
[1]It becomes ☹ once a primary cell is considered.

thus includes battery chemistry and electrochemistry, typical applications by sector, and the forecast of use, including potential market share.

4.2 Electrochemistry in Brief

A battery typically is composed of one or more electrochemical cells placed in series or in parallel. The cell in turn is an electrochemical device that produces an electric current by spontaneous redox reactions at the negative and positive electrode. These conductive electrodes are the anode and the cathode respectively, where the oxidation and reduction occur. The anode donates the electrons, whereas the cathode accepts them. This means that the anode and cathode change sides as the battery is charged and discharged. However, in the battery world, the anode is often erroneously referred to as the negative electrode, and the cathode as the positive electrode.

In the simple case of the Daniell cell (see Fig. 4.1), the galvanic cell uses two different metal electrodes, zinc and copper, immersed in a sulfate salt, referred to as the electrolyte. Both compartments are then connected

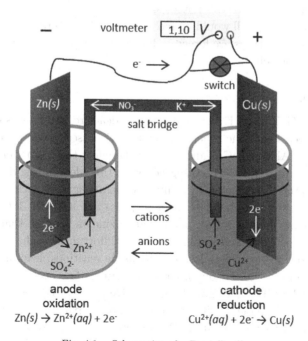

Fig. 4.1. Schematics of a Daniell cell.

via a so-called salt bridge that allows ion transport, so as to electrically balance the compartments once electrons start to flow from the anode to the cathode via an external electrical circuit doing work.

The actual redox reactions — **red**uction and **ox**idation — usually referred to as half reactions, are given below for the Daniell cell:

Zinc electrode (anode, negative electrode): $Zn_{(s)} \rightarrow Zn^{2+}_{(aq)} + 2e^-$

Copper electrode (cathode, positive electrode): $Cu^{2+}_{(aq)} + 2e^- \rightarrow Cu_{(s)}$

In order to simplify the cell notation, a cell diagram is constructed where the electrons spontaneously flow from left to right, meaning that the anode is on the left side, when the cell is discharging, i.e. doing work. This order is shown below:

$$Zn_{(s)}|Zn^{2+}(1M)||Cu^{2+}(1M)|Cu_{(s)}$$

where a vertical line represents the phase boundaries between the solid and the liquid phase and the double vertical line represents the salt bridge of the cell.

The measured electrochemical cell voltage is often referred to as electromotive force (emf) or cell potential.

In the case, this cell would be rechargeable, the spontaneous reaction is reversed when electrons are forced to flow from the positive to the negative electrode of the cell.

4.3 History

When Luigi Galvani in 1780 was studying a dead frog which was lying on a copper alloyed bench, he touched it with his iron scalpel and suddenly its leg twitched. Although Galvani ascribed the effect to "animal electricity", it was Alessandro Volta who disagreed with this idea and he discovered a few years later that placing two different metals against each other separated by a wetted intermediate layer provides a voltage. In 1800, he then constructed his famous voltaic pile based on zinc and copper that provided continuous electricity and a stable current. Unfortunately, the battery was subject to severe electrolyte loss and parasitic reactions, leading to failure of the system after about one hour's use. After a few decades the British chemist Frederic Daniell figured out to solve those problems and in 1836 he presented the Daniell cell, which uses two compartments of the metal, each soaked in its metal sulfate. Both compartments were separated by a earthenware container, which was thus porous for ions to move through, but prevented mixing of both solutions. Soon it became the industry standard, important for the new telegraph networks. The system became even more popular after the Frenchman, Monsieur Callaud, improved the system by reducing the internal resistance. Hence "the gravity" battery was born and survived till the 1950s. In 1842, the German scientist Johann Christian Poggendorff had already further improved the system in terms of voltage up to 1.9V by modifying the electrolyte composition and changing the positive electrode material to a carbon plate. Clearly the chemistry was different from the actual Daniell cell, and therefore it became later known as the "chromic acid cell" or "bichromate cell". Similarly, in 1839 the Welshman, William Robert Grove, replaced the copper electrode with platinum soaked in nitric acid. The "Grove cell" provided a high current and almost doubled the voltage compared to the Daniell cell. Unfortunately, platinum was and is expensive and the cell produces poisonous nitric oxides. Already the importance of materials plays a role as well as safety issues. Therefore, rechargeable systems would have been of definite interest. In 1859, it was Gaston Planté who invented the first rechargeable system, namely the lead acid battery. It consists of lead as the negative electrode and lead dioxide as the positive electrode both immersed in sulfuric acid. Today, this cell is still used in the transport industry as the Start, Lighting and Ignition (SLI) battery as well in stationary storage systems. Nevertheless, at that time, the weight of this rechargeable battery was enormous and was not acceptable for consumer goods. In 1866, Georges Leclanché showed a battery — the Leclanché cell or the zinc-carbon cell — that consisted still of a zinc negative electrode, but now with a manganese dioxide positive electrode, where both electrodes were immersed in the same ammonium chloride solution. In a later stage, the ammonium chloride solution was "solidified" or immobilized so as to make a dry cell, by e.g. plaster of Paris by Carl Gassner in 1886. This most famous primary battery is still manufactured in huge quantities today, mainly in the alkaline form. However, this battery was not the first cell using an alkaline based electrolyte. In 1899, the Swedish scientist, Waldemar Jungner, invented the nickel–cadmium battery. This rechargeable battery uses nickel and cadmium electrodes in a concentrated potassium hydroxide solution. Commercialization started in Sweden in 1910, and was adopted in the U.S. in 1946. The cells were superior to the lead-acid cells, but much more expensive. In order to reduce the cost, Jungner replaced nickel with iron, but it was Thomas Edison who further developed it to certain success, e.g. in rail transport. Another attempt to replace the toxic cadmium was achieved via the nickel metal hydride

(NiMH) battery. Here cadmium was substituted by an alloy of various metals, including rare earth ones. The NiMH batteries were introduced in 1989 for the consumer, although it had already been shown to work back in 1970. Still the quest was to find an improved energy density rechargeable battery to assure its use, e.g. in electric vehicles. Lithium, being the lightest metallic element, therefore was always of interest, also because it produces a high voltage. Finally in the 1970s, the first lithium batteries were introduced on the market, with a 3V primary lithium coin cell, a device still on the market. In 1980, a boost in rechargeable lithium batteries arose when the American chemist John B. Goodenough and the Moroccan research scientist Rachid Yazami proved that lithium cobalt oxide as the positive electrode and graphite as the negative electrode can both reversibly accept lithium. In 1991, Sony introduced the first commercialized rechargeable Li-ion battery.

Missing from the list are the primary batteries zinc-air, silver oxide, mercury oxide, nickel oxyhydroxide, and lithium-thionyl chloride batteries. The first four still use zinc as the negative electrode, whereas the lithium-thionyl chloride cell uses lithium. This last battery has not been generally released to the public for safety reasons, but found its use, e.g. in medical applications such as automatic external defibrillators (AEDs). The most important missing secondary batteries are the sodium-sulfur batteries — used in stationary electricity storage systems, (redox) flow batteries, lithium sulfur — developed by Sion Power in 1994[2] — and lithium air batteries. These will however be discussed in the following sections with respect to their chemistry.

4.4 Battery Composition and its Chemistry

In this section, the chemistry of the various batteries will be briefly touched upon. This is important to show which materials are relevant for the system under study. However, it does not immediately say something about the non-active components, i.e. the casing, binders, current collectors etc. These materials will be discussed in section 4.5. Furthermore, this section has been split up into primary (non-rechargeables) and secondary cells (rechargeables).

4.4.1 *Primary batteries*

Primary cells today find their place mainly in consumer goods, but also in medical devices such as implants. A striking observation is that these two sectors are both important, both for reasons of cost and recyclability. This immediately reflects the importance of the materials used.

4.4.1.1 *Alkaline batteries*

The alkaline battery has a zinc negative electrode and a manganese dioxide positive electrode. Potassium hydroxide is used as the alkaline salt in the cell, and is relevant for the two half reactions, but is not consumed in the end.

Hence, the two half-reactions at the negative and positive electrodes are:

$$Zn_{(s)} + 2OH^-_{(aq)} \rightarrow ZnO_{(s)} + H_2O_{(l)} + 2e^-$$

$$2MnO_{2(s)} + H_2O_{(l)} + 2e^- \rightarrow Mn_2O_{3(s)} + 2OH^-_{(aq)}$$

With the overall reaction:

$$Zn_{(s)} + 2MnO_{2(s)} \rightarrow ZnO_{(s)} + Mn_2O_{3(s)}$$

4.4.1.2 *Mercury batteries*

Mercury batteries have mercury(II) oxide as the positive electrode and zinc as the negative electrode. Potassium hydroxide or sodium hydroxide is used as the alkaline salt in the cell, and like in the alkaline cells it is used at the electrodes but overall is not consumed. The two half-reactions at the negative (two steps) and positive electrodes are:

$$Zn + 2OH^- \rightarrow ZnO + H_2O + 2e^-$$

$$\text{Step 1}: \ Zn + 4OH^- \rightarrow Zn(OH)_4^{-2} + 2e^-$$

$$\text{Step 2}: \ Zn(OH)_4^{-2} \rightarrow ZnO + 2OH^- + H_2O$$

$$HgO + H_2O + 2e^- \rightarrow Hg + 2OH^-$$

With the overall reaction:

$$Zn + HgO \rightarrow ZnO + Hg$$

4.4.1.3 *Silver oxide batteries*

Silver-oxide batteries are quite similar to the earlier mentioned mercury batteries, but have silver oxide as the positive electrode and also zinc as the negative electrode. For the electrolyte, aqueous solutions of sodium hydroxide or potassium hydroxide are used. The half reactions at the negative electrode (two steps) were shown earlier; the half reaction at the positive electrodes occurs also in two steps according to:

$$AgO_{(s)} + H_2O_{(l)} + 2e^- \rightarrow Ag_{(s)} + 2OH^-_{(aq)}$$

$$\text{Step 1}: \; 2AgO_{(s)} + H_2O_{(l)} + 2e^- \rightarrow Ag_2O_{(s)} + 2OH^-_{(aq)}$$

$$\text{Step 2}: \; Ag_2O_{(s)} + H_2O_{(l)} + 2e^- \rightarrow 2Ag_{(s)} + 2OH^-_{(aq)}$$

With the overall reaction:

$$Zn + AgO \rightarrow ZnO + Ag$$

4.4.1.4 *Zinc-air batteries*

Zinc-air batteries use a zinc negative electrode and a positive electrode where air can penetrate the cell. These cells are typically used for hearing aids, where you have to remove a sticker before use, so the positive electrode can be exposed to air. The electrolyte used is similar to that used for the alkaline cell, and the half reaction of the zinc electrode as well. The positive electrode (air) half reaction then is:

$$\frac{1}{2}O_2 + H_2O + 2e^- \rightarrow 2OH^-$$

With the overall reaction:

$$2Zn + O_2 \rightarrow 2ZnO$$

4.4.1.5 *Lithium batteries*

There exist many variations of lithium batteries, but all of them have a lithium negative electrode. The positive electrodes, however, vary from system to system. A brief overview is given in Table 4.2. The table includes the material and the electrolyte composition used. The half reaction at the positive electrode is somewhat complex for certain systems and will not be given here., However, the half reaction at the negative (lithium) electrode is given by:

$$Li_{(s)} \rightarrow Li^+_{(aq)} + e^-$$

Table 4.2: Composition of Commercial Primary Lithium Batteries.

Material	Typical electrolyte composition[1]
Metal oxides[2]	Lithium perchlorate in PC/DME
Metal sulfides[3]	PC, dioxolane, DME
iodine	Lithium iodide
Silver chromate	Lithium perchlorate solution
Carbon monofluorde	Lithium tetrafluoroborate in PC, DME, or Υ-BL
Copper chloride	Lithium tetrachloroaluminate in inorganic liquid SO_2
Thionyl chloride[4]	Lithium tetrachloroaluminate in thionyl chloride
Sulfur dioxide	Lithium bromide in SO_2

[1] PC = propylene carbonate, SO_2 = sulfur dioxide, Υ-BL = gamma-butyrolactone, DME = dimethoxyethane.
[2] manganese dioxide, silver oxide, vanadium pentoxide, bismuth trioxide, cobalt oxide, copper(II) oxide.
[3] iron sulfide, iron disulfide (Pyrite), copper sulfide, lead sulphide.
[4] sulfuryl chloride systems do exists as well.

4.4.2 Secondary batteries

4.4.2.1 Lead-acid batteries

Lead-acid batteries use lead, lead oxide and lead sulfate for their electrodes. The positive and negative electrodes are both immersed in concentrated sulfuric acid, which plays a crucial role in the working of the cell. The electrode reactions on discharging are:

$$Pb_{(s)} + HSO_{4(aq)}^- \rightarrow PbSO_{4(s)} + H_{(aq)}^+ + 2e^-$$

$$PbO_{2(s)} + HSO_{4(aq)}^- + 3H_{(aq)}^+ + 2e^- \rightarrow PbSO_{4(s)} + 2H_2O_{(l)}$$

With the overall reaction:

$$Pb_{(s)} + PbO_{2(s)} + 2H_2SO_{4(aq)} \rightarrow 2PbSO_{4(s)} + 2H_2O_{(l)}$$

Since the system is rechargeable, on charging, the above reaction will be reversed.

4.4.2.2 Nickel-cadmium batteries

Once the nickel-cadmium cell has been charged, the negative electrode is cadmium and the positive electrode is nickel(II) oxide hydroxide. The electrodes are separated by a porous membrane, referred to as the separator, which is soaked in an alkaline solution of potassium hydroxide.

The electrode reactions on discharging are:

$$Cd_{(s)} + 2OH^-_{(aq)} \rightarrow Cd(OH)_{2(s)} + 2e^-$$

$$2NiO(OH) + 2H_2O_{(l)} + 2e^- \rightarrow 2Ni(OH)_{2(s)} + 2OH^-_{(aq)}$$

With the overall reaction:

$$2NiO(OH) + Cd_{(s)} + 2H_2O_{(l)} \rightarrow 2Ni(OH)_{2(s)} + Cd(OH)_{2(s)}$$

4.4.2.3 *Nickel-metal hydride batteries*

The nickel metal hydride (NiMH) cell has a metallic-like material (M) as the negative electrode and nickel(II) oxide hydroxide as the positive electrode. The electrodes are separated by a porous membrane, referred to as the separator, which is soaked in an alkaline solution of potassium hydroxide. During discharging the electrode reactions are:

$$MH_{(s)} + OH^-_{(aq)} \rightarrow M_{(s)} + H_2O_{(l)} + e^-$$

$$NiO(OH) + H_2O_{(l)} + e^- \rightarrow Ni(OH)_{2(s)} + OH^-_{(aq)}$$

With the overall reaction:

$$MH_{(s)} + NiO(OH)_{(s)} \rightarrow M_{(s)} + Ni(OH)_{2(s)}$$

The metallic-like material at the negative electrode is a so-called intermetallic compound, a crystalline metallic structure of various metals with regular order and distinct composition, hence it is not an alloy, because an alloy typically allows all compositions of metal mixtures. Several compounds have been identified for this NiMH battery according to the following composition, AB_5, with A being a mixture of various rare-earth metals (lanthanum, cerium, neodymium, praseodymium) and B a transition metal (cobalt, nickel, manganese, cerium, neodymium, praseodymium,) or aluminum. Besides, AB_2 compounds have been used as well, where A is either vanadium or titanium and B is nickel or zirconium. It is further stressed that modifications of these materials was done with several other elements and added to the compounds as impurities.

4.4.2.4 *Lithium ion batteries*

Typically the lithium ion (Li-ion) battery uses a carbonaceous material (graphite, hard carbons, etc) at the negative electrode and a transition metal oxide, phosphate or silicate at the positive electrode (see Table 4.3).

Table 4.3: Various Positive Electrode Materials Used for Li-ion Batteries.

Material	Composition	Crystal structure
Lithium cobalt oxide, LCO	$LiCoO_2$	Layered, $LiCoO_2$
Lithium nickel manganese cobalt oxide, NCM	$LiNi_aCo_bMn_cO_2$	Layered, $LiCoO_2$
Lithium nickel cobalt aluminum oxide, NCA	$LiNiCoAlO_2$	Layered, $LiCoO_2$
Lithium manganese oxide, LMO	$LiMn_2O_4$	Cubic, spinel
Lithium nickel manganese oxide, LNMO	$LiNi_{0.5}Mn_{1.5}O_4$	Cubic, spinel
Lithium iron phosphate, LFP	$LiFePO_4$	Olivine
Lithium iron silicate, LFS	Li_2FeSiO_4	β-Li_3PO_4

The electrode half reactions during discharge are simplified to the following reactions:

$$LiC_{6(s)} \rightarrow Li^+_{(aq)} + 6C_{(s)} + e^-$$

$$2Li_{0.5}CoO_{2(s)} + e^- + Li^+_{(aq)} \rightarrow 2LiCoO_{2(s)}$$

With the overall reaction:

$$LiC_{6(s)} + 2Li_{0.5}CoO_{2(s)} \rightarrow 6C_{(s)} + 2LiCoO_{2(s)}$$

Besides the standard carbonaceous materials, novel types of Li-ion cells may use titanium oxides, tin compounds or silicon.

4.4.2.5 *Lithium air batteries*

The lithium air (Li-air) battery is not yet commercially available. Still this system is regarded as the holy grail. The material used for this system is lithium metal or similar negative electrodes as used for Li-ion batteries. For the positive electrode, usually carbon mats are used together with typically a transition metal oxide catalyst. The reactions at the negative and positive electrodes are respectively:

$$Li_{(s)} \rightarrow Li^+_{(aq)} + e^-$$

$$O_{2(g)} + 2e^- \rightarrow 2O^-$$

$$O_2(g) + e^- \rightarrow O_2^-$$

$$O_2^- + e^- \rightarrow 2O^-$$

With the overall reaction:

$$2Li_{(s)} + O_{2(g)} \rightarrow Li_2O_{2(s)}$$

It must be stressed that the steps and final overall reaction is still under debate.

4.4.2.6 *Lithium sulfur batteries*

The lithium sulfur (LiS) battery is on the verge of full commercialization. It offers a high capacity and is cheap, but safety might still be an issue. The materials that are used in a Li-S cell are sulfur at the positive electrode and similar materials as mentioned in the section about the Li-air battery for the negative electrode. The half reactions at the negative electrode therefore are similar as mentioned earlier, but the half reactions occurring at the positive electrode are complex via various polysulfides:

$$S_{8(g)} + 16e^- \rightarrow 8S^{2-}$$

$$S_8 \rightarrow Li_2S_8 \rightarrow Li_2S_6 \rightarrow Li_2S_4 \rightarrow Li_2S_3 \rightarrow Li_2S_2 \rightarrow Li_2S$$

These polysulfides tend to dissolve in the liquid electrolyte and thus short the cell. Besides, the sulfur compounds have an extremely low electronic conductivity. For this reason various carbon cages are being constructed to contain the sulfur compounds and to achieve an adequate electronic conductivity for this positive electrode.

4.4.2.7 *(Redox) flow batteries*

Redox flow batteries (RFB), initially developed by NASA in the 70's for its space programm, use two liquid electrolytes containing dissolved metal ions as active species, which are thus pumped across each other along an ion exchange membrane. These electrolytes at the negative and positive electrodes are referred to as an anolyte and a catholyte, respectively, and are an important part of the cell, with respect to materials. In Fig. 4.2 an example of a vanadium redox flow battery is shown, where all chemicals remain in the solutions of either the anolyte or catholyte. Besides the vanadium redox flow batteries, a number of other flow systems exist, of which only a few are commercially feasible at the moment, see Table 4.4. These include hybrid flow batteries, where typically solid species are formed or are present as a precipitate (e.g. zinc bromine, hydrogen bromine, lithium polysulfide, iron-iron), or where solid active components are dispersed in the electrolyte solvent as a kind of viscous ink (e.g. lead-acid, Li-ion systems). This however requires a smart design of the flow battery. A new trend in flow batteries is the use of an all-organic system. However, these are still under development and in a premature phase.

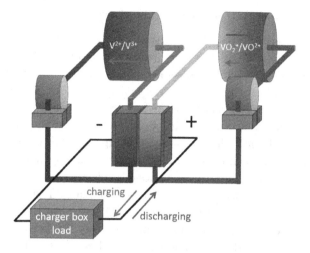

Fig. 4.2.   Vanadium redox flow battery scheme.

Table 4.4:   Material for Various Flow Batteries.[3]

Flow battery	Materials used	Reactions
Redox flow battery		
Vanadium	Vanadium, nafion	$V^{2+} \rightarrow V^{3+} + e^-$
		$VO_2^+ + e^- + H^+ \rightarrow VO^+ + OH^-$
Polysulfide bromide	Sulfur, sodium, bromine	$2Na_2S_2 \rightarrow Na_2S_4 + 2Na^+ + 2e^-$
		$NaBr_3 + 2Na^+ + 2e^- \rightarrow 3NaBr$
Aqueous redox flow	Alkalimetal, water, nafion, iron, iodine[1,2]	$Fe^{2+} \rightarrow Fe^{3+} + e^-$
		$NaI_3 + 2Na^+ + 2e^- \rightarrow 3NaI$
Non-aqueous redox flow	Organic redox couple, organic solvent[3]	Too many to present
Hybrid flow battery		
Lithium polysulfide	Lithium, sulfur	See lithium sulfur batteries
Zinc bromine	Zinc, sodium, bromine	$Zn \rightarrow Zn^{2+} + 2e^-$
		$NaBr_3 + 2Na^+ + 2e^- \rightarrow 3NaBr$
Zinc cerium	Zinc, cerium	$Zn \rightarrow Zn^{2+} + 2e^-$
		$Ce^{4+} + e^- \rightarrow Ce^{3+}$
Iron iron	Iron, graphite	$Fe^{2+} \rightarrow Fe^{3+} + e^-$
		$Fe^{2+} + 2e^- \rightarrow Fe$
Dispersed systems		
Lead-acid	See Lead-acid battery	See Lead-acid battery
Li-ion	See Li-ion battery	See Li-ion battery

[1] Only one possible example is shown.
[2] N,N,N,2,2,6,6-heptamethylpiperidinyloxy-4-ammonium chloride (TEMPTMA) is also used.
[3] This is a new class of materials, but are mainly based or organic materials.

4.5 Battery and Battery Materials Market

The global materials demand for electricity storage systems will be increasing in the next few decades. This obviously has to do with the increasing costumer interest in cellular consumer goods such as smart phones, laptops, etc, as well as (hybrid) electric vehicles. Besides, there is a trend in the market towards renewable energy systems, such as solar, wind, and hydro, that requires electricity storage devices as they are intermittent in nature. Hence, these renewable energy sources either separately or in combination require a storage system to supply continuous, reliable power, e.g. in remote areas.[1] Besides, electricity storage may well be used in parallel systems next to the power plant. The above systems thus require mainly rechargeable (secondary) batteries. Hence, the focus will be on these systems, but the primary battery market will not disappear in the next decade, and thus will be addressed as well with respect to materials demand. In this section the following key questions will be addressed:

- Which are the key end-user applications that use batteries extensively?
- Who are key players deciding on the implementation?
- How will the market for end-users grow in the next decade?
- How does the demand pattern for different battery chemistries vary?
- What are the main materials supporting the coming demand?
- What further chemistries are beyond the horizon?

Depending on their application, batteries are divided into *energy batteries* and *power batteries*, reflecting a high *specific energy* for long run-times and high *specific power* for high-current loads, respectively. Figure 4.3 shows the relationship between specific energy in and specific power for various existing energy storage systems.

4.5.1 *Key end-users and systems*

The battery market for end-users can be split into the following sectors[4]:

- Consumer goods: mobile phones, laptops/notebooks
- Transportation: automotive, locomotive, marine, and aerospace
- Utility: power plants - fossil fuels and renewables
- Industry: power tools, energy savings (*note that compared to Table 4.1, the medical part falls under Industry here, and the industry part has been split up into industry and utilities.*)

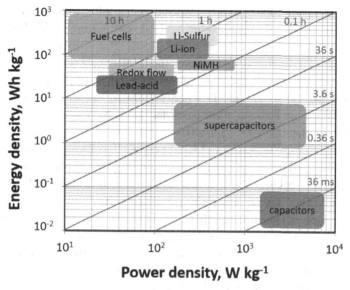

Power density, W kg⁻¹

Fig. 4.3. Specific energy *vs* specific power of rechargeable batteries. Note that the specific energy is the capacity a battery can store and is measured in watt-hours per kilogram (Wh/kg), whereas specific power is the ability to deliver the energy per time, i.e. the power, which as such is measured in watts per kilogram (W/kg).

Table 4.5: Comparison of Various Distributed Energy Storage Technologies for Utility Storage Applications (Selected Characteristics).[5,6]

	Lead-acid	Li-ion	Flow[1]
Cost	☺	☺	☺
Energy density	☹	☺	☹
Capacity	☹	☺	☺
Power density	☺	☺	☹
Installation	☹	☺	☺
Cycle life	☹	☹	☺
Depth of Discharge (DoD)	☹	☺	☺
Environment (recycleability)	☹	☹	☺

[1]Redox flow and hybrids
☹: poor ☺: average ☺: good

Since Li-ion batteries have higher energy densities compared to lead-acid or NiMH batteries, this makes the Li-ion batteries ideal for use in consumer goods, electric vehicles, energy storage for electricity utilities, and industrial applications, but at a certain cost as shown in Table 4.5. Lead acid batteries are subject to improper battery disposal in landfills, which may result in the contamination of groundwater by lead and sulfuric acid. Hence, the large scale introduction of lead acid batteries for utility services is less interesting.[1] The energy density makes these cells unattractive for consumer goods and electric vehicles. Nevertheless, due to the cost, lead-acid batteries may find their way to domestic energy storage,[5] as well as to advanced military electronic systems. Flow batteries, being huge in size, are becoming the preferred chemistry for utility-based storage, e.g. in grid energy storage or in power storage for solar and wind power. This includes load levelling, peak shaving and uninterrupted power supply (UPS) applications. The use of nickel-based rechargeable systems is expected to decline in the next few decades, and they are thus not discussed in detail.

New markets that may further boost battery growth are electric bicycles and scooters, as well as domestic/residential electricity storage.

Primary batteries will be mainly used for consumer goods, such as watches, electronic keys, remote controls, toys, flashlights, beacons, and military devices in combat, etc.[4]

4.5.2 *Key decision makers*

The four sectors addressed in the previous section (consumer goods, transportation, utility, and industry), particularly the transportation and utility sectors, are strongly dependent on political decisions. Legislation and regulation promoting the electrification of vehicles is expected to create a significant boost to (hybrid) electric vehicles sales, strongly catalyzed by the rapid development of advanced high energy density Li-ion batteries. Since renewable technology is becoming more and more mature, with the gap between conventional and renewable fuels closing, thus renewables are becoming competitive with fossil fuel power plants. Various governments encourage the use of renewable power via specific subsidies or regulations.[5,6]

The legislation, regulation promoting, and subsidies are important activities for the growth of the battery market as we will see in the following sections.

4.5.3 *Market growth by sector*

Market growth by sector — consumer goods, transportation, utilities, and industry — are discussed in several reports, for example: Fredonia,[7] Frost&Sullivan,[4] MIMB,[8] Markets and Markets,[9–11] Grand View Research.[12] It should be stressed that the focus here is not on cost, but on sales figures and market volume, so as to understand the future materials demand, rather than cost. Obviously, cost is a major driver for specific materials and is therefore the underlying reason for the selection of specific materials, and deciding the sales and volumes. Hence, market growth in cost goes hand in hand with market share in volume, and thus with materials demand. The percentage distribution of the battery market share by

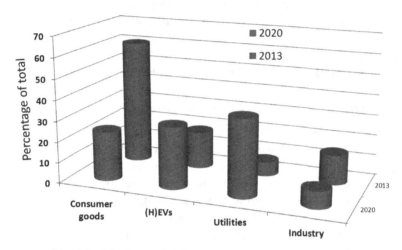

Fig. 4.4. Distribution of the battery market sales by sector.[4]

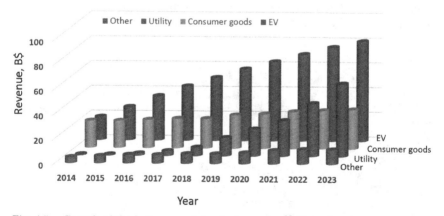

Fig. 4.5. Growth of the battery market sales by sector.[13] Note that here cost is used, but these can be just as easily seen as volume or power installed.

sector is shown in Fig. 4.4, whereas in Fig. 4.5 the annual growth is shown
by sector.[9,10]

4.5.4 *Market growth by battery chemistry*

The various applications as identified by the four sectors require different
battery specifications. Here we will distinguish between flow batteries and
other rechargeable batteries, such as Li-ion, lead-acid, and NiMH batteries.
Despite the fact that the global primary battery market is forecast to grow
at a Compound Annual Growth Rate (CARG) of 4.06% during the period
2016–2020,[14] the major battery volumes will come from the secondary battery market. This growth of the battery market will thus have an immediate
consequence on the materials demand as will be shown in section 4.5.5.

4.5.4.1 *Lead-acid, NiMH, and Li-ion batteries*

The year 2009 has been taken as a base for the distribution of the battery
market (excluding flow batteries)[4] — see Fig. 4.6. This year has been chosen
because it is foreseen that NiCad batteries and to a lesser extent NiMH
batteries will decline completely in the next decade, otherwise the market
for these battery types would not show up at all.

The growth of the primary battery market is 4.06% CAGR,[14] whereas
the growth of rechargeable batteries is about 10% CARG. The actual data
per battery chemistry are shown in Table 4.6.

Lead-acid batteries remain of interest because they are robust and find
their place in starter batteries for Start, Lighting and Ignition (SLI) and
Uninterrupted Power Supply (UPS), and deep-cycle batteries for wheeled
mobility (golf cars, wheelchairs, scissor lifts, etc.).

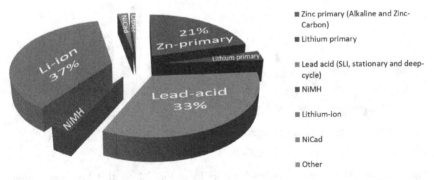

Fig. 4.6.   Battery market distribution in 2009 after Frost&Sullivan.[1]

Table 4.6:   Annual Market Growth of Various Battery Chemistries, Measured in CAGR.

	CAGR	Ref.
Primary batteries	4.06	14
NiCad	−4%	15
NiMH	−4%	15
Lead-acid	+4%	15
Li-ion	16% (3% for power tools)	15

Despite the forecasted decline in the NiCad and NiMH in total energy (Wh), on a pack level there still will be batteries available, but in the total volume of materials, this will be of minor importance. On the other hand, new systems are appearing on the horizon, such as lithium-sulfur, sodium-aqueous, magnesium-ion, and organic batteries, etc. A revival of the iron battery is also expected. These systems are still under development and will be discussed in section 4.5.6 in more detail.

4.5.4.2 *Flow batteries*

Flow batteries are often regarded as a semi-battery, a system that is a mixture of a fuel cell and a battery. However, since the waste materials after use are still stored in a container, and not disposed of to the environment, they will be considered as a battery system here.

From around 2006, Redox flow batteries (RFB) were used mainly for large scale electricity storage, e.g. for stationary applications, such as renewable energy utilization, and at the residential, industrial, and (micro-) grid level. The volume is expected to grow to a \$4bn market by 2027.[16] Although, RFBs do not deliver the same power as a Li-ion battery, they compete in terms of cycle life, safety, 25-year service life without maintenance, and thus reliability for the above mentioned application.[5] In Fig. 4.7, the expected growth of the RFB is shown. Obviously, the vanadium RFB are the ones that are most mature and have significant potential in the near future. Nevertheless, in terms of cost and environmental issues, novel systems based on less polluting materials are required. Many demonstration projects have been initiated, and several showed that RFBs are mature enough for further exploitation in China, Japan, Canada and the U.S. One example is the 15MW (60MWh) Minami Hayakita Substation in Japan for integrating photovoltaics into the grid.[17] It has further been forecasted that RFBs will compete with Li-ion and sodium-sulfur systems, which are now the two leading chemistries in this market sector.[16] Other chemistries, such

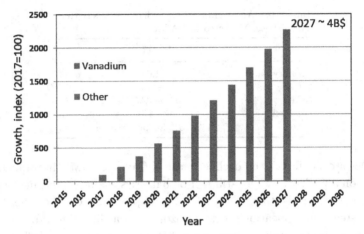

Fig. 4.7.　Market forecast for redox flow batteries.[16] Other chemistries include zinc/bromine and hydrogen/bromine.

as zinc/bromine and hydrogen/bromine, may also have the potential to gain a significant market share as shown in Fig. 4.7. Besides these mature and commercial RFBs, a number of alternatives exist as nicely described by Su *et al.*,[18] and briefly discussed in section 4.4.2.7. Several studies have appeared on the growth of flow batteries for the stationary battery storage market.[5,16,17,19,20] It was reported that approximately 72% of the stationary battery market share originated from the redox battery segment.[20] This figure will further increase, giving rise to a global flow battery market at a CAGR of 3.38% during the period 2017–2021.[20] This however, is in contrast to another study where it is believed that Li-ion batteries are going to dominate the market, i.e. 80% of the global installations by 2025.[21] Nevertheless, an increase in the use of flow batteries is expected for stationary energy storage from 145 MW in 2016 to 5,770 MW in 2025, i.e. an increase by a factor of 40. In contrast, the use of Li-ion batteries will grow only from 10,600 MW in 2016, to 19,280 MW in 2025, hence by a factor of 20.[13]

4.5.5　*Future material demand pattern by battery*

The variety of existing systems requires different battery materials, each offering their own characteristics. The demand is for low cost systems with high performance. Hence, materials and battery producers, as well as research institutions, are working to create batteries with high power density, long life cycle, low cost, and high performance that are environmentally friendly to satisfy the needs of the battery and end-user industry.[9-11] Hence, price and availability have a strong influence on battery materials use, and thus demand. The global battery materials market grew at a CAGR of 13% from 2013 to 2018 (from U$5.1 billion to U$11.3 billion),[9-11] and according to Transparency Market Research,[1] the market for battery materials will further increase at a 13.6% CAGR during the period between 2017 and 2025 (US$13.70 billion). An even more optimistic prediction was made by Fredonia,[7] where a forecast was made of 8.3% increase per year to $46.8 billion in 2019.[7]

The battery market will be dominated by secondary systems. Hence, the materials market will be dominated likewise. The relatively small primary systems market will mainly use manganese oxides, zinc, sodium or potassium hydroxide, and graphites (carbons).

The remaining usage of lead-acid batteries in the automotive sector is responsible for the growth of the materials for them. These will mainly be lead, lead oxide, graphite, arsine and other impurities, and polymers. Other rechargeable batteries, based on nickel chemistry currently on the market, will almost completely vanish in the next decade, and thus will the need for these materials, including the rare earth elements.[7] This means that for the coming decade, the Li-ion battery and its materials will dominate the materials demand. However, with increasing stationary electricity storage, the RFBs are becoming important. Since, these systems will cover enormous storage capacities, they therefore require a significant amount of materials, particularly taking the increase in renewable energy into account. The materials that are of interest are vanadium for the vanadium RFB, and zinc and bromine for the zinc-bromine RFB. Both systems require membranes which are typically of polymeric origin, and stainless steel tanks.

4.5.5.1 *Li-ion rechargeable batteries*

Li-ion batteries are more efficient as compared to others today, having a high performance with outstanding properties like high energy and power density, and a long discharge cycle. The Li-ion battery is on the market as a pouch cell — typically in mobile phones — and as a cylindrical cell — the so-called 18650 cells. These cylindrical cells are becoming more and more standard and are being used, for example, in the Tesla cars. Figure 4.8 shows the structure of such a cylindrical cell. It is constructed by coating a metal current collector with a wound laminate of positive and negative electrode materials. The positive and negative electrode materials are separated by

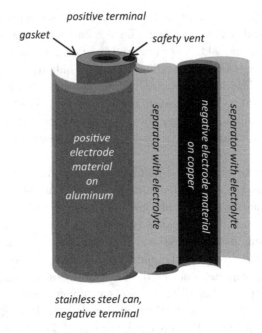

Fig. 4.8.   Cylindrical (18650) Li-ion battery showing the various coatings and foils.

a porous plastic foil soaked with the liquid electrolyte to prevent internal shorts. The electrolyte further contains a specific lithium salt.

The materials to be used in Li-ion batteries are:

- Negative electrode: lithium, carbon, graphite, silicon, lithium titanium oxide, etc.
- Negative current collector: copper
- Positive electrode: lithium cobalt oxide (LCO), lithium nickel manganese cobalt oxide (NMC), lithium nickel cobalt aluminium oxide (NCA), lithium ferrophosphate (LFP), etc.
- Positive current collector: aluminium
- Electrolyte: lithium hexafluorophosphate, ethylene carbonate, propylene carbonate, etc.
- Separator: polyethylene, polypropylene,
- Binder: PVdF, carboxy methyl cellulose (CMC)
- Casing and binder materials: stainless steel and polymers/plastics.

In Figs. 4.9 and 4.10, the forecasted amount of positive and negative electrode material are presented, respectively.[22–24] These forecasts show

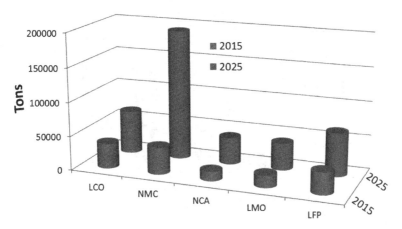

Fig. 4.9.   Forecasted market demand for selected positive electrode materials.[22-24]

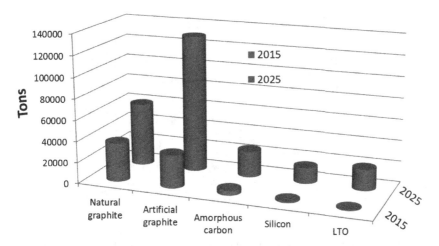

Fig. 4.10.   Forecasted market demand for selected negative electrode materials.[22-24]

that NMC and LFP and artificial graphite will be the major components for this generation of Li-ion batteries. With respect to the electrolyte (Fig. 4.11), separator (Fig. 4.12) and binder material, the demand goes linearly with the demand for the total electrode materials.[22-24] With respect to the NMC and LFP batteries, the amount of components percentage-wise per cell are presented in Fig. 4.13.[22-24] The projected electrolyte salt, binder material, and separator material are lithium hexafluorophosphate, PVdF, and polyethylene/polypropylene, respectively. This means

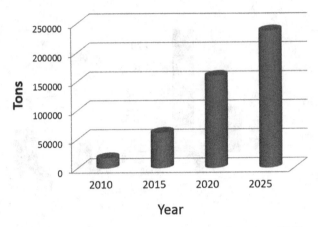

Fig. 4.11. Forecasted market demand for electrolytes.[22-24]

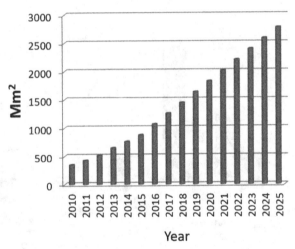

Fig. 4.12. Forecasted market demand for separator materials in surface area (Mm^2).[22-24]

that beside metal-containing and carbonaceous materials, a significant percentage of fluorine- and phosphorus-containing components are present in the battery.

4.6 Future Systems

The future systems that are of interest in terms of high power density, long life cycle, low cost, and high performance, and that are environmentally

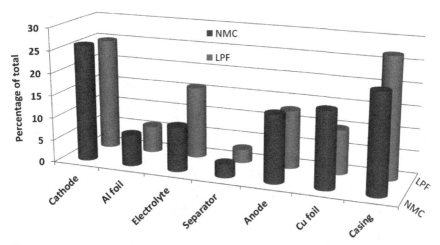

Fig. 4.13. Distribution of the various battery components in percentage for a cell made with LPF and NMC.[25]

Table 4.7: Future Rechargeable Batteries.

	Voltage V	Energy density Wh/kg	Ref.	Remark
Lithium-sulfur	2.1	500	26	Operates similarly to Li-ion
Lithium-air	1.7–3.2	13000	26	Theoretical capacity
Lithium-metal	3.6	300	26	
Solid state lithium	3.6	300	26	
Lithium aqueous	1.2–2.4	75	27	Often the positive electrode material is of the same origin as that in Li-ion batteries
Sodium aqueous	1.2–2.4	50–60	27	Often the positive electrode material is of the same origin as that in Li-ion batteries
Magnesium-ion	1.8–3.0		28, 29	Often the positive electrode material is of the same origin as that in Li-ion batteries, including sulfides

friendly to comply with the battery and end-user industry are listed in Table 4.7.

The materials that are being used in the above potentially rechargeable batteries comprise lithium, sodium, sulfur, magnesium, manganese, and eventually other transition metals such as nickel, iron and titanium.

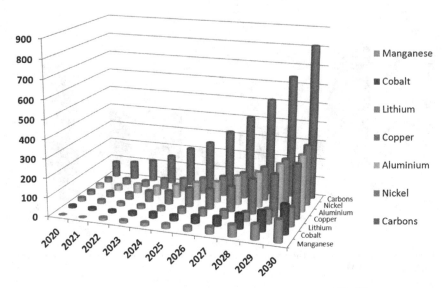

Fig. 4.14.　Forecasted market demand for metals for EVs.[30]

4.7　Discussions and Conclusions in Brief

The goal for novel types of rechargeable batteries is to find cost-effective, environmentally friendly materials delivering high performance in terms of energy density and power during a long cycle and service life. Metals will still remain the predominant type of battery material, irrespective of the choice of battery system, see Fig. 4.14 as an example for components for electric vehicles, which will be more than 50% of the total Li-ion battery demand. In summary, the materials that are expected to be needed in bulk in future batteries are:

- Negative electrode: lithium, cobalt, lead, carbon, graphite, magnesium, sodium, vanadium, antimony, etc.
- Positive electrode: (lithium/sodium) manganese oxide, lead dioxide, lithium cobalt/nickel/manganese/aluminum oxide, lithium ferrophosphate, sulfur, vanadium, etc.
- Electrolyte: ammonium chloride, sodium/potassium hydroxide, sulfuric acid, lithium hexafluorophosphate, organic solvents
- Separator: polyethylene, polypropylene,
- Casing and binder materials: stainless steel and polymers/plastics.

In order to improve the performance of the batteries, certain additives are projected, such as rare earth materials and e.g. niobium. From the

Table 4.8: Evaluation of critical elements in selected future battery materials.

	Supply risk[1]	Geopolitics[2]	Competition with food industry[3]	Ethical mining[4]
Antimony	⊗	⊗	o	o
Cobalt	⊗	⊗	o	⊗
Lithium	☺	⊗	o	o
Magnesium	⊗	☺	o	o
Niobium	⊗	⊗	o	o
Vanadium	⊗	☺	o	o
Natural graphite	⊗	o	o	o
Silicon	☺	☺	o	o
Phosphorus	⊗	o	⊗	o
Fluor	☺	☺	o	o

[1]The supply risk is measured according to the 3rd list of critical raw materials for the EU of 2017.[31]
[2]Geoplitics issues are evaluated from the same document as "Supply risks".
[3]See: .Phosphorus, Food and Our Future.[32]
[4]Ethical mining is a particular issue with respect to cobalt mining in the Democratic republic of Congo (DRC) according to many reports, particularly those from UNICEF and Amnesty International — references are not further specified.
⊗: very critical; ☺: medium critical; o: not analysed

abovementioned bulk materials and trace elements that are important for future rechargeable batteries, a number are critical in terms of supply risk — including scarcity and geopolitics, competition with the food industry, ethical mining, and carbon footprint. In Table 4.8 these materials, or actually their component elements, are thus summarised with respect to these critical aspects. In this evaluation, geopolitics is taken as an individual aspect, whereas the materials scarcity is included in the supply risk. The evaluation regarding carbon footprint has not been presented here, as it requires much more information and/or explanation to justify the qualification.

With respect to the critical elements, particularly for the ones that concern scarcity and supply risk, recycling is of utmost relevance, and needs to be regarded as an industrial sector of paramount importance.

References

1. Transparency Market Research. (2018). *Battery materials market*. Rep Id.: TMRGL16454, Albany, New York, USA: Transparency Market Research.
2. Sion Power. (n.d.). Sion Power. Retrieved from https://www.sionpower.com

3. Darling, R.M, Gallagher, K.G., Kowalski, J.A., Ha, S., and Brushett, F.R. (2014). Pathways to low-cost electrochemical energy storage: a comparison of aqueous and nonaqueous flow batteries, *Energy and Environmental Science*, 7(11), 3459–3477
4. Hofmann, S. and Höpner, A. (2015, May 5). Germany's flat battery business. *Handelsblatt Global*. Retrieved from https://global.handelsblatt.com/compa nies/germanys-flat-battery-business-209070
5. Transparency Market Research. (2017). *Vanadium Redox Flow Batteries (VRFB) market — global industry analysis, size, share, growth, trends and forecast 2017–2025*. Rep Id.: TMRGL28379, Albany, New York, USA: Transparency Market Research.
6. Energy Storage Systems. (2016). *ESS technical white paper, all-iron flow battery — overview*. Wilsonville, Oregon, USA: Energy Storage Systems. Retrieved from http://www.essinc.com/wp-content/uploads/2016/08/ESS-Technical-White-Paper-August-2016.pdf
7. Freedonia. (2015). *World battery materials*. Cleveland, Ohio, USA: Freedonia. Retrieved from https://www.freedoniagroup.com/freedonia-focus/world-bat tery-materials-FW45027.htm
8. Mordor Intelligence. (2017). *Battery market — growth, trends and forecast (2017–2022)*. Hyderabad, Telangana, India: Mordor Intelligence.
9. Markets and Markets. (2013). *Battery materials market — by types (cathode, anode, electrolyte, separator, binders, packaging material), applications and geography — trends and forecasts to 2018*. Hadapsar, Pune, India: Markets and Markets. Retrieved from https://www.marketsandmarkets.com/Market-Reports/battery-raw-materials-market-866.html.
10. Markets and Markets. (2016). *Battery energy storage system market by battery type (lithium-ion, advanced lead acid, flow batteries, & sodium sulfur), connection type (on-grid and off-grid), ownership, revenue source, application, and geography - global forecast to 2022*. Hadapsar, Pune, India: Markets and Markets.
11. Markets and Markets. (2013). *Battery materials market (lithium battery) applications & geography — 2018*. Hadapsar, Pune, India: Markets and Markets. Retrieved from https://www.marketsandmarkets.com/Market-Reports /battery-energy-storage-system-market-112809494.html.
12. Grand View Research (2017), Market research report, www.grandviewresear ch.com/industry-analysis.
13. Navigant Research. (2014). *Advanced batteries revenue and energy capacity by application, world markets: 2014–2023*. Boulder, Colorado, USA: Navigant Research. Retrieved from https://www.navigantresearch.com/research/
14. ReportsnReports. (2016). *Global primary battery market 2016–2020*. Hadapsar, Pune, India: ReportsnReports. Retrieved from http://www.reportsnrep orts.com/reports/738711-global-primary-battery-market-2016-2020.html
15. Pillot, C. (2013). *BATTERIES 2013 Conference, October 14–16, 2013, Nice, France*. [PowerPoint presentation]. Retrieved from http://www.avicenne.co m/pdf/Theworldwidebatterymarket2012-2025CPillotBATTERIES2013Nice October2013.pdf.

16. Grande, L. (2017). Redox flow batteries 2017–2027: Markets, trends, applications — large, safe, sustainable batteries for residential, C&I, and utility markets. Cambridge, UK: IDTechEx. Retrieved from https://www.idtechex .com/research/reports/redox-flow-batteries-2017-2027-markets-trends-appli cations-000531.asp, and https://www.idtechex.com/research/reports/redox -flow-batteries-2017-2027-markets-trends-applications-00010923.asp

17. U.S. Department of Energy. (n.d.). Global energy storage database. Retrieved from www.energystorageexchange.org.

18. Su, L., Kowalski, J.A, Carroll, K.J., and Brushett, F.R. (2015). Recent developments and trends in redox flow batteries. In Z. Zhang and S.S. Zhang (Eds.), *Rechargeable Batteries: Materials, Technologies and New Trends* (pp. 673–712). Switzerland: Springer International Publishing.

19. Global Market Insights. (2017). *Stationary battery storage market size — growth forecast report 2030.* Selbyville, Delaware, USA: Global Market Insights.

20. Technavio. (2017). *Redox Battery Dominates the Global Flow Battery Market.* London, UK: Technavio. Retrieved from https://www.businesswire.com/new s/home/20171222005250/en.

21. IHS Markit. (2016). Markit projects. Retrieved from www.ihsmarkit.com.

22. Pillot, C. (2016). *The worldwide rechargeable battery market 2015–2025.* Paris, France: Avicenne Energy.

23. Pillot, C. (2015). *Battery market development for consumer electronics, automotive, and industrial: Materials requirements and trends.* [PowerPoint presentation].

24. Lebedeva, N., Di Persio, F., and Boon-Brett, L. (2016). *Lithium ion battery value chain and related opportunities for Europe.* Petten, the Netherlands: European Commission.

25. Golubkov A.W, Fuchs, D, Wagner, J., Wiltsche, H., Stangl, C., Fauler, G., Voitic, G., Thaler, A., and Hacker, V. (2014). Thermal-runaway experiments on consumer Li-ion batteries with metal-oxide and olivin-type cathodes. *RSC Advances*, 4(7), 3633–3642.

26. Battery University. (2016). Summary table of future batteries. Retrieved from http://batteryuniversity.com/learn/article/bu_218_summary_table_of_f uture_batteries

27. Posada, J.O.G, Rennie, A.J.R., Perez Villar, S, Martins, V.L., Barnes, A., Glover, C.F., Worsley, D.A., and Hall, P.J. (2017). Aqueous batteries as grid scale energy storage solutions. *Renewable and Sustainable Energy Reviews*, 68(2), 1174–1182.

28. NuLi, Y. Yang, J., Wang, J., and Li, Y. (2009). Electrochemical intercalation of Mg^{2+} in magnesium manganese silicate and its application as high-energy rechargeable magnesium battery cathode. *Journal of Physical Chemistry C*, 113(28), 12594–12597.

29. Mohtadi, R. and Fuminori, M. (2014). Magnesium batteries: Current state of the art, issues and future perspectives. *Beilstein Journal of Nanotechnology*, 5, 1291–1311.

30. Burton, M. and Van Der Walt, E., (2017, August 3). Electric-car revolution shakes up the biggest metals markets. *Bloomberg*. Retrieved from www.bloo mberg.com/news/articles/2017-08-02/electric-car-revolution-is-shaking-up-the-biggest-metals-markets.

31. European Commission. (2017). Critical raw materials. Retrieved from http://ec.europa.eu/growth/sectors/raw-materials/specific-interest/critical_en.

32. Wyant, K.A, Corman, J.R., and Elser, J.J. (2013). *Phosphorus, Food and Our Future*. New York, USA: Oxford University Press.

Part II

Defining Critical Materials

Chapter 5

A Historical Perspective of Critical Materials, 1939 to 2006

David Peck

Faculty Architecture and Built Environment, Delft University of Technology, Julianalaan 134, 2628 BL, Delft, The Netherlands

Coventry University, Faculty of Engineering, Environment and Computing, 11 Gulson Rd, Coventry CV1 2JH, UK

Politecnico di Milano, MIP, Graduate School of Business, 26/A, Via Raffaele Lambruschini, 4C, 20156 Milano, Italy

The tensions around limits to growth versus tech will fix it are, today, as prominent as ever in the debates around critical materials. Changes in the demand and supply of materials has regularly led to periods of material supply problems. This chapter provides an overview of the development of critical materials from the mid-20th into the early 21st century. The overview begins with, in the U.S., the development of critical materials policy in World War II and the Cold War years, the oil crisis of the 1970s and the subsequent evolution into the early years of the 21st century. Critical materials thinking has been defined through war, the cold war and then concerns over energy availability and environmental impacts. This chapter shows how the historical military-energy framework for assessing critical materials has evolved into critical materials approaches to help address the challenges of energy, materials and the environment in the 21st century.

5.1 Introduction

Over the past 200 years, the period since the industrial revolution, there has been a tension between people who prioritize the power of an economy and people who put the care of the natural environment first. The economists' position is based upon near term models of economic growth, with productivity driven by use of natural resources. The environmentalists' position is based upon the stewardship of all living things on earth, across very long term time periods, and implies that human activity will consistently erode finite natural capital to the point of exhaustion (Tahvonen, 2000, Martin & Kemper, 2012).

Whilst the idea of economists versus environmentalists can be seen as too simplistic, in general such adversarial positions are often reasonably valid. The environmentalist has long maintained that as finite resources are depleted at an accelerated rate, with the corresponding environmental degradation, there will come a point of social, economic and environmental collapse. The economist will argue that every date proposed for the forecast collapse is never reached and that constant new technological innovations mean that the exhaustion of resources will never happen and any environmental damage can be mitigated and repaired.

These two positions can be summarized as "limits to growth" versus "tech (technology) will fix it". Whilst natural resources have not physically 'run out' there has, over the past 100 years, been periods of problems with material availability, which have been developed into definitions of critical materials. The actors involved have not been limited to economists and environmentalists and the motivations for developing critical materials thinking vary from situation to situation. One consistent factor in the story of critical materials has been the development of new materials together with new technologies.

The development of new materials is so significant that historians have taken the periods of material change to define ages in human history. This is shown in Fig. 5.1 below, which outlines in general terms the 'ages' of materials. The timeline in Fig. 5.1 is not equally spaced because the development and consumption of materials has increased rapidly over the 20th and 21st centuries when compared to past centuries.

From the mid-19th century onwards the field of materials, as a science, has developed rapidly. This evolution was driven by the demands of the industrial age. The development of steam power required new materials, whilst at the same time the availability of plentiful energy, through coal fired steam power, facilitated new technologies, which assisted the discovery, recovery and production of new materials. This industrial demand and scientific development have gone hand in hand, facilitating the introduction of new materials and their transformative effects into society.

The deepening understanding of the periodic table of elements and the corresponding development of new materials, coupled with the evolution

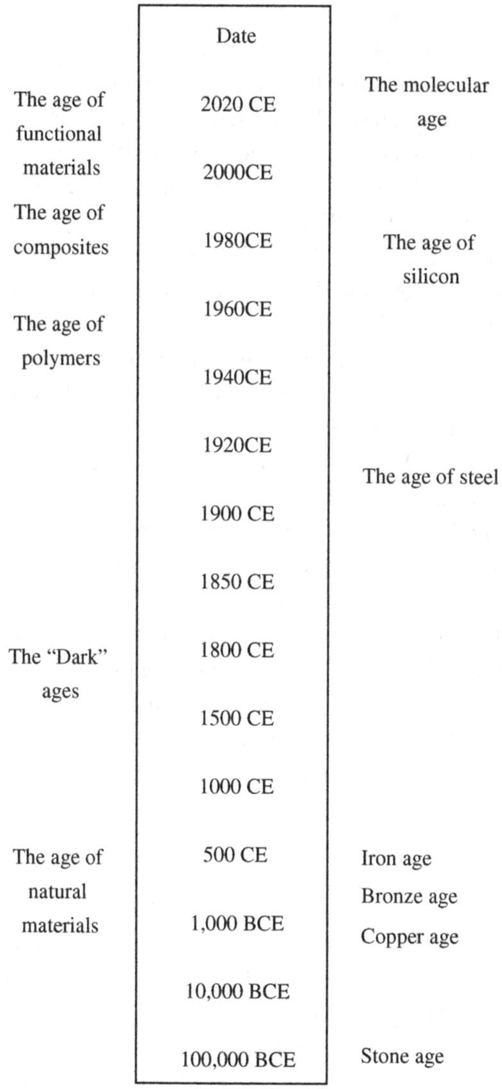

	Date	
The age of functional materials	2020 CE	The molecular age
	2000CE	
The age of composites	1980CE	The age of silicon
The age of polymers	1960CE	
	1940CE	
	1920CE	The age of steel
	1900 CE	
	1850 CE	
The "Dark" ages	1800 CE	
	1500 CE	
	1000 CE	
The age of natural materials	500 CE	Iron age
		Bronze age
	1,000 BCE	Copper age
	10,000 BCE	
	100,000 BCE	Stone age

Fig. 5.1.   A materials time line. Adapted from Ashby 2016.

of new technologies, has also driven changes in the demand and supply of materials, especially metals.

Changes in the demand and supply of materials have regularly led to periods of material scarcity, or shortage (Tilton, 2003, Johnson, *et al.*,

2007, Ashby, 2013, Ashby *et al.*, 2016). Significant problems in the supply of materials have often arisen during wartime. Conversely, risks concerning the control of material supply have often led to increases in geo-political tensions, and at worst, conflict breaking out. Wars have been won, or lost, due to control, or loss of control, of material supplies. An example of material supplies playing a significant role in the outcome of a war can be seen in World War One (WWI). By 1918, even though the Imperial German Army was still a formidable fighting force in the field, a collapse on the home front led to a rapid German defeat. This political and societal collapse was primarily the result of the severe resource restrictions within Germany. Conversely the Allies were able to overcome unrestricted German U boat action and provide for not only the rapidly growing Allied forces, but the home front too (Eckes, 1979, Stevenson, 2012).

This experience in WWI led to nations in the 1930s feeling they were threatened by lack of direct control of their own resources. These nations included Germany, Italy and Japan and were termed the 'have not' nations. Against them were the 'have' nations such as Britain and the U.S. (Eckes, 1979).

Governments have often used criticality assessments to determine the military material stockpiles in response to the risks to material supply. Such an approach can be seen in the late 1930s, when Britain began to hold stock of raw materials for the expected conflict with Germany. The British termed this activity as 'purchasing war reserves of materials' (Postan, 1952). The terms used often change but from the mid-20th and into the 21st century, material scarcity has evolved into the current, widely used, term, critical materials.

This chapter provides an overview of the development of critical materials from the mid-20th into the early 21st century. The overview begins with late 1930s uses of the term 'critical' in the U.S., the development of critical materials activity in the Cold War years, the oil — energy crisis of the 1970s and the subsequent evolution into the late 1990s and into the early years of the 21st century. This chapter will show how the background of critical materials thinking has been defined through war, the cold war and then concerns over energy availability and more recently, environmental impacts. At the same time, rapidly increasing economic growth, fuelled by technological developments, coupled with an increasing global population, has led to ever increasing material demands and corresponding losses due to waste. These demands and wastes have raised questions over risks to economic growth. Most importantly, the need for a complex range of advanced, complex, technology materials to provide solutions to climate change drive the need for critical materials action.

The tensions around limits to growth versus tech will fix it are, today, as prominent as ever in the debates around critical materials. This chapter shows how the historical military-energy framework for assessing critical materials has evolved into critical materials approaches to help address the challenges of energy, materials and the environment in the 21st century.

5.2 1939 to 1945: the Second World War and Critical Materials

This period can be viewed as one in which concerns over 'limits to *military* growth' took precedence, and this gave rise to the first uses of the term 'critical materials'. One of the first uses of the term 'critical materials' was by the U.S. government in the late 1930's. With a Second World War (WWII) looking increasingly likely, the U.S. government enacted the *"Strategic and Critical Materials Stock Piling Act of 1939"*. This act provided funding to purchase and stockpile strategic and critical materials deemed essential for military production. The list of critical materials was determined by the degree to which the U.S. was import dependent and the risk of a material shortage occurring (Eckes, 1979) (National Academy of Sciences, 2008). Other countries developed similar stockpile strategies in this period, notably Britain, but they did not use the term 'critical materials'. The approaches taken can be seen as a form of 'limits to *military* growth' as all activity was primarily aimed at winning the war.

The material needs were assessed from the material demand arising from the planned requirements of equipment, such as aircraft, ships or vehicles, etc. Also assessed was the potential of domestic material production and the substitutability of a material. The methodology was built upon the experiences of the Allies in World War One (1914–1918), in particular in 1918, when the ability to supply the armies in the field and societies at home became a significant deciding factor in the outcome of the war. In Table 5.1 it can be seen that widely used materials were, in terms of elements, less complex in WWII compared to today.

Table 5.1: The Increasing Diversity of Elements Used in Materials over the Period from WWII to 2016. Table Adapted from Ashby 2016.

	Typical elements in the materials from WWII to 2016	
Materials	WWII	2016
Iron based alloys	Fe, C	Al, Co, Cr, Fe, Mn, Mo, Nb, Ni, Si, Ta, Ti, V, W
Aluminum Alloys	Al, Cu, Si	Al, Be, Ce, Cr, Cu, Fe, Li, Mg, Mn, Si, I, V, Zn, Zr
Nickel Alloys	Ni, Cr	Al, B, Be, C, Co, Cr, Cu, Fe, Mo, Ni, Si, Ta, Ti, W, Zr
Copper alloys	Cu, Sn, Zn	Al, Be, Cd, Co, Cu, Fe, Mn, Nb, R Pb, Si, Sn, Zn
Magnetic materials	Fe, Ni, Si	Al, B, Co, Cr, Cu, Dy, Fe, Nd, Ni, Pt, Si, Sm, V, W

Note: Data from the composition fields of records in the CES EduPack '14 Level 3 database, Granta Design, 2014, (Ashby, *et al.*, 2016).

5.3 Cold War Material Stockpiles, the 1950s and 1960s

Concerns over 'limits to *military* growth' continued in the global materials chaos of the immediate post WWII period, which meant the U.S. continued their wartime critical materials stockpiling policy. By the late 1940s geopolitical tensions began to dominate, with western governments' concerns over the expansion of communism leading to the Cold War. This resulted in the U.S. intensifying their material stockpile actions. The U.S. planning frame was based on a scenario of a 3 year industrial/military mobilization period followed by a 5 year conventional (non-nuclear) war, a scenario based on the experience of WWII.

In 1950 the Cold War turned hot with the outbreak of the Korean War. The U.S. stockpiling budget increased to \$2.9 billion in just 6 months. By end of 1952 the material stockpile value was \$4.02 billion and by 1956 it was \$10.9 billion (National Academy of Sciences, 2008).

During this period the U.S. set up the President's Material Policy Commission, which was tasked to assess if global resources could meet future U.S. demand. The subsequent report, called the 'Paley report', predicted significant shortages based on estimations of future resource use, essentially concerns over 'limits to *military* growth'. The report proposed a range of technology innovations, exploiting domestic reserves and continued stockpiling to overcome material constraint problems, effectively proposing a 'tech will fix it' solution (Paley, 1952). Figure 5.2 shows a figure from the Paley report demonstrating the increase of material quantities over a seven-year period. The P-80 aircraft shown was the

NEW WEAPONS USE MORE MATERIALS

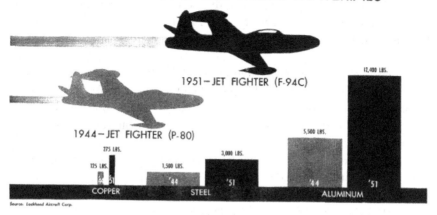

Fig. 5.2.   The increase in quantity of materials over a seven-year period (Paley, 1952).

U.S.'s first jet fighter and the F-94C was a later derivative of that same aircraft.

By the mid-1950s the Soviet Union had developed their own nuclear arsenal, including the much more powerful hydrogen bomb. This led military planners to consider material requirements in the event of a nuclear war. They decided that this scenario would mean the war would last only three years and therefore in 1958 it was decided to dispose of a portion of the critical materials stock. The reason was that a nuclear war was assumed to need less material as the fighting would not last as long. In the early 1960s the critical materials stockpile was further reduced by nearly $3.4 billion, from a figure of $7.7 billion (National Academy of Sciences, 2008).

As the global economy grew through the 1960s, the material demand increased accordingly, which in turn placed a strain on material supply. As a result, U.S. industry, which supplied the military, was experiencing material supply problems, so the government released critical material stocks into the U.S. economy, to help overcome material shortages.

5.4   Critical Materials, the Energy Crisis and Rising Environmental Concerns through the 1970's

In the 1970s, the U.S. approach to critical materials was further developed and other limits to growth factors came into play. Conflict in the Middle East led to a global oil crisis in 1973, with the U.S. and the Netherlands both experiencing a total Middle Eastern oil embargo. Through the 1970s the prices of materials continued to rise due to continued economic growth in Japan, North America, and Europe. From 1947 to 1971, raw materials prices had increased 21%. From 1971 to 1973 prices increased by 46% (Eckes, 1979).

In parallel with the energy and materials challenges, Donella H. Meadows and colleagues, based in MIT, published their 1972 book, *The Limits to Growth*, where they forecast dramatic population collapse due to exhaustion of resources resulting from rising demand from a growing global population (Meadows *et al.*, 1972).

In 1976 the U.S. re-introduced the requirement for civilian industrial material needs to be considered alongside military production needs, when considering the materials stockpile (National Academy of Sciences, 2008).

In 1977 the U.S. Congress published the proceeding of a conference entitled *Engineering implications of chronic materials scarcity: . . . Engineering Foundation Research Conference on National Materials Policy*. This was part of a conference series on the topic (Huddle and Promisel, 1977). This publication demonstrates the more complex and comprehensive approach the U.S. was developing towards critical materials. This approach laid the foundations of the critical materials approaches in the 21st century.

An example of this approach can be seen in Fig. 5.3, where material substitution analysis is conducted. This highlights areas of policymaker focus.

What was not a focus for the critical materials approach in the 1970s were materials used in low carbon renewable energies, materials used in digital technologies or the growing materials demands of emerging economies such as China and India. These considerations would come to the fore in the late 20th and early 21st century.

5.5 Critical Materials and Energy Nexus in the 1970s

In parallel with the policy focus concerning wider engineering applications for critical materials, the USA developed policies to ensure that energy independence could be achieved over a 15 year period, up to 1990. The assessment of the non-fuel minerals needed for the planned transition was led by the U.S. Geological survey (USGS). The USGS developed the Minerals for Energy Production (MEP) program (Albers *et al.*, 1976).

The MEP program aimed to determine first the materials and quantities needed to achieve full energy independence. The program had a focus

A. DESIGN REQUIREMENTS

Customer Acceptance Esthetics

Personal Bias

Market Acceptability Performance

Criteria Materials Performance

Mechanical Properties Chemical

Properties Physical Properties

Fabricability

Machineability

Toxicity

Ease of Joining Corrosion, Oxidation,

and Fire Resistance Compliance with

Specifications and Code Protection

Against Misuse Vandalism Protection

Reuse/Recyclability/Disposal

Compliance with Specifications and

Codes Reliability and Maintainability

B. ECONOMIC CONSIDERATIONS

Material Cost

Cost/Price Stability

Transportation Cost

Marketing Costs (to use substitute)

Production Costs

Investment Required to Incorporate Life

    Costs Tariffs and Taxes

C. PRODUCTION CONSIDERATIONS

Availability of Fabrication Faculties

Availability of Labor (specific skills)

Production Rates Achievable

Time Required to Incorporate Substitute

Use of Existing Faculties and Labor

Energy Requirements

D. MATERIALS

SUPPLY/AVAILABILITY

CONSIDERATIONS

Supply – Present and Future, Current and

Potential

Resources / reserves

Stockpile level

Imports / exports

Defense allocation

Inventories

Supply Assurance (Including Trade

Agreement)

Identity and Location of Supplies Forms

of Materials Available Delivery Time

(Lead Time)

E. END-USE PATTERNS Historical and

Projected Supply-Present and Future,

Current and Potential

Resources/Reserves

Stockpile Level

Imports/Exports

Defense Allocation

Inventories

Supply Assurance (Including Trade

Agreement)

Identity and Location of Supplies Forms

of Materials Available Delivery Time

(Lead Time)

F. RISK CONSIDERATIONS

Regulatory Agency Compliance (Federal,

State, Local) Environmental

Health/Safety

Energy

Economic Impacts of Using Substitute

Fig. 5.3. Engineering implications of chronic materials scarcity report.

on seeking to ensure as much of the key materials as possible would come from U.S. primary mining. In addition, the program sought to understand the foreign dependence and determine any possible alternatives, including material substitution.

Importantly the MEP program sought to understand the 'most stressed materials' and ensure, if possible, the domestic U.S. materials primary supply. The term 'most stressed' would by the late 1990s become 'critical materials'. Of note is the 1976 report by Albers, *et al.*, which does not use the term scarcity. The position of the authors of the report was that if sufficient investment and attention were to be given to exploiting primary resources, the prospect of scarcity or 'running out' would not arise.

The MEP program aimed to use early computer databases for materials stocks and flows and to use this database to understand materials 'stresses' as the U.S. developed total energy independence. The use of computers was seen as essential to manage the complexity of materials needs as part of the energy independence transition.

The report was divided into two parts; Part I covered demand and Part II supply. Part I of the report lists 31 materials essential for technologies to provide energy supply. The energy generation technologies listed cover fossil and non-fossil energy generation from coal and oil, plus nuclear as well as wind and solar. Part II of the report highlights the fact that materials independence cannot be reached and imports of some materials will always be needed. The report's Part II highlights the need for imports of, for example; cobalt, manganese, niobium, aluminum, nickel and tungsten. The report also highlights that the U.S. materials stockpile has been used extensively to address shortages in some materials.

The report does not highlight concerns about any environmental impacts of either increased material production or of increased burning of fossil fuels. Climate change concerns were not a widespread policy issue at this time. Nor were there any health and safety issues raised concerning the expected increased use of asbestos and other hazardous materials.

The conclusions of the report begin by making the statement that energy generation is essential and must always have first call on materials. The authors proposed that supply shortages could occur for other sectors as the energy sector material demand increased. The report does however highlight that there was extreme uncertainty with the supply and demand data they had available and to address this estimations were used. Interestingly, the rare earths were assessed, but not for energy production, and the USA was deemed to be self-sufficient. This position on rare earths

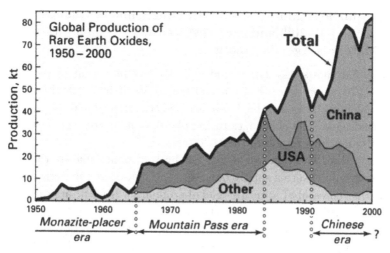

Fig. 5.4.   Global rare earth element production ($1\,kt = 10^6$ kg) from 1950 through 2000.
Source: Haxel, et al., USGS, 2002.

can be seen in Fig. 5.4, where later work by the U.S. Geological Survey in 2002 shows the period up to 1976 and beyond.

The report's conclusions highlight that for many materials significant increases in domestic supply would be needed to reach energy independence by 1990. The releases of the material stockpiles were only buying time and would not be a long term solution. The report ends with an urgent call for action to understand the scope and scale of the materials problems. The authors voice deep concern at the lack of understanding of the problems of supplying industry with the materials needed to ensure energy independence by 1990. They highlight how quickly the situation of an oil embargo impacted the U.S. and that on the topic of materials there was widespread complacency.

The solutions proposed were mainly around new geological assessments to ensure supply. Again the urgency of this work was stressed as any delay would reduce the U.S.'s flexibility to meet new challenges. There was scant mention of secondary materials supply via recycling or of reuse of materials, as part of a resource efficient approach. The authors felt confident the U.S. could become both energy independent and significantly more materials independent by 1990. They warned against reliance on 'crash programs', or quick fixes, as the lead times for geology and mining programs can be up to 20 years (Albers et al., 1976).

5.6   Technology Materials and Tackling Climate Change in the late 20th and early 21st century — the Critical Materials Agenda Emerges

As Fig. 5.4 shows, the hopes and calls for urgent action of the Albers, et al., 1976 report were far from realized through the 1980s. In turn, the technology revolutions of the digital age and economic globalization had not been envisaged. The challenges to the planet of man-made climate change were also not understood in the 1976 report.

From the 1980s onwards it was felt by U.S. policy makers that global markets would ensure a supply of energy and non-energy materials. Essentially this represented a 'tech + global economy will fix it' approach. This situation started to change from the late 1990s onwards with the rise of new technologies, globalized economic growth, increasing levels of waste, concerns over energy & climate, resource nationalism, and shifts in geopolitics, which have combined to drive the need for the development of a more sophisticated approach to critical materials. This development has ensured that around 2006, *critical materials* had become a distinct field.

The growth of new technologies has made the critical materials challenge even more complex. The number of elements needed in modern

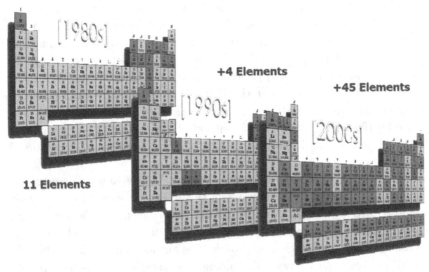

Source: T. McManus, Intel Corp., 2006

Fig. 5.5.   An Intel corp. internal image of the growing complexity of elements in electronic chips. McManus, Intel Corp., 2006.

Table 5.2:   The Shift from Material Scarcity towards Current Understanding of Critical Material Approaches (Peck, 2016).

Title	Author	Term	Descriptor
1. Depletion and the Long-run Availability of Mineral Commodities **2001**	Tilton J E, Report published by IIED for WBCSD, Washington D.C, USA	Mineral commodities depletion	Mineral resource availability Mineral depletion Shortages and scarcity = opposite of availability Excess of demand over supply Declining availability
2. On Borrowed Time? Assessing the Threat of Mineral Depletion **2003**	Tilton J E, RFF Press, Washington, D.C. USA	Mineral depletion	Mineral resource availability Mineral depletion Shortages and scarcity = opposite of availability Excess of demand over supply Declining availability
3. Scarcity and Growth in the New Millennium: Summary **2004**	R. David Simpson, Michael A. Toman, and Robert U. Ayres Discussion Paper 04-01 Resources for the Future, USA	Scarcity New Scarcity	Resource limits to growth
4. Minerals, critical minerals, and the U.S. Economy, **2008**	Eggert R. G. *et al.*, Minerals, Critical Minerals, and the U.S. Economy, National Research Council, USA	Critical minerals; Critical materials	The two dimensions of criticality are: 1. importance in use; 2. availability
5. Methodology of Metal Criticality Determination, **2009**	Thomas E. Graedel, *et al.*, Yale University, USA	Metal criticality	A critical metal involves three dimensions: 1. supply risk; 2. environmental implications; 3. vulnerability to supply restriction

societies covers much of the periodic table. As an example Fig. 5.5 shows that between the 1980s and the 2000s electronic chip manufacture required 49 more elements as complexity increased.

The U.S. policy of holding large strategic material stockpiles came to an end in 1992. Between 1993–2005 the stockpile was reduced by 75%.

Politically, the view was that markets could always supply what was needed (Abraham, 2015). Whilst this is a further example of the 'tech + global economy will fix it' approach, concerns persisted over the rate of material use, increasing complexity and global dependence. This shift is highlighted in examples of literature from 2001 to 2009, which is shown in Table 5.2.

The European Union (EU), USA, Japan, and other countries, came together at times to collaboratively develop their critical materials approaches. Their thinking not only saw material criticality in military terms but also from economic and geo-political aspects, taking energy and environmental concerns into account.

The full involvement of the EU in addressing critical materials was an important step in the history of critical materials. The EU consulted other nations, especially the U.S., to help them develop their own distinctive European critical materials strategy.

A key important realization began in the early 21st century as it was recognized that geologists & miners could not alone provide all the solutions to critical materials. Conversely some sustainability thinkers and actors realized that geologists & miners are essential to understanding critical materials and finding solutions. The age of multi-disciplinary thinking along the entire materials value chain had begun.

5.7 Conclusion

This chapter shows that the development of approaches towards critical materials began during times of geo-political tensions and conflict in the 20th century. Actions around material criticality were significant throughout the 1940s, 1950s and 1960s, mainly focused on the stockpiling of materials deemed essential to military equipment production. The main driver was concerns over 'limits to military growth' because of concerns over the availability of materials, driven by geo-political concerns and conflicts.

The development of more contemporary critical materials approaches began in the 1970s, with the combined effects of limits to growth concerns and with the energy/oil crisis. At the same time, materials supply/price problems served to intensify action. Low carbon renewable energies, advanced technology materials or emerging economies requirements were not a feature of critical materials thinking at this time. The solutions proposed to material criticality in the 1970s built upon previous decades, with calls to increase understanding of domestic primary resources, increase exploitation of those resources, increase understanding of the stock flows of materials, seek greater economic control, and stockpile where needed. This represents a 'tech will fix it' approach which has remained in place.

Through the 1980s, 1990s and into the 2000s continued growth in overall material demand was coupled with the emergence of new technologies which resulted in a much more varied palette of elements and materials emerging. The approach of governments in this period was that globalised materials markets would supply the materials needed and also stabilise prices if they spiked. Most of the policy proposals of the 1970s were forgotten as markets in the post-Cold War era opened up.

As the 21st century began it was starting to become clear that revised methodologies to assess critical materials were needed and by 2006 the development of these approaches was well underway. The work of the past decades has been revised and many of the historic policy recommendations restated. One significant difference of the 2006 approach compared to the past was that of the environment. The limits to growth frame has been adjusted to focus on the effects of pollution and climate change. By 2006 data was beginning to show that the tech will fix it approaches to tackling climate change (low carbon energy and mobility, etc.) will require significant increases in materials, which in turn will increase challenges in materials criticality.

The tensions around 'limits to growth' versus 'tech will fix it' are a feature in discussions concerning critical materials. The 'running out' or 'mineral exhaustion' concerns of finite materials are referred to, but this thinking, over the near term, relies on technologies not changing. However, technology does constantly change. At the same time rapidly increasing material consumption, plus the corresponding waste and energy/pollution drives a revised 'limits to growth' framework. The tensions are often contradictory and positions along national political lines as well as geo-political tensions are often firmly taken.

Some features from the reports arising out of the energy–materials crisis of the 1970s are consistent. Time is urgent, but the world moves at a slow pace and manages somehow. Risks are high and increasing but then they either disappear or shift and multiply. Energy requirements and technology changes in energy generation are essential but never taken seriously enough. The need for much improved material demand–supply data is urgently needed but never fully developed. The market meets needs and manages each crisis but concerns linger, sometimes it seems more by luck than judgement. 'Wolf' has been cried often but the wolf never really came, whilst at the same time knowledge that the wolf is nearby in the forest increases.

A sensible way forwards is to recognize the importance of both 'limits to growth' together with 'tech will fix it' thinking and collaboratively seek to find solutions across the entire materials value chain, be they critical or not. As so often, we have faced similar challenges in the past, and an improved understanding of our history could help us.

References

1. Abraham. D. S. (2015). The elements of power: Gadgets, guns, and the struggle for a sustainable future in the Rare Metal Age. New Haven, Connecticut, USA: Yale University Press.
2. Albers. J.P., Bawiec, W.J., Rooney, L.F., Goudarzi, G.H., and Shaffer, G.L. (1976). *Demand and supply of nonfuel minerals and materials for the United States energy industry, 1975-90 — a preliminary report.* Washington D.C., USA: United States Geological Survey.
3. Ashby, M.F. (2013). *Materials and the environment — eco-informed materials choice* (2nd ed.). Waltham, Massachusetts, USA: Butterworth-Heinemann.
4. Ashby. M.F., Balas, D.F., and Coral, J.S. (2016). *Materials and sustainable development.* Waltham, Massachusetts, USA: Butterworth-Heinemann.
5. Eckes. A.E. (1979). *The United States and the global struggle for minerals.* Austin, Texas, USA: University of Texas Press.
6. National Research Council. (2008). *Minerals, critical minerals, and the U.S. economy.* Washington D.C., USA: National Research Council.
7. Graedel, T. (2009). Defining critical materials. In R. Bleischwitz, P.J. Welfens, and Z. Zhang (Eds.), *Sustainable growth and resource productivity — economic and global policy issues.* Sheffield, UK: Greenleaf Publishing.
8. Haxel, G.B., Hedrick, J.B., and Orris, G.J. (2002). *Rare earth elements — critical resources for high technology.* Reston, Virginia, US: United States Geological Survey. Retrieved from https://pubs.usgs.gov/fs/2002/fs087-02/.
9. Huddle, F.P. and Promisel, H.E. (Eds.). (1977). *Engineering implications of chronic materials scarcity.* Washington D.C., USA: US Government Printing Office.
10. Johnson, J., Harper, E. M., Lifset, R., and Graedel, T. E. (2007). Dining at the periodic table: Metals concentrations as they relate to recycling. *Environmental Science and Technology, 41*(5), 1759–1765.
11. Martin, R.L. and Kemper, A. (2012, April). Saving the planet: a tale of two strategies. *Harvard Business Review.* p. 49.
12. Meadows, D.H., Meadows, D.L., Randers, J., and Behrens, W.W. III. (1972). *The limits to growth: A report for The Club of Rome's project on the predicament of mankind.* Washington D.C., USA: Universe Books.
13. McManus T. (2006). *Internal report.* Santa Clara, California, USA: Intel. In Graedel, T.E. (2006). *Determining the criticality of materials.* [PowerPoint presentation].

14. National Academy of Sciences. (2008). *Managing materials for a twenty-first century military.* Washington D.C., USA: National Academy of Sciences.

15. Paley, W.S. (1952). *Resources for freedom.* Washington D.C., USA: The President's Materials Commission.

16. Peck, D.P., Bakker, C.A., Kandachar, P.V., and de Rijk, T. (2017). Product policy and material scarcity challenges : The essential role of government in the past and lessons for today. Amsterdam, the Netherlands: IOS Press.

17. Postan, M.M. (1952). *British war production.* London, UK: Her Majesty's Stationery Office.

18. Simpson, R.D., Toman, M.A., and Ayres, R.U. (2004). *Scarcity and growth in the new millennium: Summary.* Washington D.C., USA: Resources For The Future.

19. Stevenson, D. (2012). With our backs to the wall: Victory and defeat in 1918. London, UK: Penguin Books.

20. Tahvonen, O. (2000). Economic sustainability and scarcity of natural resources: A brief historical review. Washington D.C., USA: Resources For The Future.

21. Tilton J.E. (2003). On borrowed time? Assessing the threat of mineral depletion. Washington D.C., USA: Routledge.

22. Tilton J.E. (2001). *Depletion and the long-run availability of mineral commodities.* Washington D.C., USA: International Institute for Environment and Development.

Chapter 6

Defining the Criticality of Materials

T.E. Graedel and Barbara K. Reck

Center for Industrial Ecology
Yale University
195 Prospect Street, New Haven, CT 06511, USA

Criticality can be defined as "the quality, state, or degree of being of the highest importance", but how can we understand what is meant by "highest importance"? In this chapter we define and describe a multi-parameter approach to the criticality issue that involves (as do the efforts of other researchers and governments) a variety of geological, economic, technological, environmental, and social concerns. Our results suggest that the highest level of concern should be for metals whose processing and use involves extensive separation from parent ores, high levels of embodied energy, little opportunity for substitution, and low levels of recyclability. Improved approaches to material use should thus involve the preferential utilization of non-critical materials, attention to the potential for material reuse at the design stage, and a focus on increasing the efficiency of recycling.

6.1 Introduction

In 2006, the United States National Research Council (NRC) undertook a study to address the lack of understanding and of data on nonfuel minerals important to the American economy. The report, titled *Minerals, Critical Minerals, and the U.S. Economy*,[1] defined the criticality of minerals as a function of two variables, importance of use and availability, in which the vertical axis reflected importance in use and the horizontal axis was a measure of availability. Since that original work, other researchers and organizations have addressed the same topic, although from a variety of methods and approaches.[2-6]

6.2   The Methodology of Criticality

A major focus of criticality studies has been on how best to quantitatively determine "importance" and "availability". Our research group at Yale extended the two-axis concept of the National Research Council[1] to three dimensions (Fig. 6.1), each dimension comprising one axis of "criticality space" — Supply Risk (SR), Environmental Implications (EI), and Vulnerability to Supply Restriction (VSR). Utilizing this methodology is an exercise in both data acquisition and expert judgment. For many of the geologically scarcer "specialty" metals, data are in short supply. In developing the methodology, a balance was sought between analytical rigor and data availability in order to evaluate the criticality of as many metals as possible, and to draw attention to cases for which data are simply not adequate. Additionally, efforts to explore the criticality of metals generally consider only the global level, but organizational differences make a uniform analytical approach for all organizational levels impractical. Our methodology was thus developed at three organizational levels (corporate, national, and global).

A suitably comprehensive assessment of criticality involves incorporating information from widely disparate specialties and data sources, from geology, technology, economics, human behavior, expert assessment, and many more. Some useful data sets are quantitative while some are

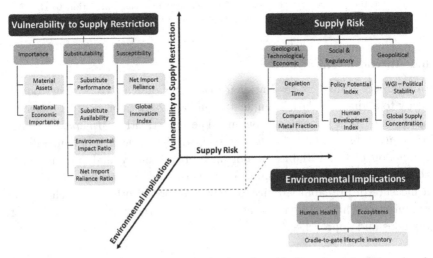

Fig. 6.1.   The three-axis approach to evaluating the criticality of metals. (Reproduced with permission from Graedel et al.[7]).

qualitative; some are well defined, others less so. In response to this complexity, it is important to present the methodology in highly transparent fashion, and invite users to redefine aspects of the work as may be most useful to them.

6.2.1 Temporal perspective

A detailed discussion of the temporal complexities that emerge when evaluating criticality can fill many pages, but in brief, no single approach is suitable for all time scales or all interested parties. What we describe below, and in more detail in Graedel *et al.*,[7] is a methodology to prescribe criticality as a snapshot in time, but one that inherently incorporates some time-dependent considerations. Other assessments of criticality, cited above, share many of these characteristics.[8]

6.2.2 Supply risk

Because the different temporal perspectives suggest that no methodology focused on a single time scale can adequately serve the complete spectrum of interested parties, it is useful to determine Supply Risk (SR) for both the medium term (5–10 years) and for the longer term (a few decades). The former is likely to be most appropriate for corporations and for governments, while the latter will perhaps best serve long-range planners, futurists, and the community of scholars dealing with sustainability.

Because our medium-term methodology is of particular relevance to corporations and nations that utilize materials rather than, or in addition to, supplying them, our focus was on *using* entities rather than *sourcing* entities (i.e., manufacturing firms rather than mining firms). The methodology evaluates SR for using entities on the basis of three components: (1) Geological, Technological, and Economic, (2) Social and Regulatory, and (3) Geopolitical (Fig. 6.2). The first of these components aims at measuring the potential availability of a metal's supply, including both primary and secondary (recycled) sources, while the latter two address the degree to which the availability of that supply might be constrained. Each component is evaluated on the basis of two indicators, as shown in Fig. 6.2. All indicators are scored on a common 0 to 100 scale with higher values suggesting a higher level of risk.

The most obvious questions related to a metal's availability in the ground are "How much is there?", "Is it technologically feasible to obtain?", and "Is it economically practical to do so?" It is generally surprising to the

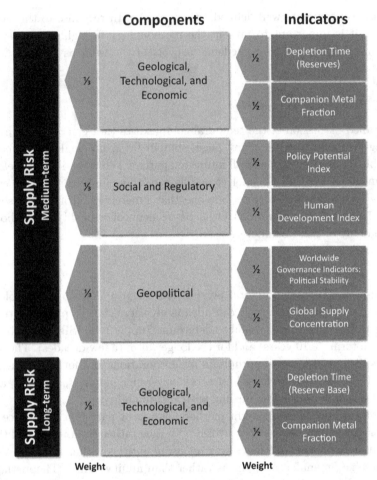

Fig. 6.2. Diagram of the Supply Risk axis, its components, and its constituent indicators for the medium-term perspective (used mainly in conjunction with the corporate and national level assessments), and for the long-term perspective (used mainly in conjunction with global level assessments). (Reproduced with permission from Graedel et al.[9]).

non-geologist that these simple questions are very challenging to answer in any useful way.

The U.S. Geological Survey (USGS) characterizes the Reserves for each metal: the amount that has the potential to be extracted within the next few years. To estimate the relative availability of the metal, we determine the amount of supply being met by recycling, and then calculate the amount of time it would take to deplete the geological reserves at the current rate of demand.

The metals that have been in common use for millennia are those that can be found in relatively high concentration (a few weight percent) in good-sized deposits. Generally speaking, these are also elements whose abundances in the continental crust are relatively high. In contrast, where the crustal concentration of a metal is less than about 0.1%, it will seldom form usable deposits of its own, but occurs interstitially in the ores of metals with similar physical and chemical properties. Such metals, if recovered, are termed "companion metals", and the principal metals in the deposits "host metals". The availability of the companions therefore depends on not only whether they are recovered but also upon the magnitude of the mining of the relevant host metal. To express the potential for SR related to the host-companion relationship, the percentage of a target metal that is extracted as a companion is used as the relevant metric.

Two indicators — the Policy Potential Index (PPI) and the Human Development Index (HDI) — are employed to quantify the social and regulatory component of the SR evaluation. Detailed information about each may be found in the Supporting Information of Graedel *et al.*[7] Each index is comprised of multiple variables that are aggregated into a single score for individual nations and, in some cases for sub-national jurisdictions. The final PPI and HDI metal indicator scores are obtained by weight-averaging each jurisdiction's transformed index score by its annual production for the metal being studied, with the transformations discussed in the mentioned Supporting Information. For the HDI, the production quantities used in the weighting should be either the metal's mining, smelting, or refining production, whichever yields the highest risk score. The rationale for this approach is to emphasize the highest risk in the supply chain, as the process step that has the highest risk is the "bottleneck" most likely to cause the supply constraint. This selection of the highest risk production weighting is not used for the PPI, because the PPI is inherently based on mining factors and should thus only be based on mining considerations. For companion metals, it is often the case that no mining production data are available. In such cases, the mining production of the host metal is used in the calculation.

Nations that are politically unstable pose a higher risk of mineral supply restriction than those that are not. The WGI is utilized to quantify this risk, and has been used in previous criticality assessments.[2,10] The index encompasses national social, economic, and political factors that are associated with underlying vulnerability and economic distress. A number of specific criticisms of WGI have been answered by the WGI researchers. We recognize these challenges, but nonetheless feel that the WGI is a satisfactory indicator for our purposes based upon its use in previous criticality assessments.

Mineral deposits are not equally or randomly distributed on Earth. Some minerals are predominantly found in only a few countries while others have more widely dispersed ore deposits. In general, the more concentrated the mineral deposits, the higher the risk of supply restriction. HHI is a metric commonly used to measure market concentration. Its first noted use for the purpose of evaluating the availability of mineral resources is in a recent article by Rosenau-Tornow et al.,[10] where it is used to measure the concentration of mining production at both the national and corporate levels. (It is also used by the European Commission to weigh SR). HHI is utilized in this study to quantify the risk of having "all of your eggs in one basket" by examining the degree of production concentration.

Each component score is the average of its indicator's scores, and the final SR score, calculated by averaging the three component scores, locates the metal under study on the SR axis in criticality space.

6.2.3 *Environmental implications*

Metals can often have a significant environmental impact as a result of their toxicity, the use of energy and water in processing, or emissions to air, water, or land. We designate an axis on the criticality diagram to depict the environmental burden of the various metals, thus moving from a criticality matrix to a "criticality space". There are two components of the analysis, as shown in Fig. 6.1.

The Environmental Implications (EI) evaluation included in our methodology is not intended to be viewed as the regulatory measures that may restrict one's ability to obtain mineral resources; that issue is addressed in the Social and Regulatory component of SR. Rather, it should be viewed as indicating to designers, governmental officials, and nongovernmental agencies the potential environmental implications of utilizing a particular metal. For this evaluation, the inventory data from the ecoinvent database[11] is utilized because of the breadth and depth of that database. From the ecoinvent inventory data, the damage categories Human Health and Ecosystems are calculated according to the ReCiPe Endpoint method, with "World" normalization and "Hierarchist" weighting. The third damage category according to this method, Resource Availability, is not incorporated in the EI evaluation because it is addressed in the SR methodology. The summation and subsequent scaling of the two damage category evaluations provides a single score on a common 0 to 100 scale for a cradle-to-gate (from the unmined ore to the manufacturing front gate) environmental impact assessment.

6.2.4 *Vulnerability to supply restriction*

No single approach is appropriate for evaluating VSR at each of three organizational levels (corporate, national, and global). For example, a particular metal may be crucial to the product line or operations of some corporations, but of little or no importance to others. Similarly, countries with a strong industrial base will value certain metals more than may technologically depauperate countries. As it happens, there are some indicators in common among the various organizational levels but other indicators that may be specific to only one or two. As a consequence, we have developed three distinct, yet often overlapping, methodologies for the three organizational levels. The methodologies utilize indicators adjusted to a common 0 to 100 scale. In several of the cases, in which a qualitative assessment is thought to be the most desirable approach, we provide a scoring rubric in which the 0 to 100 range is divided into four equal "bins". Each bin has a range of 25 points to represent the level of uncertainty in the assessment, and the middle score for each bin is utilized as the default score for those cases in which specifying an exact number proves too great a challenge.

A complication in assessing the VSR is that, unlike assessing the level of SR, it is important to evaluate each significant end-use application of a metal separately. This is because the degree of importance and the substitutability of the metal in question generally vary from one end-use application to another.

The VSR is dependent on the importance of the metal in question and the ability to find adequate substitutes if the metal is unavailable. Quantifying the VSR is thus conducted by evaluating two components, Importance and Substitutability, using several indicators assessed independently for each end-use application of the metal. The corporate-level assessment is directed to a corporation's current and anticipated product line, with special emphasis paid to economic considerations. A third component, Ability to Innovate, is included at this organizational level because more innovative corporations are likely to be able to adapt more quickly to supply restrictions.

Assessing the VSR on a national level differs from the corporate assessment in several ways. The importance of the element in question is again a central component, but one in which the indicators relate to domestic industries and the country's population. Importance and Substitutability are retained, but are evaluated somewhat differently. Importance here is composed of two indicators — National Economic Importance and in-use stock. National level substitutability is identical to that described for the corporate-level assessment, except that it substitutes Net Import Reliance Ratio for Price Ratio. Ability to Innovate is comprised of Net Import Reliance and a measure of innovation, measured by INSEAD's country-level Global Innovation Index.[12] Additional details are given in the Supporting Information. (The details of sourcing-nation evaluations differ somewhat from those of using nations; this methodology is currently under development.)

6.3 Comparing the Criticality of Metals

We have applied the Yale methodology to 62 metals and metalloids (hereafter termed "metals" for simplicity of exposition)[9] — essentially all elements except highly soluble alkalis and halogens, the noble gases, nature's "grand nutrients" (carbon, nitrogen, oxygen, phosphorus, sulfur), and radioactive elements such as radium and francium that are of little technological use. An overview of the findings is provided by plotting the three-axis results for each metal in criticality space (Fig. 6.3). A number of metals of quite different criticality properties are concentrated at the middle of the diagram; they may rank moderately high on one or two of the axes, but not extremely high on any. Another group of metals is concentrated toward the lower left front corner. For those metals, criticality concerns are relatively low on all three axes. A third group is located toward the right side of the diagram. For those metals, the concern is largely related to supply risk. Only platinum and gold appear toward the upper left back corner. Gold has large geological reserves and a low companion fraction. As a consequence the supply risk is low, but gold's high cradle-to-gate environmental impacts per kilogram of metal (related to extraction and processing from ore deposits) and its high vulnerability to supply restriction (related to its near-universal use in electronics, jewelry, and investments and its lack of available suitable substitutes) render it of special interest. Platinum has

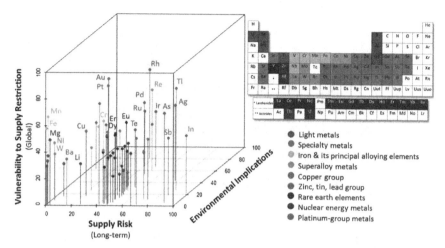

Fig. 6.3. The global assessment for 62 metals, 2008 epoch, global level, in criticality space. The highest level of criticality is at 100, 100, 100 (back right top). The metals assigned to the several groupings are indicated in color on the periodic table. (Reproduced with permission from Graedel *et al.*[9]).

similar energy and environmental challenges and, in addition, its deposits are geopolitically highly concentrated. It is notable that the rare earth elements (dark blue) form a pattern of medium supply risk but of sequentially increasing environmental implications; this reflects the fact that the rare earth elements are sequentially separated, with additional energy required for each separation.

A comprehensive perspective on global-level criticality is provided by displaying the values of the three criticality space axis variables on the periodic table, as shown in Fig. 6.4. Fig. 6.4(a) illustrates that the highest values for supply risk are concentrated in groups 13–16/periods 4–5. The figure demonstrates that the metals so important for high-tech applications, such as electronics and thin-film solar cells, are most crucial from a supply risk perspective. For environmental implications (Fig. 6.4(b)) the highest values are in groups 8–11/periods 5–6. In the case of vulnerability to supply restriction (Fig. 6.4(c)), the highest values include thallium, lead, arsenic, rhodium, and manganese.

There are only a few metals that have an overall high score along the supply risk dimension (i.e., the metals that have small geological resources relative to their current demands and that are mainly recovered as byproducts of other metals, with byproducts called companions in our analysis).

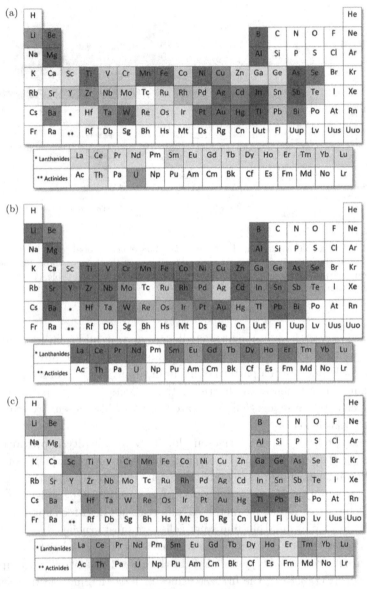

Fig. 6.4. Two-axis depictions of criticality for 62 metals, 2008 epoch, global level for
(a) supply risk, (b) environmental implications, and (c) vulnerability to supply restriction. (Reproduced with permission from Graedel *et al.*[9]).

These include indium, arsenic, thallium, antimony, silver, and selenium, metals important in modern electronics and thin-film solar cell technology. From an environmental implications perspective, the most concern rests with precious metals (gold and the platinum group metals, in particular), because of environmental impacts related to extraction and processing. On the vulnerability to supply restriction dimension, the degree to which suitable substitutes are unavailable is a signal of concern. That parameter singles out magnesium, chromium, manganese, rhodium, yttrium, and several rare earths for attention. All of the elements mentioned above should thus be targeted for special consideration in any general effort to minimize the use of metals that are more problematic from various criticality perspectives.

Unlike many research results in the physical sciences, a criticality of metals assessment should not be regarded as static, but as a result that will evolve over time as new ore deposits are located, political circumstances change, and technologies undergo transformation. This dynamic characteristic of metal criticality requires that evaluations such as that done in the present work be periodically updated. However, data revisions are not frequent, and major transformations in technology and society often occur slowly. A recent study looked at the criticality of six metals in the US, Australia, and at global level in 2008 and 2012 and found no major changes in criticality over this four-year time period.[13] We thus regard criticality reassessments on perhaps five-year intervals as both practical and perfectly adequate for most uses.

6.4 Discussion

The assessment of criticality is not purely of academic interest, but also of significant value to industrial product designers and to national policymakers. Designers are already advised to choose materials so as to minimize embodied energy and energy consumption during use. The present study adds an additional dimension to materials choice: that of minimizing criticality in material choices. For designers, the criticality designations are surely relevant to efforts that seek to minimize corporate exposure to problematic metals in product design, especially for products expected to have long service lives. Perhaps more important to designers than the aggregate assessments, however, are those for individual indicators, because manufacturers may be able to minimize or avoid some risks if those risks are recognized, especially if current designs involve metals in or near problematic regions of criticality space. For example, efforts can be made to find secure sources of supply, to increase material utilization in manufacturing, to reduce the use of critical metals, or to increase critical metal recycling.[14] Cross-metal analyses of specific criticality indicators can also reveal properties of individual metals or metal groups, as we have shown in the cases of potential substitutability and environmental implications. Considerations such as these extend the product designer's remit from a sole focus on materials science to consideration of corporate metal management as well. In the case of supplier nations or user nations, recognizing the regions of opportunity and of danger in connection with their own resources and industries can minimize risk going forward.

A final point of discussion relates to the relevance of the present work to national and global resources policy. Whether or not individual products or corporate product portfolios are designed with metal criticality in mind, it is indisputable that the world's modern technology is completely dependent on the routine availability of the full spectrum of metals, now and in the future. Tomorrow's technology cannot be predicted with much confidence, especially in the longer term, but it would be quite short-sighted were one or more metals to be depleted to the extent that their use in new technologies could not be confidently assumed. Such occurrences would be less likely to happen if metal criticality were routinely considered by industries and governments. In any case, metal availability in perpetuity should not be taken for granted.

References

1. National Research Council. (2008). *Minerals, critical minerals, and the U.S. economy*. Washington D.C., USA: National Research Council.
2. European Commission. (2010). *Critical raw materials for the EU*. Brussels, Belgium: European Commission.
3. European Commission. (2014). *Report on critical raw materials for the EU*. Brussels, Belgium: European Commission.
4. European Commission. (2017). *Study on the review of the list of critical raw materials*. Brussels, Belgium: European Commission.
5. Hatayama H. and Tahara K. (2015). Criticality assessment of metals for Japan's resource strategy. *Materials Transactions*, 56(2), 229–235.
6. Morley N. and Eatherley D. (2008). *Material security: Ensuring resource availability for the UK economy*. Aylesbury, UK: Oakdene Hollins.
7. Graedel, T.E., Barr, R., Chandler, C., Chase, T., Choi, J., Christoffersen, L., et al. (2012). Methodology of metal criticality determination. *Environmental Science and Technology*, 46(2), 1063–1070.

8. Graedel, T.E. and Reck, B.K. (2016). Six years of criticality assessments: What have we learned so far? *Journal of Industrial Ecology*, 20(4), 692–699.

9. Graedel, T.E., Harper, E.M., Nassar, N.T., Nuss, P., and Reck, B.K. (2015). Criticality of metals and metalloids. *Proceedings of the National Academy of Sciences of the USA*, 112 (14), 4257–4262.

10. Rosenau-Tornow, D., Buchholz, P., Riemann, A., Wagner, M. (2009). Assessing the long-term supply risks for mineral raw materials — a combined evaluation of past and future trends. *Resources Policy*, 34(4), 161–175.

11. Frischknecht, R., Jungbluth, N., Althaus, H.J., Doka, G., Dones, R., Heck, T., Hellweg, S., Hischier, R., Nemecek, T., Rebitzer, G., Spielmann, M. (2005). The ecoinvent database: overview and methodological framework. *The International Journal of Life Sciences Assessment*, 10(1), 3–9.

12. INSEAD. (2010). *Global innovation index 2009–10*. Fontainebleau, France: Confederation of Indian Industry, INSEAD.

13. Ciacci, L., Nuss, P., Reck, B.K., Werner, T.T., and Graedel, T.E. (2016). Metal criticality determination for Australia, the US, and the planet — comparing 2008 and 2012 results. *Resources*, 5(4), 29.

14. Duclos, S.J., Otto, J.P., and Konitzer, D.G. (2010). Design in an era of constrained resources. *Mechanical Engineering*, 132(9), 36–40.

Chapter 7

Identifying Supply Chain Risks for Critical and Strategic Materials

James R. J. Goddin

Granta Design, Cambridge, UK

For many companies understanding the environmental impacts of their products and operations is steadily rising upwards in their business agenda.

Common business drivers include:

— Legislation on energy consumption, hazardous substances and conflict minerals.

— Volatile material and energy prices.

— Product marketing, brand value and Corporate Social Responsibility (CSR)

— Stimulus for product innovation.

Despite these significant and growing pressures, many companies have not yet been able to effectively implement systems or tools to manage the environmental issues associated with the products they develop.

Approaches such as Life Cycle Assessment (LCA) have been suggested. However, these generally require significant knowledge and expertise, both to perform the initial analysis and to understand the results. Furthermore, the LCA approach is substantially divorced from product development activities as it is generally applied at the end of the product development process. This results in poor engagement of designers and engineers with environmental issues and a general lack of support (or capability) to address these issues within the organization at an early enough stage to implement effective design decisions.

This chapter introduces the current best practice employed within industry when attempting to overcome these implementation barriers, focusing upon the adoption of risk-based approaches to the management of product sustainability issues and the utilization of materials information that is available during the product design process.

The approach aims to integrate product sustainability into the strong culture for business risk management that already exists within most advanced manufacturing organizations.

By presenting product environmental sustainability issues in terms of the business risks they engender and by integrating with existing business risk management systems, a business risk-based approach will result in a wider understanding of environmental issues and ultimately result in these issues being managed as a normal part of the engineering design process.

In this chapter, we focus specifically on risks associated with Critical and Strategic Materials. However, the approach outlined has already been very effectively applied to restricted substances (such as those impacted by REACH legislation) and more traditional LCA based indicators such as Energy, CO_2 and water which also comprise areas of legitimate business risk when considered appropriately.[1]

7.1 What is a 'Critical Material'?

There has been much discussion over the past few years about the issues surrounding the supply of certain 'critical' or 'strategic' raw materials. For the most part these discussions have focused upon the supply of elements, including rare earths such as neodymium and dysprosium, which are important for the permanent magnets used in many electric vehicles and wind turbines. Elements also commonly include those used in other high value or specialist alloys such as those used in the challenging environments experienced in automotive and aerospace applications or in electronic components. In some more recent reviews the materials included within the definition of Critical Materials has been extended to include biotic resources such as timber and palm oil and it has also been argued that certain engineered materials such as long strand carbon fiber should also be included on the basis of intellectual property which limits the number of commercially licensed manufacturers.

Definitions for what comprises a 'critical material' abound[2] but the definition which the authors have settled on, and which generally appears to be accepted within industry, is:

Critical Material: An element, composition, biotic or man-made material which, by virtue of its properties, enables a product to deliver value-added functionality, wherein the ability to substitute that functionality using an alternative material is limited or would incur significant penalties in terms of cost, performance or safety and for which one or more of its constituents or pre-cursors is at risk of experiencing a supply disruption.

In 2010 the 'Critical Raw Materials for the EU' report highlighted fourteen elements or groups of elements that were considered both economically significant to the EU and for which an elevated risk of supply disruption was perceived to exist.[3] These included antimony, beryllium, cobalt, fluorspar, gallium, germanium, graphite, indium, magnesium, niobium, platinum group metals, rare earth elements, tantalum and tungsten.

Since 2010 the EU list has been updated twice and a variety of other reports have highlighted elements of concern to other economic regions or in some cases to specific sectors or technologies of economic importance within those regions — see Fig. 7.1 below for example.

In most lists, criticality is represented as an indication of Supply Risk on one axis against an indication of Economic Importance on the other. In the case of the analysis by Graedel *et al.*, environmental impacts have been used as an additional third axis.[4]

Following the publication of the EU 2010 list, concern grew within industry that such lists were starting to be used as de-facto resources for the assessment of business risk and that their use in this way might inadvertently lead to false negatives when these regional or sectorial lists were applied at a business level.

It is important to understand that, with limited exceptions, lists of critical materials such as those produced by the EU were never intended to be used in the context of business specific risks but rather were intended to guide policy decisions at the level of the economic region in question. In short, the lists were intended to inform foreign policy and domestic investment decisions rather than guide engineering decisions or business level risk assessments.

It is also very important to remember that, in most cases, elements such as those highlighted in the EU list of critical materials are very rarely used in isolation but rather as part of a broader composition of elements. The ability to substitute an 'at risk' element therefore needs to be assessed in the context of that composition in the specific application in which it is being applied by the business. In many cases this also mandates that the broader design of the system is considered, including any knock-on effects that potential substitutions may have. This in turn leads to a system based upon a rational materials selection approach as long advocated by Prof. Mike Ashby.[5]

7.2 Critical Materials — Assessment of Business Risk

In order to implement a business centric approach to critical materials, businesses therefore need to have an understanding, at a sufficiently detailed level, of the elements extracted and refined by international mining

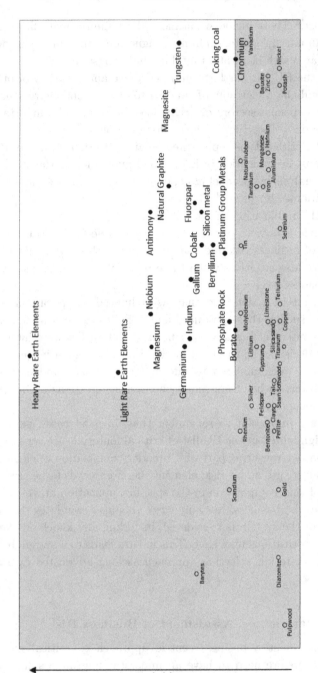

Fig. 7.1. Critical Raw Materials for the EU, 2010.

operations, the alloys they are used in and the properties those alloys have as a result of their composition or processing.

Combining this information with appropriate risk data for the supply chain of each element then serves to highlight the relative supply risks of the materials that the business uses.

Many of the 'economic area' critical material lists aggregate and weight different classes of risk. This leads to a single metric which, whilst easy to plot, also masks the origin of the risk. In order for a business to understand how to mitigate its risks and to understand the severity, it is important to retain traceability over the source of each risk and we have found that by keeping each risk separate and distinct the user can pick and choose which risks they are most concerned about and apply the information in the correct context.

When considering supply risk it is important to consider the lifetime of the product being assessed. If the product has an extended lifetime (in aerospace 20+ years is normal) and may require replacement during that lifetime, then the likelihood that supply risks will change needs to be considered. During such periods the concentration of world supply can change significantly, new legislation can be enacted and countries descend into conflict — any of which may increase the risk of availability of the material required.

7.2.1 *Business impact*

In order to understand criticality however we need to extend beyond this to consider the impact a supply risk would have on the business if it led to a restriction in access to the resource.

As previously stated, the best way to approach this is from an understanding of the properties that are required by the product and the ability of other materials to be used which have a lower supply risk associated with them. The combination of this information will then serve to inform the user of the level of risk that exists and the severity of the impact of that risk if it happens. A common approach within industry is to use a risk matrix such as the one illustrated in Fig. 7.2, below.

This information can then be used to inform appropriate risk ranking and from there the definition of mitigation strategies.

When considering Impact, whilst some attempt has been made to consider this for regional lists of critical materials, we have observed that industry prefers to retain this assessment for themselves and to make their own

SUPPLY RISK	**Very High**	9	14	19	24	29
	High	7	12	17	22	27
	Medium	5	10	15	20	25
	Low	3	8	13	18	23
	Very Low	1	6	11	16	21
		Very Low	**Low**	**Medium**	**High**	**Very High**
		IMPACT				

Fig. 7.2. Illustration of a risk matrix for considering severity as a function of risk and business impact.

judgements. The selection of a material for a specific application is often a highly proprietary mixture of technical, legal, economic and supply chain driven factors. The impact of a critical materials risk is therefore equally specific to the company involved and needs to consider each of these factors and the relative risks of possible alternative materials that could be used and the knock-on effects to the overall systems design such substitutions often entail.

In our experience, companies are typically very well suited to make these assessments and typically have systems in place to consider and manage such impacts. Our assessment therefore focuses on providing sufficient information to enable companies to quickly identify and assess the supply chain risks associated with the materials used in their products.

7.2.2 *Supply risk*

The challenge of assessing supply risk quickly and easily at the business specific level has been addressed to a significant degree by Granta Design and Rolls-Royce in a U.K. funded project called SAMULET and the approach has since been expanded further in close consultation with industry, in particular the Environmental Materials Information Technology (EMIT)[6] consortium and the ADS Design for Environment working group.

This type of bottom-up assessment of supply risk is now favored by many in industry as a more accurate means of assessing just how significant supply risks may be.

A significant part of the challenge is that manufactured parts, assemblies and products necessarily comprise many materials each of which will contain multiple elements in varying proportions. And as if it were

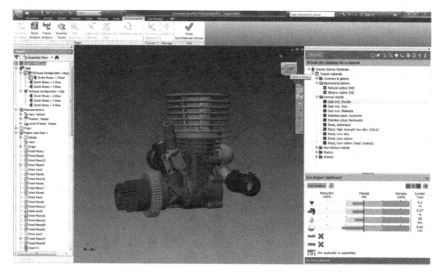

Fig. 7.3.   Example of a Computer Aided Design (CAD) tool from Autodesk incorporating the Granta Design plug in Eco-Materials Advisor,[7] enabling the assignment of materials and the subsequent evaluation of critical materials, restricted substances and environmental risks.

not hard enough to simply assess which elements are in a product, the relative proportions of each element can also have a significant impact, especially when considering the effects of material performance and price volatility.

The ability to subsequently substitute materials for which a supply risk exists for one or more elements demands an additional level of information, that relating to the properties of the material and its potential alternatives. It also requires tools that enable users to effectively screen alternative materials for risks so as to avoid substituting one risk for another, potentially more costly risk.

In the approach summarized here, Granta Design compiled data[8] to enable the evaluation of risk metrics for 65 elements in the periodic table. These risks were subsequently mapped to the compositions of ~1,800 commercially available alloys.

Tools utilizing this data and embedded within web or computer aided design (CAD) tools enable the evaluation of complex designs embodying many of these materials with dynamic feedback to the user of the risks present in the design — see Fig. 7.3, above.

7.3 Risk Metrics

In their approach, the EU as well as many other authors have aggregated various types of risk using weighting to reflect perceived importance. Whilst this approach has the benefit of providing a single number for comparison it fails in two significant respects.

1. Aggregated risks mask the reason why the risk exists.
2. The aggregation approach adopted doesn't consider materials as compositions of multiple elements, each of which carries a different risk.

Both of these issues are significant when considering critical materials from a business context. In the industry-based approach individual risk metrics are reported individually but can be aggregated if required considering only the risks considered relevant to the business. This is enabled by maintenance of a pedigree for each material linking the material to its composition and the elements within that composition to the supply risks specific to that element.

At the present time the number of risks considered is relatively small but comprises risks for which data is readily available and which have a direct bearing upon the accessibility or economics of sourcing and using materials in products or on a variety of legal or reporting requirements that companies face in relation to their products.

The approaches used however are remarkably flexible regarding the source indicators that are used and the authors consider that further indicators will emerge in time to consider other environmental or social factors such as child mortality and child labor, education, or social equality.

7.4 Abundance

Abundance risk represents the fundamental risk to supply associated with the fact that a material is generally scarce on Earth. This is based on the value of abundance in parts per million with abundance risk levels being attributed for threshold values as illustrated in Table 7.1:

Some elements which have a low abundance (i.e. an abundance risk level of 'Very High' are still mined effectively and are sufficiently valuable to justify the cost and energy of extraction (for example, gold). Abundance risk does not therefore provide an indication of production infrastructure. Note also that the uncertainties associated with resource estimates tend to be very large. Mining companies may invest in exploration of reserves

Table 7.1:   Abundance Risk Thresholds.

Abundance in the Earth's crust (ppm)	Abundance risk level
<0.01	Very High
0.01–1	High
1–100	Medium
100–10,000	Low
>10,000	Very Low

in order to justify their commercial investment decisions, but they don't necessarily aim at proving the full extent of the ore body.

Thus, the abundance risk metric does not in itself indicate the risk that a mineral will become unavailable in future (or within the foreseeable time horizon which is typically up to 20 years), since this also depends on geopolitical and economic factors. However, it can provide a preliminary indication of other supply risks that might be associated with the mineral as a pointer to further analysis.

Abundance risk is particularly useful when used in combination with the other Critical Materials risk metrics as a means of understanding why other risks exist, and may help to guide appropriate mitigation actions.

7.5 Monopoly of Supply

This metric represents the degree of concentration of production of an element within any one country, which is used to gauge the risk of monopolistic production within the supply chain.

In general, the Herfindahl-Hirschman Index (HHI) is accepted as an indicator of the monopoly of supply that currently exists for an element, and is defined as the sum of the squares of the market share for the producers of that element, where the market shares are expressed as fractions.

In Granta's Critical Materials data module, the market share is taken to be the total annual production of that element by country. The production data used to calculate the Herfindahl-Hirschman Index (HHI) has been compiled from a range of sources and is expressed as a value between 0 and 10,000 as illustrated in Fig. 7.4, below:

An HHI value approaching 0 would represent a very diverse supply with no single country of production dominating. This would equate to a very low-risk supply in terms of monopolistic production.

An HHI value approaching 10,000 (the maximum possible value) would indicate a significant degree of monopoly in the supply of that element,

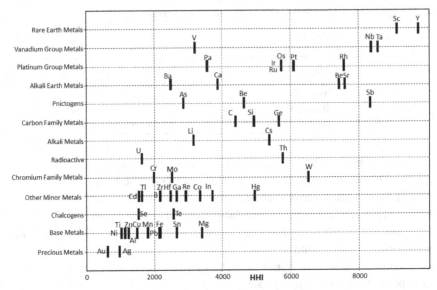

Fig. 7.4. Illustration of the Herfindahl-Hirschman Index for materials based on 2010 production data from the British Geological Survey and the U.S. Geological Survey.

and a Very High risk of monopolistic production. A similarly high risk of potential supply disruption would also then exist should that country wish to exert control over that supply (for example, for domestic use) — other indicators, such as price volatility, might point towards this type of behavior.

Note that this metric focuses specifically on the country of production, and does not consider monopoly by a company or organization. For instance, one company could produce 100% of the world output of an element, but if that production takes place in many different countries the HHI value will be low.

When interpreting the business risk associated with a monopolistic supply, it is also important to consider:

- The country which dominates that supply (if this is domestic to you, then this will probably pose a lower risk).
- The existence and reliability of supply agreements and ownership of non-domestic resources, which are also options for mitigating this risk.
- The export restriction history of the country that dominates the supply. For example, China was subject to a successful dispute settlement initiative at the World Trade Organization level following its imposition of

export restrictions (including quotas and export duties) on other WTO members.

7.6 Sourcing and Geopolitical Risk

This metric is based on a modified and scaled version of the Herfindahl-Hirschman Index (HHI) that embodies the geopolitical risk of the producing countries, as well as the degree of monopoly in the supply of a material.

The measure used for geopolitical risk is the World Bank's Worldwide Governance Indicator (WGI)[9] which represents six dimensions of governance:

- Voice and Accountability
- Political Stability and Absence of Violence
- Government Effectiveness
- Regulatory Quality
- Rule of Law
- Control of Corruption

These have been aggregated to provide a single indicator (WGI) which is expressed for 213 economies. To calculate the index, the Worldwide Governance Indicator for each producing country is combined with production data for each element so that the index value calculated for an element best reflects the WGIs of the countries which produce the majority of that element. The approach used to calculate monopoly of supply is the same as for the Herfindahl-Hirschman Index (HHI) and uses the same production data.

Sourcing and geopolitical risk is also attributed with threshold values to express this as a qualitative risk level, using the thresholds shown in Table 7.2, below.

The Sourcing and geopolitical risk metric reflects the premise that an element for which production is dominated by one highly stable country

Table 7.2: Sourcing and Geopolitical Risk Thresholds.

Sourcing and geopolitical risk HHI	Sourcing and geopolitical risk level
>4	Very High
3–4	High
2–3	Medium
1–2	Low
<1	Very Low

may not necessarily pose as much of a risk to supply disruption as an element which is produced by 10 countries each of which is politically unstable.

A Sourcing and geopolitical risk HHI value of less than 1 (i.e. a risk level of Very Low) represents a diverse supply with the bulk of this supply being produced by politically stable countries. The risk of a supply disruption occurring for such materials as a result of political or civil unrest is expected to be Very Low.

At the other end of the scale, a Sourcing and geopolitical risk HHI value of greater than 4 (i.e. a risk level of Very High) represents a more monopolistic supply with those producing countries being subject to greater political instability. The risk of political or civil unrest disrupting the supply of such materials is considered to be Very High.

In between these two endpoints, the ranking of an element or material depends on the interaction between the two factors.

Sourcing and geopolitical risk is therefore typically considered to be a more reliable metric than the simple HHI for judging whether a significant supply disruption is likely to occur as a result of political instability or civil unrest in one or more of the more significant producing countries.

7.7 Environmental Country Risk

This metric is also based on a modified and scaled version of the Herfindahl-Hirschman Index (HHI) that embodies the risk of the producing countries limiting production of a resource on account of environmental legislation, as well as the degree of monopoly in the supply of the material.

The measure used for the environmental risk associated with each country is the Environmental Performance Index (EPI) produced by Yale University.[10] The EPI encompasses the two primary environmental drivers of environmental health and ecosystem vitality. Each of these objectives is represented by a number of environmental indicators, grouped into policy categories (as shown in Table 7.3, below), which are weighted and then aggregated to give the EPI score for each country. Thus, a country's EPI score reflects its degree of environmental quality and maturity.

To calculate the Environmental country risk HHI, the EPI scores are combined with production data for each element so that the index value calculated for an element best reflects the environmental maturity of the countries producing that material. The approach used to calculate monopoly of supply is the same as for the Herfindahl-Hirschman Index (HHI) and uses the same production data.

Table 7.3: Environmental Performance Indicators Developed by Yale and used for the Calculation of the Environmental Performance Index.

Objective	Policy category	Indicators
Environmental health	Air pollution (effects on human health)	Indoor air pollution
		Particulate matter (air)
	Water (effects on human health)	Access to drinking water
		Access to sanitation
	Environmental burden of disease	Child mortality
Ecosystem vitality	Air pollution (effects on ecosystem)	SO_2 emissions per capita
		SO_2 emissions per \$ GDP
	Water resources (effects on ecosystem)	Change in water quantity
	Biodiversity and habitat	Biome protection
		Marine protected areas
		Critical habitat protection
	Forests	Forest loss
		Change in forest cover
		Change in growing stock
	Fisheries	Coastal shelf fishing pressure
		Fish stocks overexploited
	Agriculture	Agricultural subsidies
		Pesticide regulation
	Climate change and energy	CO_2 emissions per capita
		CO_2 emissions per \$ GDP
		CO_2 emissions per KWh of electricity
		Renewable electricity

While the EPI indicator for a region is not solely driven by mineral extraction, by its nature the mining and the refinement of some elements tends to be damaging or at least disruptive to the environment. If not properly controlled, extraction of a material can involve significant deforestation, airborne pollution, the use of highly toxic substances and significant amounts of energy, and downstream impacts to human health, fisheries, and habitats.

- A material with an Environmental country risk HHI value greater than 4 (i.e., a risk level of Very High) suggests that countries producing that material are generally exercising a lower degree of care for the environment and for human health and well-being. This means that production is likely to be more damaging to the environment and to the native population than it might otherwise be. It may also mean that those producing countries will be under greater pressure to improve their environmental performance, and less able to make these improvements without potentially significant disruptions to supply.

- A material with an Environmental country risk HHI value lower than 1 (i.e., a risk level of Very Low) suggests that countries producing that material are generally exercising a much higher degree of care for the environment and for the health and well-being of their population. This means that actions have probably already been taken to reduce the damage that production poses to the environment and to minimize the impacts on human health and well-being. It may also mean that those producing countries are also already well equipped to respond to future environmental legislation and pressures to improve, and that supply disruption is likely to be minimal.

The production of some materials will naturally center on countries which exercise lower levels of environmental and legislative control, as production in these countries can be more readily established and material produced at a lower cost in the absence of any company-specific controls. This metric alone does not provide business assurance that a company's own supply is not from a country with a lower level of environmental control, but does provide a fair indication of the likelihood that their supply is from such a country.

7.8 Conflict Mineral Risk

This metric currently indicates the risk that a material's sourcing or production may have helped to finance conflict in the Democratic Republic of Congo, either directly or indirectly — a definition that is likely to be expanded to include other minerals and other regions once relevant EU legislation comes into force in the coming years.

The implication of this risk is a legal requirement to provide an independent traceability audit and report of the material's supply chain and to disclose information that may be perceived by the consumer as being detrimental to brand value.

In general terms, a conflict material is an element which has been obtained from a mineral whose production or trade has financed conflict. At the moment, this definition is applied only to minerals sourced from the Democratic Republic of Congo (DRC), or from the adjoining countries of Angola, Burundi, Central African Republic, Congo Republic (a different nation than DRC), Rwanda, Sudan, Tanzania, Uganda, and Zambia, through which minerals might otherwise conceivably be traded legitimately.

The concept of a conflict mineral was enshrined under the US Conflict Minerals Law (section 1502 of the Dodd-Frank Act),[11] which was signed into law in the US on the 21st July 2010 and included:

- columbite-tantalite (coltan)
- cassiterite
- gold
- wolframite

or any derivative of these; or any other mineral or derivative determined by the US Secretary of State to be financing conflict in the Democratic Republic of Congo.

In terms of this law, the label 'conflict mineral' is applied as a blanket term to all minerals named within the law and their derivatives, regardless of their origin or whether trade in any specific unit of them has financed conflict. For instance, anything containing tin (a derivative of cassiterite) is considered to contain a conflict mineral, regardless of where or how the tin was obtained.

Under the US Conflict Minerals Law, manufacturers were required to investigate and report on their use of conflict minerals (as defined within the law), and to provide an independent third-party audit of the supply chain and full disclosure of the source for each conflict mineral used. This requirement necessarily extended outside the US, due to the nature of the global supply chain involved and the requirement to trace the material back to its source. It therefore potentially carried a significant administrative burden.

In the Granta database each element is assigned to one of three risk categories, with a risk level of 'High', 'Caution', or 'None', as shown in Table 7.4, below.

Table 7.4: Conflict Material Risk Thresholds.

Risk category	Conflict material risk level
Conflict mineral as defined in US Conflict Minerals Law	High
Other elemental minerals reported to be sourced from DRC and adjoining regions	Caution
Absence of reliable information	None

A risk level of High indicates that the composition includes a 'conflict mineral' — an element which is known to be produced by the DRC or one of its adjoining countries (but could also be produced elsewhere) and which, according to US law, would require independent third-party supply chain traceability audits and reporting to verify the source of the actual supply used.

A risk level of Caution was introduced to indicate that the composition includes an element which is not formally defined as a conflict mineral under US legislation, but is otherwise known to be produced by the DRC or an adjoining country.

A risk level of None indicated that there is no current evidence that the element has a source in the DRC or any of its adjoining countries.

7.9 Price Volatility Risk

Price volatility reflects historic fluctuations in the price of a material as sold on the open market, and is calculated as the percentage difference between the maximum and minimum price (in USD/kg) over the past five years, relative to the minimum price.

Considered simplistically, prices for materials tend to increase when demand for that material accelerates faster than supply. How much prices increase is a reflection of what the market will bear in terms of price, how valuable that material is for the particular application, the availability of more economical and perhaps more abundant substitutes, the value of competing technologies, and how rapidly supply can be scaled to meet demand (including potential barriers such as monopolies of supply).

Price fluctuations in a material can also arise due to fluctuations in another commodity, in particular the cost of energy — as the extraction and refining of materials sometimes requires significant amounts of energy, in some cases as much as 580,000 MJ/kg (e.g. for Rhodium). For such materials, it is perhaps to be expected that a comparatively small increase in the price of energy will have a significant impact on the downstream price of the material to the consumer (unless other mechanisms exist to prevent it). This is typically reflected also by a higher price for that material.

While historic price fluctuations do not provide any certainty over future price variations, they may provide an insight to the ability that global production has to scale supply with demand, and may also reflect the exposure the production of that material has to increases in the cost of energy.

Table 7.5:   Price Volatility Risk Thresholds.

Price volatility (%)	Price volatility risk level
>400	Very High
300–400	High
200–300	Medium
100–200	Low
<100	Very Low

Price volatility risk level expresses this as a qualitative risk level, using the thresholds shown in Table 7.5, above. In setting these thresholds Granta Design adopted the approach suggested by the EU Report of the Ad-hoc Working Group on defining critical raw materials (2010).

As price volatility is calculated from historic market rates, it will not necessarily reflect any business-specific contracts a company may have which enables them to buy materials at a different or preferential rate, nor will it reflect any market trading practices. This metric is unable to provide any guidance as to future price fluctuations, and neither does it provide any business assurance that a company's own supply is immune from future price changes. It does, however, provide a fair indication of the ability the supply-chain of that material has to react to changes in demand (which may arise due to the emergence of a new technology, or from legislative requirements forcing the adoption of alternate materials for existing technologies), and to other more common market forces such as the cost of energy.

7.10   Application of Risk Metrics at a Business Level

Through the Granta Design software, the risks associated with each of these elements have now been linked to the 1,800+ alloys contained in Granta's Material Universe database based on the composition of elements in each alloy. This database already contains significant volumes of material property information across a wide range of attributes and is supported by tools for materials selection, materials substitution, and materials optimization.

Through interfaces embodied within a variety of popular CAD interfaces or through web-based applications, reports can be generated which summarize each of the elements contained in a product and the level of each risk category. Similar reports can also be generated for substances restricted by legislation such as RoHS or REACH which are linked to materials and processes in a similar manner and also for Energy and CO2 footprint which are also significant business drivers.

Reports on critical materials are available at different levels. At the top-level reports indicate the highest value within each risk category for each of the elements within each part. This top-level analysis enables specific assemblies to be identified as a potential source of risk within a product that may contain many different materials, see Fig. 7.5 for an example top level report from the Granta Product Risk package.

At lower levels, risks are identified within each assembly and then for each element within that assembly, thus enabling hot-spots to be identified quickly and effectively and for risk mitigation strategies to be appropriately developed, see Fig. 7.6 for an example detailed report from the Granta Product Risk package.

Summary for all parts

Part	Part count	Mass (kg)	Abundance risk	Conflict material risk	Environmental country risk	Price volatility	Sourcing and geopolitical risk	Monopoly of supply (HHI)	Max. price variation (USD)
Part 1	1	5.0	Low	None	High	Very high	High	6300	2.7
Part 2	1	12	Low	None	Medium	Very low	Medium	4100	54
Part 3	1	4.0	Medium	Caution	Very high	Very high	Very high	7600	37

Fig. 7.5. Top-level summary of critical materials and conflict minerals risks for a made-up product.

Part Summary: Part 1

Material	Mass (kg)	Abundance risk	Conflict material risk	Environmental country risk	Price volatility	Sourcing and geopolitical risk	Monopoly of supply (HHI)	Max. price variation (USD)
Cast iron, gray, flake graphite, EN GJL 100	5.0	Low	None	High	Very high	High	6300	2.7
Overall for part	5.0	Low	None	High	Very high	High	6300	2.7

Details: Elements in Part 1

Material: Cast iron, gray, flake graphite, EN GJL 100

Element	Mass (kg)	Abundance risk	Conflict material risk	Environmental country risk	Price volatility	Sourcing and geopolitical risk	Monopoly of supply (HHI)	Max. price variation (USD)
Carbon	0.18- 0.20	Low	None	High	Low	High	6300	0.17
Iron	4.6- 4.7	Very low	None	Low	Very high	Low	2500	2.3
Manganese	0.025- 0.035	Low	None	Very low	Very low	Very low	1700	0.018
Silicon	0.12- 0.13	Very low	None	Medium	Very low	Medium	4800	0.16
Overall for material	5.0‡	Low	None	High	Very high	High	6300	2.7

Fig. 7.6. Lower level summary indicating the critical materials and conflict minerals risks for a made-up part and for the elements contained in the materials of that part.

7.11   Case Study

As an example of how this bottom-up analysis might be used, we consider a high temperature nickel-based alloy such as the single crystal superalloy CMSX-4, typically used by a wide range of aero engine manufacturers for components such as combustors, turbines, vanes, and blades. The composition of the alloy is illustrated in Fig. 7.7, below.

If the risk profile of this alloy had been assessed according to the original list of 14 elements highlighted by the Critical Raw materials for the EU report, we should primarily have been concerned about the risks represented by Cobalt, Tungsten and Tantalum each of which represents between 6% and 10% of the composition by mass.

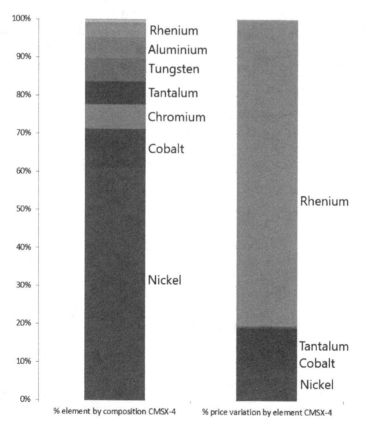

Fig. 7.7.   Composition of CMSX-4 and the relative price variation of the elements factoring in composition and the maximum elemental price variation over a five-year period.

   Looking at the five-year price variation for each element however and
scaling for the composition of the alloy using the Granta Design tools we get
a very different picture. Although Rhenium comprised only 3% of the composition of the alloy, it represented the biggest risk for the alloy as a whole
in terms of price variation, accounting for over 80% of the compositional
price variation (over 5 years) with Nickel and Tantalum making up the bulk
of the remaining 20%. Rhenium and Nickel both have comparatively low
supply risks according to the EU report and so a risk assessment based
upon these elements of regional concern might not have highlighted the
most significant economic risks to a business using this alloy.

   Figure 7.8 illustrates the geopolitical supply risk for the elements contained in CMSX-4, Tungsten and Cobalt appear to represent the most
significant risks in terms of the potential for geopolitical supply disruption,
with Tantalum rating fifth, below chromium and Molybdenum — neither of
which were included in the original list of 14 elements of concern to the EU
as a region. Of these, cobalt and tungsten are considered to be of Medium
to High risk and so these are probably of greatest concern, the supply of

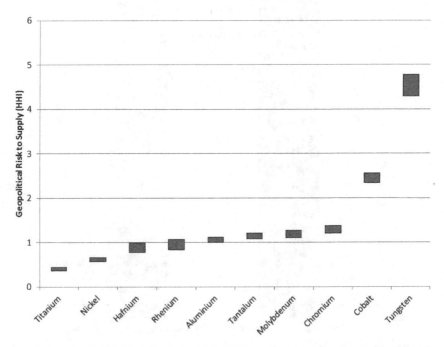

Fig. 7.8.   Modified Herfindahl-Hirshman Index representing the geopolitical supply risk
for the elements in CMSX-4.

these elements being dominated by the DRC (53% of supply of Co) and China (85% of supply of W).

The benefits of applying critical materials risk metrics in a business-specific environment are that this therefore highlights the most relevant risks for a product and by presenting these risks individually in a readily accessible form it is possible to identify assemblies, materials and elements that may be of concern and to use these to investigate possible business risk mitigation strategies.

7.12 Responding to Supply Risks

If we consider monopoly of supply against sourcing and geopolitical risk we can produce Fig. 7.9 below, where we have illustrated four distinct zones for the 65 elements under consideration.

Politically Stable, Low monopoly:

Elements which have a low monopoly of supply and for which the countries producing them are typically geopolitically stable.

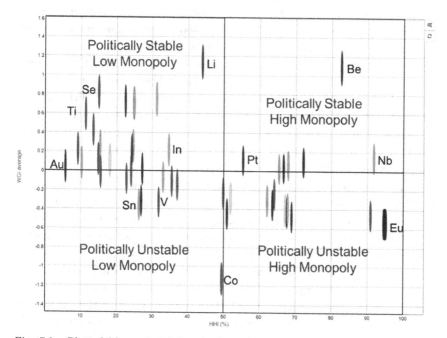

Fig. 7.9. Plot of Monopoly of Supply (x-axis) vs. Sourcing and Geopolitical Risk (y-axis), for 65 elements.

Politically Stable, High monopoly:

Elements where a monopoly in production does exist but where the countries producing the element are geopolitically stable and therefore less likely to enact supply restrictions.

Politically Unstable, Low Monopoly:

Elements which are commonly produced in countries that are geopolitically unstable but where a monopoly doesn't exist. For these elements supply from a different country is likely to be possible is supply is constrained by one region.

Politically Unstable, High Monopoly:

These are the elements most of concern as the supply is monopolized by countries that may be more likely to undergo regime change, political or other disruptions which could lead to reduction in supply through various mechanisms.

This is a simple illustration of the complexity contained within some of the more simple materials supply risk metrics and highlights why being able to quickly identify risks is important and also why being able to classify the nature of the risk matters to industry. How a company would respond to an element they use in one of the zones listed above is inherently different to how they might respond to an element in another zone.

In some cases, a simple supply agreement may be sufficient to mitigate the risk, in other cases a much more comprehensive response might be required which might include relocation of manufacturing facilities, stockpiling of materials or investment in mining activities to secure supply.

Being able to correctly judge the correct response requires data that can be used easily and which can be drilled into reliably to provide the insight needed to inform business decisions.

Although often touted as a solution, many businesses are very reluctant to substitute materials use in their products. In many cases, materials have been used for many years and changing them takes a considerable investment which is only typically worthwhile if there is a significant economic benefit that can be reaped through enhanced sales resulting from improved performance or from a long term and reliable decrease in material costs.

One potential disruptor in this space is the concept of a Circular Economy put forward by the Ellen MacArthur Foundation.[12] In theory, leasing based approaches may enable higher value materials to be used in applications which may not be economically viable within a linear (i.e. take-make-dispose) economic setting. Such approaches however are only likely to be viable if significant performance benefits exist and where the lifetime of the product can be suitably extended through the use of these higher value materials to the extent that long term savings in manufacturing and transportation can offset the additional materials costs.

The viability of such a circular economy approach can now be assessed using circularity metrics developed by the Ellen MacArthur Foundation and Granta Design in combination with critical materials risks so as to manage the tradeoff between product performance, cost and risk.[13]

7.13 Materials Analysis and CSR

Having shown an example of the risks of considering a material's criticality at a broad regional level rather than on the basis of the material's composition and within the context of a specific business application, might there be any risks to consider with the application of conflict minerals legislation such as the Dodd-Franck act or the legislation currently being considered in the EU?

Currently the Dodd-Frank act considers materials sourced from the Democratic Republic of Congo or from the nine surrounding regions of Angola, Burundi, Central African Republic, Congo Republic, Rwanda, Sudan, Tanzania, Uganda, and Zambia. For this legislation there is an established causal link between mining activities and the fueling of war and human rights violations. These are not however the only regions noted to be in a state of conflict. There are multiple sources and definitions of conflict, and the following are useful resources when considering the broader scope of conflict regions:

- The Uppsala Conflict Data Program (UCDP)[14]
- The PRIO Centre on Culture and Violent Conflict[15]

Whilst there may be no direct causal link between mining activities and conflict evidenced in other regions as there is in the Democratic Republic of Congo, the presence of conflict in these regions nevertheless poses a potential risk to the supply of materials from these regions, as evidenced by the geopolitical supply risk for materials associated with these countries in the Granta Critical Materials data module.

In addressing such risks should there perhaps be a concern that an unintended consequence of these legislative measures may be a reduction in trade with these regions which might in turn have unintended consequences for the economic development of the producing region?[16]

If we consider a scenario that represents most end user companies, one in which there is little or no direct traceability of raw materials back to the specific country of origin, we might extend the analysis already presented to instead consider a scenario in which we assume that production of the material used in our products is represented by the global production figures for the elements contained within it.

Considering again a component manufactured from CMSX-4 superalloy, the countries from which our material might be sourced can be represented in Fig. 7.10, below:

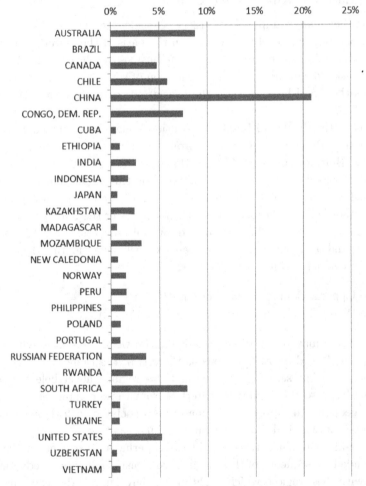

Fig. 7.10. Percentage production of CMSX-4 elemental components by mass, assuming sourcing follows global production for each element (2011 production figures).

If we then consider some of the socio-economic factors influencing these countries, Fig. 7.11 illustrates two of the many social metrics contained in the Granta Design Sustainability database. These highlight an elevated infant mortality and a lower human development index in many of the countries from which materials for this alloy may be sourced.

Thus the Granta Design Sustainability database can be used to consider other factors that might not be connected directly to materials, but rather the communities they are sourced from. Whilst in some cases there is no direct causal link between mining activities and quality of life in a particular region, the presence of social factors within regions where materials are being sourced should nevertheless be of interest when considering measures that may be taken as part of a corporate social responsibility program. With changing perceptions of company responsibility widening to that of the communities they operate in or source materials from, it is important to consider how corporate social responsibility activities could contribute to improved development of those communities.

Importantly, companies should not just cease sourcing from a conflict region or one with low quality of life scores, but rather encourage the development of conflict-free mines or initiatives to improve quality of life in those regions and work with governments and NGOs to improve the stability of the country. Leaving such areas could be a missed opportunity to use industry's buying power to bring about changes to a region, but also could allow other companies from countries with lesser social obligations to take their place.

The broader consideration of regions of conflict is a potential requirement for future ethical sourcing policy. As such, work is being done on this metric to understand and evaluate the social, economic and environmental standards within a given supply region for the purposes of determining any potential procurement risks to a business's ethical standards or policies.

7.14 Elements or Compositions

In much of the literature on critical or conflict minerals, criticality is considered at the elemental level. However, as we have already stated, elements are rarely used in isolation but rather within a composition such as an alloy, where it is the composition in combination with the manufacturing process and the product and system design than yields the performance that the application demands. Dealing with compositions is not however reflected well in any of the existing literature.

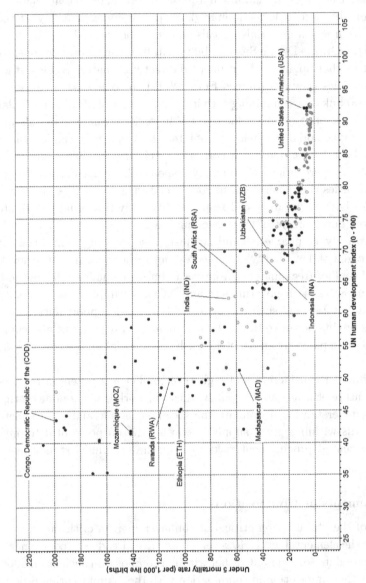

Fig. 7.11. Two socio-economic indicators for the countries producing elements contained in CMSX-4. X-axis — UN Human development Index. Y-axis — Under 5 mortality rate (per 1,000 live births).

For example, if we look at the property data within the Granta Materials Universe database and consider the maximum service temperature of the 1,800+ alloys contained within it we can create Fig. 7.12 where we can see that for alloys with a maximum service temperature above 1,400°C we will always have a material with a 'Very High' sourcing and geopolitical risk.

This plot however only indicates that alloys contain at least one element that has a 'Very High' sourcing and geopolitical risk, but it doesn't reflect how much of the alloy has that risk. The alloys classed at Very High risk may contain 100% or 0.01% and still fall within this classification.

One method to illustrate this difference has been explored by Granta Design and considers the aggregate risk of the material composition. By this method we effectively attribute the supply of the alloy itself to the countries that produce the elements it contains while weighting the allocation by each country's production of each element and then following the same methodology for calculating risk.

This 'Compositionally Aggregated' indicator then reflects not just the presence of a high-risk element but the concentrations of each element and their relative risk. Figure 7.13 reflects how Sourcing and Geopolitical risk differs by adopting this approach by comparing this compositionally aggregated metric against the original 'worst-case' methodology that was based purely on the presence of high risk elements.

From this figure we can immediately identify that the commercial purity rare-earth elements are considered Very High risk within both methodologies. Within the compositionally aggregated methodology, these materials represent a high proportion of a high-risk element.

Below this however we have a range of alloys, including magnesium alloys such as WE43A which would have been considered very high risk by the worst-case methodology but which according to the compositionally aggregated methodology are only marginally higher risk than some of the alloys that would have been considered Medium or High risk.

This compositionally aggregated indicator is important as it highlights two distinct cases:

In the first case, an elevated risk exists because of a trace element. In these cases, the material may not function as needed without that element but the ability to source enough of that element might not be significantly impacted — depending upon relative market demand for the element. For example, it may be possible to source enough of the element to produce the quantity of alloy needed without a very significant increase in cost.

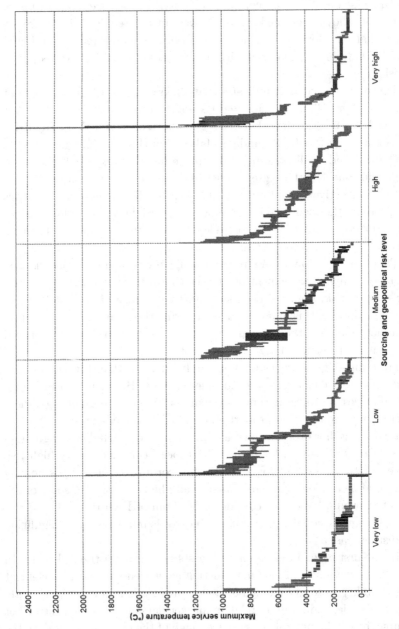

Fig. 7.12. Plot courtesy of Granta Design of Maximum service temperature against Sourcing and Geopolitical Risk.

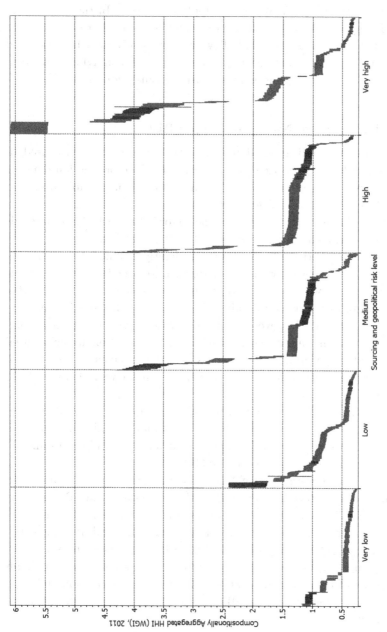

Fig. 7.13. Comparison of compositionally aggregated sourcing and geopolitical risk and the worst-case approach adopted by most reports.

In the second case, an elevated risk exists because a significant proportion of the composition is at risk of supply disruption. In these cases, the economic impacts of a supply disruption may be more severe.

The figure also illustrates that alloys which may otherwise be considered only at medium risk today because of the risk of their individual elements may, on aggregate, represent as high a risk as some higher risk alloys.

7.15 The Temporal Nature of Supply Risks

With each of the indicators discussed in this chapter, the data used to calculate risk has always been historic in nature. The production data for each country is typically at least 1–2 years old at point of publication and similar delays exist for WGI and EPI data.

What we are in effect looking at is the history of the risk of our materials and not the risk as it exists today or will exist in the future.

In this context the data may appear slightly redundant, as what we would all like to know is what will happen next over the lifetime of our products, so that we can act accordingly to mitigate our risks. Unfortunately none of us has a crystal ball with which to predict the future.

What we can do however is consider critical materials risks beyond a single point to highlight trends. Taking the compositionally aggregated sourcing and geopolitical risk we discussed in the previous section, we can look at how this metric has changed over the fifteen years between 1996 and 2011.

In Fig. 7.14, we plot the compositionally aggregated risk for each material using data from 1996 against the same data for 2011. Each material is reflected as a range because of compositional variation and uncertainty in both the production figures and the WGI itself.

In this plot, materials which have not changed in risk lie along the black line, some materials (such as Palladium-silver alloy and Ruthenium) fall below this line and have in effect decreased in risk over this period.

Many other materials however fall above this line and have increased in risk. These materials include the rare earth elements such as Dysprosium, but also include the majority of steels, magnesium and aluminum alloys. Some of these materials were relatively low risk in 1996 and were only marginally higher risk in 2011. Others have increased in risk more significantly.

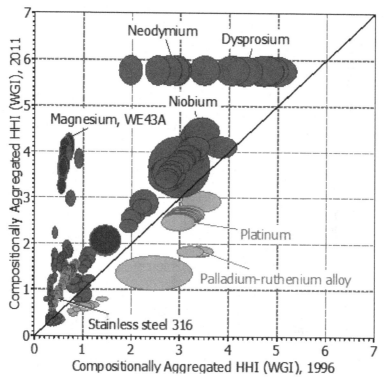

Fig. 7.14.   Compositionally aggregated sourcing and geopolitical risks for 1996 and 2011 for 1,800+ alloys from the Granta Design Material Universe database.

We can reflect this increase by dividing the geopolitical risk score in 2011 by the score in 1996, Fig. 7.15 illustrates this metric against density for materials in Granta Designs database.

From Fig. 7.15 we can observe that steels have experienced a modest increase in risk over this period, whereas many aluminum and magnesium alloys have experienced a more significant increase in risk.

Digging into the data behind this, the increase in aluminum and magnesium risks result from a significant increase in production of aluminum and magnesium by China — a country considered by the World Bank as being at an elevated geopolitical risk.

The small increase in the risk of steel might not be considered of particular concern. However, steel tends to be a material used in significant

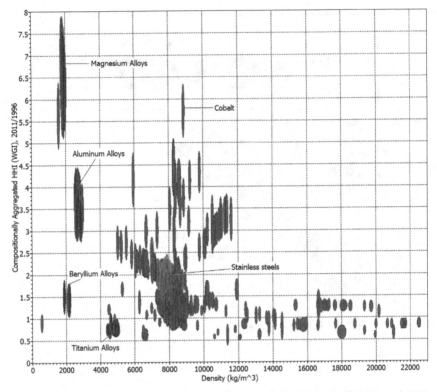

Fig. 7.15. Relative increase in sourcing and geopolitical risk between 1996 and 2011 against material density, from Granta Design Materials Universe database.

quantities and a comparatively minor increase in production costs can have a significant economic impact on the businesses producing or consuming it.

7.16 Conclusions

The assessment of risks associated with critical materials and conflict minerals is an emerging field and one which places a strong requirement on the understanding of what our products contain and what economic, geopolitical, environmental and sociological factors exist that might restrict supply and production.

Addressing supply risks at a regional or sectorial level, whilst beneficial for the purposes of defining policies or broad research agendas, does not always yield the correct information when considering supply risks at a business or product level where a broader range of risks needs to be considered and presented either without aggregation or at least with the ability to drill-down to the assemblies, materials and elements that may be of greatest concern.

The ability of businesses to mitigate risks to supply also requires an understanding of materials and the properties that make them suitable for use as well as the lifecycles of the products themselves. These influence our ability to reuse, remanufacture or recycle these materials — which is particularly important in cases where no suitable substitute currently exists. In the Western world in particular, the concept of the Circular Economy is increasingly seen as a means of reducing our reliance on the continued import of virgin materials from areas impacted by conflict or in which there is an emerging middle class who will increasingly require their materials for the legitimate benefits of their own industries and populations.

Responding to materials supply risks may also inadvertently lead to changes in supply chains which have the potential to detrimentally impact the economic, environmental and social progression of the countries in which those supply risks exist. The potential to use supply-based information not just as a driver to alter supply chains to avoid risk but instead as an incentive to drive investment into social and environmental welfare programs through corporate and social responsibility commitments and ethical sourcing agendas — even when no established causal link exists between the material production and the deficiencies of the producing region — certainly warrants further investigation.

References

1. ADS. (n.d.) Design for Environment (DFE): Position papers on access to resources, energy consumption, recyclability potential, waste and water consumption. Retrieved from https://www.adsgroup.org.uk/membership/gr oups-committees/design-for-environment-dfe/
2. Peck, D. (2016). *Prometheus missing: Critical materials and product design* (Doctoral thesis). Delft University of Technology, the Netherlands. Retrieved from https://repository.tudelft.nl/islandora/object/uuid%3Aa6a69144-c78d -4feb-8df7-51d1c20434ea
3. European Commission. (2010). *Annex V to the report of the ad-hoc working group on defining critical raw materials*. Brussels, Belgium: European Commission.
4. Graedel, T.E., Harper, E.M., Nassar, N.T., Nuss, P. and Reck, B.K. (2015). Criticality of metals and metalloids. *Proceedings of the National Academy of Sciences of the USA*, 112(14), 4257–4262.
5. Ashby, M. and Johnson, K. (2002). Materials and design: The art and science of material selection in product design. Oxford, UK: Butterworth-Heinemann.

6. Granta. (n.d.). The Environmental Materials Information Technology (EMIT) Consortium. Retrieved from http://www.grantadesign.com/emit/index.htm.
7. Autodesk. (n.d.). Eco materials advisor. Retrieved from https://www.autod esk.com/developer-network/certified-apps/eco-materials-adviser.
8. Granta. (n.d.). Granta design critical materials data module. Retrieved from https://www.grantadesign.com/products/data/critical.htm.
9. The World Bank Group. (n.d.). World Bank governance indicators. Retrieved from http://info.worldbank.org/governance/wgi/#home.
10. Emerson, J., Esty, D.C., Levy, M.A., Kim, C.H., Mara, V., de Sherbinin, A. and Srebotnjak, T. (2010). *2010 environmental performance index.* New Haven, Connecticut, USA: Yale Center for Environmental Law and Policy.
11. US Conflict Minerals Law. (2010). In the Dodd-Frank Wall Street Reform and Consumer Protection Act (Section 1502).
12. Ellen MacArthur Foundation. (2013). Towards a circular economy. Cowes, UK: Ellen MacArthur Foundation.
13. Ellen MacArthur Foundation. (2015). Circularity indicators: an approach to measuring circularity. Cowes, UK: Ellen MacArthur Foundation. Retrieved from https://www.ellenmacarthurfoundation.org/assets/downloads/insight/Circularity-Indicators_Methodology_May2015.pdf.
14. Uppsala University. (n.d.). Uppsala Conflict Data Program conflict ency-clopaedia. Retrieved from http://www.pcr.uu.se/research/ucdp/ucdp-confl ict-encyclopedia.
15. PRIO. (n.d.) PRIO Centre on Culture and Violent Conflict. Retrieved from https://ccc.prio.org/.
16. Propper, S. and Knight, P. (2013, December 4). 'Conflict free' minerals from the DRC will only be possible if companies stay. *The Guardian.* Retrieved from https://www.theguardian.com/sustainable-business/conflict-free-mine rals-drc-companies-stay.

Chapter 8

In Search of an Appropriate Criticality Assessment of Raw Materials in the Dutch Economy

Elmer Rietveld and Ton Bastein

The Netherlands Organization of Applied Science (TNO), unit Strategic Analysis & Policy unit Anna van Buerenplein 1, the Hague, Netherlands

Past events and predictions suggest the need for a methodology to assess the criticality of raw materials to national economies. Existing criticality methodologies were combined to develop a raw materials criticality methodology for the Dutch economy, considering materials embedded in intermediate or finished goods as well. Indicators are described according to their relevance for assessment of risk for supply security, financial damage or reputational damage for companies. The impact aspect is based on value added in exported domestically produced goods.

8.1 Introduction to the Necessity of Raw Material Criticality Assessment

Introductions explaining the need for raw materials criticality assessment often cite the expected developments in population size, affluence and technology.[1] During the 20th century, population and wealth growth led to an increase in the extraction of construction materials by a factor of 34, ores and industrial minerals by a factor of 27, fossil fuels by a factor of 12 and biomass by a factor of 3.6.[2] The global population is expected to reach 9 billion by 2050 and 10.1 billion by 2100.[3] A strong increase of resource consumption in emerging and non-western economies can be anticipated, where GDP growth rates of over 5% year are predicted for the coming decades.[4] Consequently, a tripling of the global consumption of materials has been predicted.[5] Foresight studies into the impact of technology developments do not offer a prospect of a stable market that guarantees that the pace of R&D can match these population and affluence developments.[6] Furthermore, the energy transition, widely regarded as an urgent global challenge, is strongly dependent on raw material supply.[7,8]

Disruptions to supply are perhaps the most drastic problem for economic activities based on manufacturing, as opposed to private or public services. In a manufacturing sector study commissioned by the Dutch Employers Organization for the High-tech equipment FME,[9] a large proportion of the companies interviewed indicated that they had experienced problems as a result of disruptions to the supply of their most critical materials. These were mostly related to problems of their intermediate suppliers and were thus seldom related to genuine problems in the supply of raw materials.

The use of raw materials in modern economies therefore deserves the same careful consideration that has been given to labor, capital, energy and R&D. They are a critical ingredient in maintaining the desired quality of life as (mostly Western societies) have been fortunate to enjoy in recent decades. The link between raw materials and societal impact induced the Dutch national government to commission a study to get a clear picture of both the direct and indirect dependence of the Dutch economy on raw materials. Hence the Netherlands economy serves as case study for the criticality methodology.

8.2 Common Aspects of Criticality Assessments

The assessment of risks and vulnerabilities of raw materials seem to have three common elements. These are: documenting the resources and reserves, identifying the risks (supply disruptions combined with the impact of these disruptions) that prevent the access to these resources and the impact in case mitigating these risks is ineffective.

"Classical" risk analysis is principally concerned with investigating the risks surrounding a plant (or some other object), its design and operations. Such analysis tends to focus on causes and the direct consequences for the studied object. Vulnerability analysis takes a broader perspective and focuses both on consequences for the object itself and on primary and secondary consequences for the surrounding environment. It also concerns itself with the possibilities of reducing such consequences and of improving the capacity to manage future incidents".[10,11] In general, a vulnerability analysis serves to "categorize key assets and drive the risk management process".[12,13]

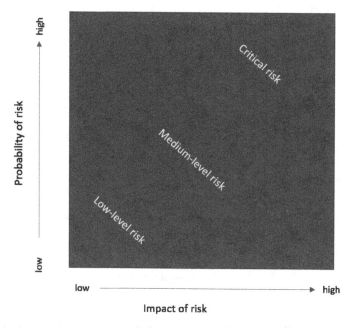

Fig. 8.1.   Classical risk analysis plot.

The outcome of a risk analysis takes the shape of a vulnerability diagram, such as the one depicted in Fig. 8.1.

The criticality assessment of the EC (revised in 2016 by the JRC) at first glance follows this general line of thinking: the assets chosen are (biotic and abiotic) raw materials, the criticality investigated is the probability for a supply disruption of a specific raw material, and the consequences of that supply risk for the potential damage for the European economy. The schematic vulnerability plot used by the EC is given in Fig. 8.2.

In this Fig. 8.2, the term chosen on the y-axis "A measure of risk of supply shortages" mixes the terms for risk and probability.[14] The supply risk combines elements relevant for probability assessment (production concentration for instance) with elements that are related to potential mitigating measures (such as substitution and recycling). However, the substitution options of a given material do not interfere with the probability of a supply disruption of the material. To discard substitution as an element of the y-axis of the EC-methodology[15] may therefore be considered. The y-axis could then be interpreted as a measure of the probability of supply disruptions.

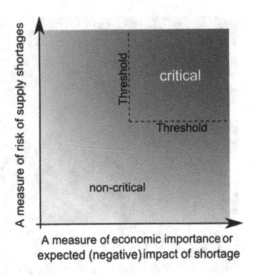

Fig. 8.2.   Vulnerability plot.[16]

The "measure of economic importance or expected negative impact of shortage" is a measure that is well in line with the Impact measurement that is an implicit aspect of any risk analysis. In the current EC/JRC-methodology schematically shown in Fig. 8.2, substitution is introduced here as well. It may be argued that the search for substitutes is a possible subsequent action in case a risk is deemed to be critical, and that therefore substitution may not be part of the EI-component of the EC-methodology. However, the presence of a likely substitute for a given material may indeed alleviate the economic impact of supply shortage, and therefore constitutes a relevant element of the EI-determination.

As stated in another chapter, it can be interesting for further development of criticality assessments of raw materials supply to study and interpret indicators that have not yet been employed in these EC studies. In studying other approaches, it is irrelevant for our current purposes to comment in detail on the mathematics that are chosen to combine various indicators to a common composed measure, as long as the method and its underlying data are reported in a transparent way.

Various insightful comparisons between raw material criticality methodologies and their outcomes have been published before.[17–20] An observed overlap between methods was reported (Fig. 8.3). In the subsequent sections, these review studies have been used implicitly.

Indicator	Dimension	Data type	
• Geopolitical concentration	SR	●	
• Static reserve range	SR	●	
• Mine Production	SR	●	Frequency
• Economic Relevance	VU	⊙	of criteria
• Supply & demand trends	SR	●	in
• Strategic relevance	VU	⊙	
• Recycling rates	SR	●	criticality
• Substitutability	VU / SR	⊙	studies
• Production as by-product	SR	⊙	
• Political conditions	VU	⊙	
• Company concentration	SR	●	
• Emerging technologies	VU / SR	⊙	
• Production costs	SR	●	
• Functionality & Technology	VU	⊙	
• Ability to drive through price incr.	VU	⊙	
• Damage Potential	ER	●	
• Impact on climate change	SR / ER	⊙	
• Exploration budget & investment	SR	●	

⊙: Qualitative data ● : Quantitative data VU : Vulnerabilty SR : Supply risk ER : Environmental risk

Fig. 8.3. Often mentioned criticality criteria.[20]

They turned out to be the most appropriate set of indicators for assessing critical raw materials in the Dutch economy.

8.3 In Search of Appropriate Supply Risk Factors for The Netherlands

In the Dutch situation the Ministry of Economic Affairs was interested in two aspects: (i) an assessment of the vulnerability of Dutch industry to the secured supply of (abiotic) raw materials, and (ii) a strong support for Dutch companies enabling them to assess their own vulnerability vis-à-vis these supply issues.

Based on this, stakeholder consultation with the Dutch industrial community[9] resulted in a broader set of indicators, not solely focusing on security of supply, but also on potential risks for profitability and company reputation.

The vulnerability of the Dutch economy overall was assessed by calculating the value added as a result of the use of a certain material. The results of that assessment are briefly discussed in section 5.

8.3.1 *The security of supply perspective*

The security perspective takes the view of governments, operating in an increasingly tense geopolitical theatre.[21] Global population growth and the increasing prosperity of the world's population go hand in hand with a strong increase in the demand for a wide range of raw materials. This need is growing most rapidly in emerging economies, where raw materials are required for both building basic infrastructure as well as to meet the growing demand for consumer goods. Partly as a result of the changing geopolitical relations caused by this, the degree of certainty regarding the supply of raw materials for economies which are net importers of materials (such as the EU-28 countries), is decreasing.

Numerous governments around the world are responding to the increasing pressure on the stability of raw material supply by participating more in primary mining, focusing on their own mining industry, stock piling materials, committing more resources to research and development into alternative materials, more efficient use of materials, and intensifying recycling. In addition to this transparent trade in raw materials, a vital arbitrating role for the WTO is becoming increasingly important. EU policy currently focuses on these last two elements (efficient use and recycling) in addition to the intensification of mining in the EU itself.[22]

The following set of indicators was introduced to the Dutch governmental and industrial users to inform them about the security of supply perspective.

8.3.1.1 *Geo-economic factors: The R/P ratio*

In principle, supplies of fossil and mineral raw materials are finite. However, the quantities of mineral raw materials that could still theoretically be extracted are not (yet) relevant to this discussion. What is important is whether the combination of available exploration and extraction technologies on the one hand and economic reality on the other, allow the extraction of sufficient quantities of the minerals in question per unit time. Estimating worldwide reserves is therefore a complex and dynamic activity. Adjustments to estimates of reserves (such as those published in the USGS Mineral Commodity Summaries) appear to be carried out on the basis of administrative actions rather than on the basis of an analysis of new proven reserves. An estimate of future consumption plays no part whatsoever in the determination of reserves.[23]

To say the least, the use of the R/P-ratio, the number of years of production with the currently published reserves, is highly questioned as a relevant indicator. Still, it was chosen as one of the indicators for long term supply of security, with the observation that a proven reserve of more than a thousand years (such as for rare earth elements) versus a proven reserve of less than 20 years (as is the case for antimony, strontium, zinc, gold, tin and silver), leads to relevant awareness in designers of new applications.

8.3.1.2 *Geo-economic: Companionality*

Another important geological-economic characteristic of mineral raw materials is the degree to which they are extracted as the main product (or co-product) of mining operations or as a by-product. Many mineral raw materials are only extracted as by-products ("companions") of other raw materials (the so-called "hosts"). In such cases, the profitability of the mine will not depend on the extraction of the companion. Such dependence can lead to a lack of market elasticity: a sudden increase in demand (for example, as a result of technological innovation) will not — if it concerns a by-product or "companion" — immediately lead to an increase in production or the establishment of new mining operations, unless the process efficiency of the companion extraction increases or if full use is not yet made of all the companion raw material that can be extracted.[24] A further consequence is that, if global demand for a host raw material stabilizes or even decreases (as is the case with lead), the extent to which companions can be extracted will decrease, even when demand increases.

A high level of companionality can therefore be considered a supply risk for the longer term. Materials for which companionality is high are almost all rare earth elements (except yttrium and cerium), all platinum group metals (except platinum), indium, rhenium, tellurium, zircon and cobalt.

8.3.1.3 *Geopolitics: Concentration of (production and reserves of) materials (HHI) in source countries*

Many authors point to the influence of changing balances of power in the world and the risks associated with these, in combination with the fact the extraction of many mineral raw materials is limited to just a few source countries.[25] The concentration of production in source countries is generally considered a relevant indicator of supply insecurity in the short term.

The degree of monopoly forming is expressed in most studies using the so-called Herfindahl-Hirschman Index (HHI), which is composed of the total sum of squares of the extraction concentrations per source country. This is an accepted standard for concentrations in a sector (in this case, source countries). The maximum value is therefore 10,000 (one country produces 100% of the total volume). The EU study into critical materials subsequently weighs the contributions to this HHI per country to the World Governance Index (WGI). This increases the contribution of unstable countries to this risk factor. In our assessment for Dutch stakeholders we considered the separate reporting of this World Governance Indicator worthwhile and relevant for industry and therefore the HHI was not weighted by a countries WGI. An HHI of over 2500 is considered to represent a situation of risky monopoly formation. Most raw abiotic materials (in the Dutch study 18 out of 64 studies materials) fall into this category.

The same comparison can also be incorporated into a risk analysis for the long term: in determining the concentration of the geographical distribution of economically viable reserves (as reported in the USGS Mineral Commodity Summaries: this could be labelled HHI_{res}. A high reserves concentration may indicate a future supply monopoly. Striking examples of materials for which the reserve concentration is reported to be much higher than the production concentration are the platinum group metals (South Africa claiming high reserves) and phosphate rock (Morocco claiming high reserves).

8.3.1.4 *Geopolitics: Existing export restrictions (OECD data)*

An interesting indicator for use in relation to a dominant position is the extent to which export restrictions are imposed by a source country. The data held by the OECD covers 72 countries (the EU is considered as one region) for the period 2009–2012 and 80% of the global production of minerals, metals and timber. The measures cover prohibitions on export and export restrictions, export duties, licensing requirements and obligations in relation to the local market. There is a strong dynamic growth in such measures: 75% of all the measures that were in effect in 2012 had been introduced after 2007.[26] The fact that China, holding strong positions in many mineral markets, has proven to be willing to exert their power and impose export restrictions, justifies that the supply risk for minerals from China is considered higher than that for minerals from elsewhere on the basis of their market concentration alone.

8.3.1.5 *End-of-life recycling input rate*

In this chapter we focus on the problems associated with the supply of raw materials. The import of materials and goods leads to the formation of a so-called "urban mine" in our society, a supply of raw materials stored in our infrastructure, capital investments and the products we consume. In the coming decades, recycling will become an ever more important source of materials and must ensure that the depletion of resources proceeds at a slower pace.[26,a]

Harvesting those materials may make Europe less dependent on the import of raw materials. In essence, 'urban mining' leads to an increase in the number of source countries (decreasing HHI) and generally occurs in well-governed countries. This indeed reduces supply risk. Since well documented knowledge about the volumes and origins of recycled metals is difficult to retrieve, the end-of-life recycling rate was used in the Dutch study as such.

8.3.2 *The financial perspective: Price volatility and value chains*

Businesses often have an elaborate system in place to secure their supply chain. At the same time, these supply chains are not very transparent. Beyond a first-tier supplier the actual knowledge about the composition of components is evaporating quickly. In the Dutch situation, many industries are back-end assemblers or OEM (Original Equipment Manufacturers) who indeed have a limited knowledge about the chemical, elemental composition of their delivered processed materials (such as chemical compounds and base metals), components and other intermediate products (intermediates), as schematically shown in Fig. 8.4 below.

Companies may be involved at each of these levels and obtain their materials and goods from each of these higher levels. Therefore an analysis of the vulnerabilities of raw materials only (opposed to all type of goods) provides a limited assessment of the overall vulnerabilities found in the supply chain. A wide availability of basic raw materials, combined with a

---

[a]It is important to note that the role of recycling is important here because of the large proportion of metals in the list investigated. Thanks to good process technologies, metals often retain their quality. For biotic materials, this situation is more complex. Where recycling does take place, it will often lead to a decline in quality and hence value. The recycled material is in such cases no substitute for "virgin" material.

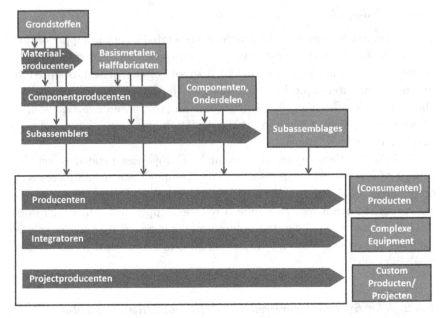

Fig. 8.4.　Illustrating different stages of use of raw materials in the value chain.

bottleneck in the supply chain (e.g. the presence of a very limited number of suppliers of intermediate products) can cause a vulnerable situation. That study[9] reported an abundance of acute supply disruptions which were unrelated to raw material supply disruption, but in fact were connected to other problems found further up the value chain.

Besides a limited knowledge about the impact of supply disruption of a single raw material, this lack of knowledge also prevents a company from assessing the impact of price increases and price volatility on its business.

8.3.2.1　*Operating profit: Price volatility of (raw) materials*

Increasing and varying raw material costs affect operating profits and – particularly in the case of an uneven playing field – competitiveness. Concerns about operating profits are therefore important at both the corporate and national level.

It is a fact that the price volatility of mineral resources is high and has increased since the turn of the century. Price volatility can have various causes. It may arise as a result of an imbalance between supply and (for some applications rapidly) increasing demand, export restrictions or speculation on the commodities market.[28]

The effects on the supply side include uncertainty about the profitability of mining investments, which leads to shortages in the long term. In this sense, price volatility could be an indicator of the risk of supply uncertainty in the long term. The same applies to the phenomenon of "supply shocks" (moments where a sudden decline in production leads to an immediate price increase).

On the demand side, price volatility may lead to problems when prices cannot be passed on to customers and where a "level playing field" for producers in different countries does not exist. The influence of this depends strongly on the contribution made by the cost of this raw material to the cost of the final product.

To determine effects on operating profits one needs to know the price volatility per raw material and an estimation of the quantities of a raw material that are used. Price volatility can be expressed in different ways. In a TNO study,[29] MAPII, the Maximum Annual Price Increase Index was introduced, as a measure of the maximum relative price increase that has occurred during the past 20 years. The MAPII represents the highest price increase per year in that period, divided by the price of raw materials at the beginning of the year with the highest price increase. A MAPII of 1.0 means that the price rose by 100%, i.e. doubled, during a given year during this period. Using the MAPII, the impact of price volatility on a product or product group can be determined as follows:

$$\sum_{x=n}^{m} ((\text{MAPII}_x \cdot \text{P2011}_x) \times \text{TS}_x) \times \frac{\text{W(import)}}{\text{V(import)}}$$

In this formula, MAPII_X is the maximum annual percentage increase in the price of a raw material (determined for the period 1990–2011), P2011_x is the price level in 2011 of the raw material, TS_x is the characteristic proportion of a raw material in a particular product group, W(import) is the weight of the volume of imports of all products within a product group and V(import) is the value of imports of all products within that product group. The important parameter TS_x is further elaborated in section 4.

The price developments are based on fragmentary data gathered from the USGS Mineral Commodity Summaries. The price developments reported there are based not only on a variety of sources but also on different

product qualities. Notwithstanding these limitations, however, it is possible
to generate a clear picture of the extent to which prices may fluctuate from
year to year in the worst-case scenario.

8.3.3 *The reputation perspective: CSR and externalities*

Even when supply is guaranteed and price volatility has no major impact on
business operations, conditions in source countries may still have an adverse
effect on business in case a company's reputation is at stake. This is particularly an issue when the primary mining operations (including financing
local conflicts, poor working conditions and local environmental pressure)
can cause a negative image of companies using those materials, even when
the business involved is much further down the value chain, and is not
primarily involved in mining or processing of these raw materials. The regulation issued by the European Parliament is just one example of the widely
felt need to initiate supply chain due diligence.[30] Especially the increasing
impact of social media makes companies increasingly vulnerable for such
upstream issues. In the Dutch context it was felt desirable to define and
present a number of indicators that provided some insight in these reputational aspects.

8.3.3.1 *Environmental impact of resource extraction*

Awareness of the environmental impact of the mining and refining of raw
materials can be important in — for example — being prepared for criticism
and in seeking possible alternatives that are less damaging to the environment. Since the results of a study performed in The Netherlands links raw
materials with their use in products (including cases when an individual
company may not be aware of this), such information with regard to raw
materials at the product level will be included in the self-assessment tool
to be made available to companies.[31]

A methodology that could be employed is the use of midpoint indicators (following the ReCiPe method and the EcoInvent databases) for raw
material production (indicators for a wide variety of environmental impacts
can be retrieved with this method). In terms of environmental impact, gold
and the platinum group metals stand out far beyond other raw materials
(caused by greenhouse effect but also particulate matter formation during
mining).

8.3.3.2 *Performance of source countries in terms of human development (Human Development Index HDI)*

One of the factors indicating the relationship between potential social problems and raw materials is the human development index (HDI).[b] The HDI is roughly composed of: life expectancy, average years of schooling, expected years of schooling and gross national product per capita.

When including this parameter, the potential reputational damage for using tantalum and cobalt are prominent: the important role of the African Great Lakes Region and the extremely low HDI in that area means that these materials stand out negatively.

8.3.3.3 *Regulations pertaining to conflict minerals*

A factor with a particular influence on corporate reputation, with repercussions for the entire supply chain, is the debate on the import of conflict minerals. The European Commission has designed a system that should lead to an end of imports of certain minerals (tin, tantalum, tungsten, gold (TTTG)) from conflict areas ('conflict-affected and high-risk areas' means areas in a state of armed conflict, fragile post-conflict as well as areas witnessing weak or non-existent governance and security, such as failed states, and widespread and systematic violations of international law, including human rights abuses") by European refiners and smelters. These regulations are similar to those adopted by the US government via the Dodd-Frank Act.[32] The Dodd-Frank Act imposes specific requirements for the traceability of tin, tantalum, tungsten and gold with respect to the export of products containing these materials to the United States.[33] That means that information about these four materials is relevant not only to reputation but also as regards the export situation.

When such materials become available through recycling processes, the dependence on source countries will be reduced, provided that recycling and any subsequent processing of the recycled materials takes place locally. The precise details of the nature and above all the location of recycling are currently very unclear.

---

[b] "The Human Development Index (HDI) is a summary measure of average achievement in key dimensions of human development: a long and healthy life, being knowledgeable and have a decent standard of living." The HDI is compiled and reported by the UN Development Programme.

8.4 Placing Value Added on the X-axis

The impact of a supply chain disruption of raw materials is based on an economic perspective that directly relates it to themes such as competitiveness, the labor market and the trade balance. The previous section focused on the criticality indicators for the y-axis. This section deals with the way the economic impact on the Dutch society has been addressed on the x-axis.

It is commonly accepted that[15] the impact of raw material supply disruptions are best expressed on the x-axis by the value added in an economy. To do this in the best possible way, a relatively detailed relation (CN 6-digit and CPA 6-digit product groups) was determined between raw materials and their application in processed materials, intermediate and final products. This allows the impact in value added per sector and domestic export per product group to be expressed, clarifying the impact on the labor market and global competitiveness respectively. The TS_x parameter discussed in the subsection about operating profit represents the "typical share" of a raw material in the total net weight of a product group (for methodological detail, see[29]), illustrating the need to link the 6-digit product groups to raw materials.

Information linking products and raw materials are available from literature[c] and auxiliary sources[d] for most products. Such shares can be checked or corrected by more detailed but limited databases such as EcoInvent.[e] However, the analysis is initially concerned with the qualitative link between raw materials on the one hand and products on the other. When estimating the economic impact, we assume that the amount of a specific raw material in a given product is irrelevant, but that each material is essential for the quality of the delivered product and therefore the related competitiveness of the company concerned. Such allocation of raw materials to products is of course not unique to the case of the Netherlands. Therefore, this assessment could easily be extended to an analysis on European scale, thereby providing an (more detailed and accurate) alternative for the EC-analysis based on the rough economic impact using NACE2 data and share of application of a material expressed as TS_x.

---

[c]For a full list, see TNO (2015).
[d]Such as trade databases like e-to-china.com and werliefertwas.de
[e]https://www.ecoinvent.org/

8.5 An Appropiate Criticality Assessment Applied to the Dutch Economy

The Dutch case study results are shown in Fig. 8.5. In this plot the short term criticality was expressed as:

Criticality = HHI * (WGI$_{\text{weighted}}$+ OECD restrictions$_{\text{weighted}}$) * (1-%EOL-RR)

The short-term security of supply of raw materials was considered low in the Dutch study where the source country concentration is high AND where the source countries have a mediocre World Governance Index and have proven willing to impose export restrictions AND where recycling of end-of-life products is low.

The importance of iron, copper and aluminum exceeds that of other raw materials. These materials are used in a large number of products which have added value in almost all sectors. This makes them the most important materials in our economy.

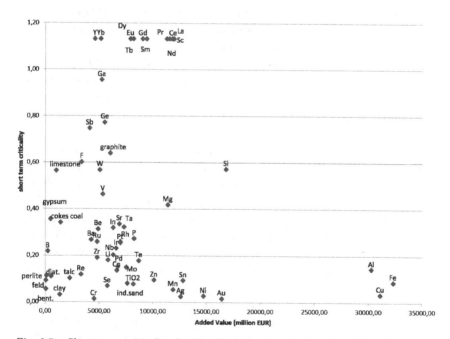

Fig. 8.5. Short term criticality for The Netherlands: security of supply in relation to added value per raw material.

Furthermore, the significant importance of silicon, gold, silver and important alloying elements such as nickel, tin, magnesium and zinc stands out. In addition, a group of rare earth metals (lanthanum, cerium, neodymium, praseodymium and scandium) have been identified as important to the Dutch economy.

8.6   Challenges Ahead: Biotic Materials, Future Supply and Substitution

Three important challenges that could shape the analytical face of criticality assessments in the coming years are discussed in this section.

8.6.1   *Should substitution be a part of vulnerability assessments?*

As was discussed in section 2, substitution is a rather common factor in criticality analyses. It is debatable whether this factor is relevant in assessing supply risk, but its use in assessing vulnerability and impact is rather straightforward: the availability of a readily available and decently performing substitute certainly alleviates the vulnerability of a country in case of a supply disruption. The level of detail and the type of assessment is however very different among the various reports.

Habib[34] states: *"the assessment of the substitutability of a given resource can in our view benefit from a more specific and technology-oriented perspective. Often, substitutability is assessed at the elemental level, i.e., substitution of one element by another based on key physical/chemical properties of the element as such, e.g. substitutability of copper by silver or aluminum based on their conductivity properties or the like, and mainly qualitatively assessed based on expert judgements. We argue that assessing the vulnerability of being dependent on a specific resource including the options for, and ease of, substituting one resource by another, needs a more comprehensive assessment. Such an assessment should rely on a holistic understanding of product development and technology development and address ways to substitute not only the elemental resource as such, but the complex technological solutions to creating the features and functionalities of the products and systems of concern."*

This implies a meaningful broadening in the scope of substitution. The example used in their paper (alternative technological solutions for direct geared wind turbines without needing permanent magnets based on neodymium and dysprosium) indeed suggest that the substitution options for NdFeB-magnets are rather large and thus the resulting impact low. This means that from a societal point of view, one may indeed choose a broad scope for substitution and easy substitution indeed may alleviate risks for a society being deprived of essential functionalities.

In practical terms however, this use of substitutes is often difficult from the point of view of an individual company. In the example of lighting for instance, it can be suggested that all imaginable light sources (LEDs, CFLs, conventional bulbs, candles) are alternative solutions for the same function and therefore substitutes. It would however result in completely changed industrial value chains and therefore such substitute definition is too wide. The JRC methodology published in 2016[15] bases itself on a 'drop-in' technology type of substitution, enabling companies to move to a different product without significantly altering the value and production chain. Some alloying elements (such as Nb, W, V) can be interchanged in similar process equipment and deliver comparable performance. Though such examples may exist, from a company point of view substitution always leads to considerable impacts such as decreased performance, changing process parameters, and increased costs. Therefore, in the Dutch study, solely focusing on a company perspective, substitution was not incorporated in vulnerability assessments.

Helbig et al.[18] conclude in their review of criticality methods that substitution is the single most frequently applied indicator for vulnerability. The overview that they provide demonstrates that the availability of substitutes, their performance, or the share of products for which substitution is considered impossible is all assessed by expert opinion (on 3,4 or 5 point scales). Most of these expert opinions are based on the previously indicated direct substitution of materials for other materials. This consensus among cited publications is in contrast with the JRC-methodology currently employed: this search for substitutes is based on literature searches or commonly known substitutes (so as to ensure ready availability instead of potential long-term options of low TRL). In comparison with the known expert-based substitute indicators, it may be questioned whether the complex calculations suggested actually provide meaningful or desired detail.

8.6.2 *How to include future developments in an impact assessment?*

Incorporating future developments will increase the relevance and decrease the reliability of every criticality assessment. As was stated[34]: *"criticality is a situation/condition of the system under study due to some property leading to criticality. Thus, criticality is a dynamic instead of static phenomenon which is subject to change over time. This is also because the indicators used to assess criticality are dynamic in nature"*.

All existing methodologies lack elements that are either future-oriented or dynamic in nature. Several criticality reports claim to have included dynamics, i.e. changes over time, in their assessment[8,34–38] some of them related to the supply side and some of them to the (economic) vulnerability side. The suggestion made previously to include assessments about reserves and reserve concentration in countries may already be considered a primitive method to include dynamics in the supply risk assessment. Some[34] use this difference between current HHI and potential future HHI (based on reserve data) as a method to 'predict' the future development of HHI.

The reports that focus on one particular sector (for instance the defense sector, or the renewable energy sector) can use technology roadmaps for their dynamic assessment: the predicted future demand for that particular application can be compared to the current need. In case this future need is significantly higher, or even exceeds the current total production of a particular material, the probability for a future supply shortage is relatively large. Consequently, the impact on the sector itself is also large, since the predicted roadmap will face barriers in order to be executed. For a more general assessment, such as the one for the EU-28, such an approach for a future-oriented assessment can obviously not be followed. Erdmann[36] uses the known change of the share of German usage vis-à-vis global usage and the change of imports to Germany (during a 5-year period preceding the analysis) as elements (both of a 10% weighing factor) in the vulnerability assessment. In their analysis, the materials Gallium, Rhenium and the Rare Earth elements are assessed to be very critical partly because of the growing consumption as a raw material in Germany (growth rates between 50% and 80% between 2004 and 2008) in combination of course with the already high consumption share of the German industry. These data are extracted from trade statistics.

In,[16] the authors use the instrument of future price development and volatility as measures for future availability development. Based on historic trend and regression analyses the authors conclude *"that future price trends and volatility are significantly influenced by a number of current material specific and general economic indicators, such as country concentration, secondary production, or interest rate."* The relations seem to be different for different materials, especially the minor elements with specific applications and potentially high growth rates and non-transparent price formation mechanisms following different predicted routes than the major metals. All in all, the inclusion of price developments for a broad assessment of raw materials proves to be complex and partly relates to the same, rather simple indicators such as country concentration and speculation in non-transparent markets.

Olivetti[39] reviews the modelling methods that have been applied to estimate future materials use and availability-derived supply risk within studies of materials criticality. These authors point to the fact that all aspects commonly used in criticality assessments are expected to change over time, and that therefore proper future-oriented assessments require careful modelling. They observe that *"it is not uncommon to find that "importance" and "availability" characteristics of materials markets are nevertheless treated as intrinsic materials properties whose values are reducible to simple economic or geophysical accounting. Worse, this oversimplification can be an attractive short-cut in criticality policy discussions. The challenge for those who study criticality risk is to find techniques for assessing importance and availability that avoid this shortcut, giving appropriate consideration to the techno-economic dynamics of real systems."*

The paper states that flows of future consumption are often estimated using empirical models of historic consumption of the material based on simple time-series regressions. These models may be enriched by scenarios including predicted population and GDP growth. The system dynamics and agent based modelling approaches indicate nevertheless that analyses of future vulnerabilities and supply issues require in-depth research and modelling activities and display significant uncertainties.

Future-oriented assessments of supply beyond the use of R/P-ratios might be based on the relation between investments in exploration and the actual discovery of raw material.[40] When it comes to the development of the reserves there should arise a relationship between the amount of investments put into detecting reserves (exploration phase) and the extent to which significant finds are made. Richard Schodde, Managing Director, Minex Consulting, noted in his presentation to the IMARC conference in 2014[41] that this is not the case, and that gives cause for concern. An analysis of expenditure in the exploration of metal and minerals could in principle contribute to gaining a better insight into long-term availability. After all, when there is no investment in the search for new supplies then no new supplies will be found.

Due to the difficulty in predicting the relationships between exploration and eventual mining investment, and the fact that only sketchy data is available for a few commodities, the extent of investment in exploration can currently not be used as an indicator for long term supply security.

8.6.3 *How to complete the picture of raw material criticality assessment by including biotic materials?*

The topic pertaining to the difference between biotic and abiotic materials is expected to return in the coming years. Recent work has elaborated the difference in assessing them.[42,43] Looking at the indicators discussed in section 3, a bold but reasonable statement can be made that there is in fact relatively little difference in the assessment methods between the abiotic and biotic realm. It may look like a paradox to observe that there are huge differences between the required data and corresponding analysis when looking at production capacities, reserves, and end-of-life recycling-input-rates. At the same time, all the other indicators are either completely similar (concentration of source countries, MAPII, export restrictions, human development, environmental performance) or irrelevant (companionality, conflict minerals). Furthermore, the challenges faced by sustainably recycling biotic materials may be very different compared to the abiotic materials, but they are also less complex.[44] The brunt of the additional analytical work to assess materials like natural rubber, soy, cacao, tropical wood etc. instead of molybdenum or copper thus focuses on production capacity and reserves. Elements like geophysical properties, pathogens, degradation and other ecosystem issues come into play when considering biotic production capacity. However, these elements are essentially related to a particular piece of land, not to a particular material. Therefore, it may be assumed that using country concentration factors (such as HHI) encompasses all of these seemingly very different factors.[45] For a high country concentration of harvesting may lead to a high (local) impact of pathogens, local droughts, etc. The most important difference between biotic and abiotic materials ironically seems to be the difference between the crust and the surface.

8.7 Summary and Concluding Remarks

Many publications report assessments of raw materials criticality. Though most authors develop 'proprietary' assessments, the overall approach and the nature of the indicators are remarkably similar. Also for the Dutch situation a set of indicators was chosen based on these commonly adopted indicators (section 3).

The general approach of a *risk analysis (determining the probability of an event and the consequences if that event takes place)* is taken by many authors. Regarding the common elements of the assessments, there are a number of conclusions that can be drawn.

With respect to assessing the probability of supply disruption we can conclude:

- *Recycling* is used as an indicator for the supply risk axis in several studies. Though no evidence was found that recycling already impacts supply risk, it is considered relevant to include recycling because it often indicates the availability of a secondary source in consumer countries.
  It is worthwhile devoting effort to *assess production volumes and countries for secondary materials*, so that these data can be included in the generally accepted HHI indicator.
- *Distribution of reserves* over the globe (as opposed to distribution of current production) is already used in several papers, and may be considered for future use for long term risk analysis. The EU-28 is the proper podium to identify long term upcoming monopolies and consider action, given the reliability and cost of data gathering and the purpose of the analysis enabled by that data. For shorter term company actions, reserve distribution is indeed less relevant.
- *The companionality* ("by-product issues") is an indicator already used in several studies and is worthwhile considering in future vulnerability assessments, though more effort should be paid to the insight in current refining capacities and the extent to which the maximum levels of companions are currently harvested.

With respect to assessing the impact of supply disruption we can conclude:

- *Substitution options* are commonly employed as an element that has an impact on vulnerability; a debate about the level at which substitution is considered (material for material, product for product, functionality for functionality, process for process) is not conclusive which renders this indicator prone to varying interpretation. Short term substitutes of high TRL that do not significantly alter production processes may be a *narrow* but workable *definition* on a company (and thus economy and added value) level.

– The relation between raw materials and the direct impact on the economy benefits from deep knowledge about the *actual application of raw materials in products*; the estimates currently employed in the EC-assessments (gross allocation of raw material use to NACE sectors) should be refined to a great extent with some existing methods. Our study assessing Dutch vulnerability has made an attempt to allocate raw material use in products, product groups and subsequently sectors in great detail. Further refinements however are desirable to identify more in detail which economic actors face risks in times of supply insecurity.

Several papers conclude that these vulnerability assessments should pay *more attention to the time-dynamics of the raw materials market* and should provide more data about the future situation. Some methods that were discussed require deep (agent-based or system dynamic) modelling and it is obvious that such methods require further development to be meaningfully deployed for substance (i.e. elemental) criticality assessment. The use of exploration investments was also shown to be non-conclusive. However, it might be considered to use trends of production and consumption over limited historic time-series in order to highlight issues for materials that have experienced high demand growth under stagnating mining capacity or unexpected high price volatilities.

   A clear conclusion can be drawn regarding the supply and value chain of raw materials. With only a few exceptions, *none of the criticality methodologies pay attention to the potential vulnerability caused by processes in the value chain* between the actual mining process and the final consumption by a company or country. This grossly overestimates risks at the mining stage and underestimates the vulnerabilities due to production concentrations in the refining industry and the manufacturing industry further down the value chain. The emphasis in the raw materials debate may therefore in some cases focus on the wrong materials and wrong players and actions. In the current EC-methodology this is partly addressed by at least assessing whether the 'next step' in processing (i.e. refining) of materials exhibits higher country concentration than the mining stage. Ideally, for strategic value chains, such analyses should be taken beyond the point of refining and dive deeper in the value chain. An example of such an approach is given in Fig. 8.6 below, illustrating the vision of the US Department of Defence regarding the dependence on a foreign value chain.

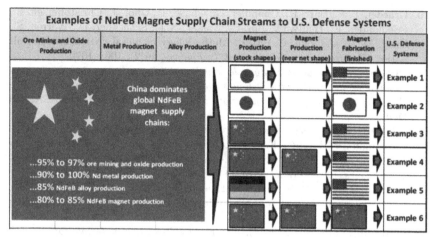

Fig. 8.6. Rare earth based permanent magnet supply chain as shown by US Department of Defense.

References

1. Alcott, B. (2010). Impact caps: Why population, affluence and technology strategies should be abandoned. *Journal of Cleaner Production*, 18(6), 552–560.
2. Fischer-Kowalski, M., *et al.* (2011a). *Decoupling natural resource use and environmental impacts from economic growth*. Nairobi, Kenya: UNEP.
3. UN Population Fund. (n.d.). World population trends. Retrieved from http://www.unfpa.org/pds/trends.htm
4. World Bank. (2011). *Global economic prospects*. Washington D.C., USA: World Bank.
5. Von Weizsäcker, E.U., *et al.* (2014). *Decoupling 2: technologies, opportunities, and policy options*. Nairobi, Kenya: UNEP.
6. Marscheider-Weidemann, F., Langkau, S., Hummen, T., Erdmann, L., Espinoza, L.T., Angerer, G., Marwede, M. and Benecke, S. (2016). *DERA Rohstoffinformationen* 28: 353. Berlin, Germany: Rohstoffe für Zukunftstechnologien.
7. Zepf, V., Reller, A., Rennie, C., Ashfield, M. & Simmons, J., (2014). *Materials critical to the energy industry: An introduction* (2nd ed.). London, UK: BP.
8. Moss, R.L., Tzimas, E., Willis, P., Arendorf, J., Espinoza, L.T. (2013). *Critical metals in the path towards the decarbonisation of the EU Energy Sector — assessing rare metals as supply-chain bottlenecks in low-carbon energy technologies*. Brussels, Belgium: European Commission.
9. Bastein, T. and Bol, D. (2012). *Critical materials — a view from the industrial-technological sector in The Netherlands*. The Netherlands: FME-CWM.

10. Graedel, T. E., Barr, R., Chandler, C., Chase, T., Choi, J., Christoffersen, L., Friedlander, E., Henly, C., Jun, C., Nassar, N.T., Schechner, D., Warren, S., Yang, M.Y., and Zhu, C. (2012). Methodology of metal criticality determination. *Environmental Science and Technology*, 46(2), 1063–1070.

11. DCGIS. (2012). *Outil d'analyse de la vulnerabilite des entreprises aux approvisionnements de matieres critiques non energetiques*. Paris, France: DCGIS.

12. US DOE. (2010). *Critical materials strategy*. Washington D.C., USA: US DOE.

13. US DOE. (2011) *Critical materials strategy*. Washington D.C., USA: US DOE.

14. Christmann P., Labbé J.F. (2015). *Notice de realisation et d'utilisation des fiches de synthese sur la criticite des matieres premieres minerales nonenergetiques*. Paris, France: BRGM.

15. Blengini, G.A., *et al.* (2016). EU methodology for critical raw materials assessment: Policy needs and proposed solutions for incremental improvements. *Resources Policy*, 53, 12–19.

16. European Commission. (2014). *Report on critical raw materials for the EU*. Brussels, Belgium: European Commission.

17. Erdmann, L. and Graedel, T. E. (2011a). Criticality of non-fuel minerals: A review of major approaches and analyses. *Environmental Science and Technology*, 45(18), 7620–7630.

18. Helbig, C., Tuma, A., Wietschel, L., and Thorenz, A. (2016). How to evaluate raw material vulnerability—an overview. *Resources Policy*, 48(June 2016), 13–24.

19. Gleich, B., Achzet, B., Mayer, H., and Rathgeber, A. (2013). An empirical approach to determine specific weights of driving factors for the price of commodities — A contribution to the measurement of the economic scarcity of minerals and metals. *Resources Policy*, 38(3), 350–362.

20. Mayer, H. and Gleich, B. (2015). Measuring criticality of raw materials: an empirical approach assessing the supply risk dimension of commodity criticality. *Natural Resources*, 6(1), 56–78.

21. AEA Technology. (2010). *Review of the future resource risks faced by UK businesses and an assessment of future viability*. London, UK: DEFRA.

22. European Commission. (2008). *Commission staff working document accompanying the communication from the Commission to the European Parliament and the Council — The raw materials initiative: meeting our critical needs for growth and jobs in Europe*. Brussels, Belgium: European Commission.

23. British Geological Survey. (2012). Risk list 2012. Nottingham, UK: British Geological Survey.

24. Hatayama, H. and Tahara, K. (2015). Criticality assessment of metals for Japan's resource strategy. *Materials Transactions*, 56(2), 229–235.

25. Frondel, M., Grosche, D., Huchtemann, D., Oberheitmann, A., Petersand, J., Angerer, G., Sartorius, C., Buchholz, P., Rohling, S., and Wagner, M. (2006). *Trends der Angebots- und Nachfragesituation bei mineralischen Rohstoffen*. Essen, Germany: RWI, ISI, BGR.

26. Kooroshy, J., Meindersma, C., Podkolinski, R., Rademaker, M., Sweijs, T., Diederen, A., Beerthuizen, M., and de Goede, S. (2010). *Scarcity of minerals: A strategic security issue*. The Hague, the Netherlands: HCSS.

27. Duclos, S. J., Otto, J. P., and Konitzer, G. K. (2010). Design in an era of constrained resources. *Mechanical Engineering*, 132(9), 36–40.

28. OECD. (2009). *A sustainable materials management case study — Critical metals and mobile devices*. Paris, France: OECD.

29. Bastein, T., Rietveld, E., Keijzer, E. (2015). *Materialen in de Nederlandse economie*. The Hague, the Netherlands: TNO. Retrieved from https://www.rijksoverheid.nl/documenten/rapporten/2015/12/11/materialen-in-de-nederlandse-economie

30. EU. (2017). Regulation 2017/821 of the European Parliament. Retrieved from http://eur-lex.europa.eu/legal-content/EN/TXT/?toc=OJ%3AL%3A2017%3A130%3ATOC&uri=uriserv%3AOJ.L_.2017.130.01.0001.01.ENG

31. WWF. (2014). *Critical materials for the transition to a 100% sustainable energy future*. Gland, Switzerland: WWF International.

32. JOGMEG. (2015). *A study of a stable supply of mineral resources* [poster]. Tokyo, Japan: The fifth EU-US-Japan Trilateral Conference on Critical Materials.

33. Bae, C. (2010). Strategies and perspectives for securing rare metals in Korea. *Proceedings of the Workshop on Critical Elements for New Energy Technologies*. Boston, MA: MIT.

34. Habib, K. and Wenzel, H. (2015). Reviewing resource criticality assessment from a dynamic and technology specific perspective — using the case of direct-drive wind turbines. *Journal of Cleaner Production*, 112(5), 3852–3863.

35. US DoD. (2011). *Strategic and critical materials: 2011 report on stockpile requirements*. Virginia, USA: US DoD.

36. Erdmann, L. and Behrendt, S. (2011). *Kritische Rohstoffe fur Deutschland*. Berlin, Germany: KfW Bankengruppe.

37. Moss, R.L., Tzimas, E., Kara, H., Willis, P., and Kooroshy, J. (2011). *Critical metals in strategic energy technologies — Assessing rare metals as supply-chain bottlenecks in low-carbon energy technologies*. Brussels, Belgium: Joint Research Centre, European Commission.

38. Nieto, A., Guelly, K., and Kleit, A. (2013). Addressing criticality for rare earth elements in petroleum refining: The key supply factors approach. *Resources Policy*, 38(4), 496–503.

39. Olivetti, E., Field, F., and Kirchain, R. (2015). Understanding dynamic availability risk of critical materials: The role and evolution of market analysis and modelling. *MRS Energy and Sustainability*, 2, E5.

40. Rosenau-Tornow, D., Buchholz, P., Riemann, A., and Wagner, M. (2009). Assessing the long-term supply risks for mineral raw materials — A combined evaluation of past and future trends. *Resources Policy*, 34(4), 161–175.

41. Schodde, R. (2014). Uncovering exploration trends and the future [presentation]. Melbourne, Australia: International Mining and Resources (IMARC) Conference.

42. Bach, V., Berger, M., Finogenova, N., and Finkbeiner, M. (2017). Assessing availability of terrestrial biotic material in the product system (BIRD). *Sustainability*, 9(1), 137.

43. Blengini, G., Nuss, P., Dewulf, J., Nita, V., Talens Peiro, L., Vidal Legaz, B., Latanussa, C., Mancini, L., Blagoeva, D., Pennington, D., Pellegrini, M.,

van Maercke, A., Solar, S., Grohol, M., and Ciupagea, C. (2017). *EU methodology for critical raw materials assessment: Policy needs and proposed solutions for incremental improvements.* Brussels, Belgium: Joint Research Centre, European Commission. Retrieved from http://publications.jrc.ec.eu ropa.eu/repository/handle/JRC105697

44. PBL. (2011). *Scarcity in a sea of plenty? Global resource scarcities and policies in the European Union and the Netherlands.* The Hague, the Netherlands: PBL.

45. Dewulf, J., Mancini, L., Blengini, G.A., Sala, S., Latunussa, C. and Pennington, D. (2015). Toward an overall analytical framework for the integrated sustainability assessment of the production and supply of raw materials and primary energy carriers. *International Journal of Industrial Ecology,* 19(6), 963–977.

Part III
Critical Material Mitigation Strategies

Chapter 9

Circular Product Design: Addressing Critical Materials through Design

Conny Bakker[*,‡], Marcel den Hollander[*], David Peck[†], Ruud Balkenende[*]

*TU Delft, Faculty of Industrial Design Engineering,
Landbergstraat 15, 2628 CE Delft, The Netherlands

†TU Delft, Faculty of Architecture and the Built Environment,
Julianalaan 134, 2628 BL Delft, The Netherlands

For product designers, the world has traditionally been one of resource abundance. Introducing them to a resource-constrained world thus requires new design strategies. This chapter explores how embedding circular economy principles into design practice and education could help product designers take critical material problems into account. We introduce four product design strategies that address materials criticality: (1) avoiding and (2) minimizing the use of critical materials, (3) designing products for prolonged use and reuse, and (4) designing products for recycling. The 'circular' strategies (3) and (4) are elaborated, as these sit most firmly within the remit of product design. This leads to a typology of circular product design that redefines product and material lifetime in terms of obsolescence, and introduces a range of approaches to resist, postpone or reverse product and material obsolescence. The typology establishes the basis for the field of circular product design, bringing together design approaches that were until this date unconnected and paving the way for the development of detailed design methods.

9.1 Introduction

"Product designers have never been taught to regard materials as anything but commodities to be employed as necessary or convenient."[1]

Product designers and engineers have traditionally focused on achieving increasingly higher performance in products, using the full range of elements in the periodic table. Whilst this has delivered an amazing range

---

‡Corresponding author: Conny Bakker, c.a.bakker@tudelft.nl

of products and technologies, it has also resulted in increasing material complexity, decreasing recycling rates and thus increasing criticality risks.[2] In this chapter we will discuss the lack of awareness of product designers with regards to critical materials and explore how embedding circular economy principles into design practice and education could mitigate some of the concerns raised. Criticality of materials is defined by their perceived supply risk, the environmental implications of mining and processing the materials, and a company's (or country's) vulnerability to supply disruption.[2]

9.2 Four Product Design Strategies to Address Criticality

Even though product designers are increasingly asked to consider energy efficiency and recyclability when making a choice of materials, criticality aspects have hardly received any attention so far.[3] This can be explained in part by understanding the way product designers choose materials: they usually base their choice on a trade-off between functionality, quality (grade) and cost. These trade-offs are done both for engineering material requirements[4] and for more subjective aspects, such as the user perceptions of material qualities and meanings.[5] When it comes to specific metal alloys or components such as printed circuit boards, designers choose the constituent materials, which often contain critical elements, only implicitly.[3]

In the case of mobile electronics, for instance, the highly competitive market seeks designs that are thinner, lighter, higher performance, more power efficient (e.g. battery) and robust (e.g. waterproof).[6] This has driven designers to select high performance components that use critical materials, and this increases the risk of critical materials problems. At the same time societies, companies and governments have rapidly become dependent on electronics and the impact of any restriction in the supply of the critical materials used in these technologies could be severe.[3,7] Duclos *et al.*, when assessing criticality risks for the company GE[8] argue: "Elements that are determined to be high in both impact to the company and in supply and price risk require a plan either to stabilize their supply or to minimize their usage." In this chapter we set out to develop such a 'plan' for product designers.

What possible strategies are there for product designers to address critical materials? Following the European waste framework hierarchy of prevent, reduce, reuse, recycle[9] there are basically four strategies: 1) avoiding the use of critical materials, 2) minimizing the use of critical materials, 3) designing products for prolonged use and reuse, which results in a decreased use of critical materials over time, and 4) designing products that are easy to recycle, which results in the recovery and reuse of critical materials. We will address these four strategies in more detail here.

9.2.1 *Strategy 1: Avoid the use of critical materials*

If product performance specifications allow, designers can specify alternative materials that do not contain critical elements. Substituting one material for another is however not always possible. Sometimes the characteristics of a product are such that material substitution results in unwanted changes in a product's properties and/or performance.[3] And for some critical elements, there simply are no known substitutes, meaning that extensive research would be needed to develop a suitable alternative.[8] Designers who want to avoid using critical materials may choose to work with material scientists and process engineers to explore material substitution options. This will require that product designers have a good knowledge and understanding of materials properties and production processes. To give an example, the production of PET (polyethylene terephthalate) plastic bottles requires the use of germanium (a critical material) as polymerization catalyst. During the production of PET the catalyst material dissipates completely into the PET and is not recovered.[10] A designer who wants to address this issue needs to work closely with material scientists and process engineers to find substitute catalyst materials.

Designers could also take a more systemic perspective and look for alternative products and technologies that could address a user's needs.[8] A case study of wind turbines was for instance presented by Habib & Wenzel.[11] The case study demonstrates, using a product design tree approach, the availability of several viable (and non-critical) alternatives to the direct-drive wind turbine technology that currently utilizes neodymium and dysprosium in its permanent magnet generator.

9.2.2 *Strategy 2: Minimize the use of critical materials*

This is a less radical strategy than avoiding the use of critical materials altogether. Greenfield *et al.*[7] describe the example of the electronic component manufacturer TDK, that managed to reduce its dependence on dysprosium by 20–50% through process redesign. In some cases, improving processes might be relatively straightforward, as the use of critical materials was never scrutinized in much detail before. For product designers to apply this strategy successfully, they again need to work closely with material scientists and process engineers. On a more systemic level, efficiency improvements can be made through a change in the basic technology, for instance the move from fluorescent light bulbs to LED lighting.[12] Nevertheless, minimizing the use of critical materials in products has potential drawbacks: even though the amount of critical materials per product may decrease, the total volume of product sales may increase. This may negate any advances made in reducing the use of critical materials. And secondly, ever smaller amounts of critical materials in complex components (such as printed circuit boards) may make critical material recovery through recycling even harder.[3]

9.2.3 Strategy 3: Design products for prolonged use and reuse

Designing products that last, and that can be repaired, refurbished and remanufactured easily and economically, will result in a decrease in the use of critical materials over time. In the words of Stahel[13] this reuse of products results in: "a slowdown in the flow of materials and goods through the economy, from raw materials production to recycling or disposal." Like in the previous strategy, creating products with a longer service-life will not avoid the use of critical materials, but it will reduce their overall consumption. Hampus *et al.*[14] calculate that in general, the extension of the use phase of electronic products (through repair) leads to a decreased use of critical materials, but they also see limitations of this approach. For instance, if the prolonged lifetime comes at the cost of an increased critical material content per product or component. This is especially problematic if high-quality recycling is lacking.[14]

9.2.4 Strategy 4: Design products for ease of recycling

Design for recycling is aimed at recovery of materials from end-of-life products. Its essential objective is "to keep the quality of the old material as high as possible".[15] Designers should therefore ensure that products consist of materials that are compatible in the recycling process, or that can be separated in compatible material fractions after manual or mechanical disassembly. In spite of the extensive body of literature on design for recycling, there is hardly any literature on design for recycling of critical materials. This may prove to be an important research gap, because the current recycling rate of most critical materials is extremely low: around 1%.[16] This is partially because they are used in very small quantities in products, and are often highly mixed with other materials (as alloys for example), which makes them difficult to separate.[3] According to Ylä-Mella & Pongrácz,[17] other reasons for the low recycling rates of critical materials are missing economic incentives for recycling, and a lack of appropriate recycling technologies and infrastructure. Complicating factors are the current trend of product miniaturization and increased integration of materials, which is good from the perspective of minimization of the use of critical materials, but which may hamper their recovery.

These four strategies are complementary and interdependent. Focusing only on one strategy may have adverse effects, as was illustrated in the examples above. Furthermore, the concept of critical materials is relative and subject to regular change.[2] Materials become more or less critical over time as geopolitical circumstances change, new mines open up or old mines close down, new technologies develop with different material requirements, or new data becomes available about physical reserves or environmental impacts of mining and processing. Designers therefore need to take a systemic perspective, and if the use of critical materials cannot be avoided (strategy 1), they should aim at using the three other strategies in conjunction, all the while monitoring the possibility of trade-offs and negative side-effects over time.

The critical materials problem puts the work of product designers in sharp perspective. Designers are familiar with the general solutions described in strategies 1 and 2: creating innovative new technologies, developing substitutions and making efficient use of resources. They have just not applied their skills to critical materials problems very often (if at all).

Designers are however quite unfamiliar with strategies 3 and 4: creating products for prolonged use and reuse, and designing for recycling. The current product design methodology, taught at design schools, has a bias towards product acquisition and first use. Strategies 3 and 4, however, deal with everything that happens, or could potentially happen, to a product (and its constituent materials) in its subsequent lives. The current product design methods that address reuse, repair, remanufacture or recycling lack a coherent framework and common language, they are severely out of date and are fragmented throughout design research literature and practice. Strategies 3 and 4 therefore need further exploration and development. As they fit well within the remit of Circular Product Design, in the remainder of this chapter we will refer to circular product design when discussing strategies 3 and 4. We will attempt to create an organisational structure and common language for circular product design that can guide designers when addressing critical materials through design.

9.3 Circular Product Design

Circular product design aims at keeping products, components and materials at their highest economic value and lowest environmental impact for as long as possible, by designing for long product life and by looping back used products, components and materials into the economic system through repair, refurbishment, remanufacture and recycling.

9.3.1 *Design for product integrity*

Instead of discussing the design of products for 'prolonged use and reuse', we introduce the term Product Integrity. This is defined as the extent to which a product remains identical to its original (e.g. 'as manufactured') state, over time. The idea of preserving as much of the original product as possible was captured by the Ellen MacArthur Foundation in their descriptions of the principles of a circular economy: "The tighter the circle, i.e., the less a product has to be changed in reuse, refurbishment and remanufacturing,...the higher the potential savings on the shares of material, labor, energy and capital embedded in the product...".[18] This echoes the ideas of Walter Stahel who introduced the Inertia Principle in his book *The Performance Economy*[13]: "The smaller the loop, the more profitable it is. Do not repair what is not broken, do not remanufacture something that can be repaired, do not recycle a product that can be remanufactured." The starting point is the original product, and the intention of the Inertia principle is to keep the product in this state, or in a state as close as possible to the original product, for as long as possible, thus minimizing and ideally eliminating environmental costs when performing interventions to preserve or restore the product's added economic value over time. Design for product integrity is a more precise way of saying 'design for prolonged product use' because it also gives guidance: there is a priority order; As the Inertia principle starts from the highest level of product integrity, moving down the hierarchy may be inevitable in the real world, but is not the preferred direction.

9.3.2 *Design for recycling*

Recycling of materials implies loss of function and value. Nevertheless, recycling is the necessary 'last resort' option when products or parts are no longer useful. This means that, in contrast to the previous strategy, recyclability is a mandatory requirement for every product.[19]

The goal of design for recycling in a Circular Economy is to create closed loop, or primary, recycling. This is mechanical, physical and/or chemical reprocessing, where the reprocessed materials have properties equivalent to the original materials. In the field of recycling, primary, secondary, tertiary and quaternary recycling is distinguished,[20] creating a priority order of decreasing materials integrity. In the case of secondary recycling, the reprocessed materials have lower properties than the original materials. This is also referred to as downcycling or downgrading.[20] Tertiary recycling (recovery of chemical constituents, or feedstock recycling) and quaternary recycling (recovery of energy) are not considered in this chapter, as in these instances materials integrity is fully lost.

9.4 Typology of Circular Product Design

In order to understand how 'Design for product integrity' and 'Design for recycling (or materials integrity)' can be used to address critical materials challenges, they need to be brought together in a coherent framework that gives guidance to product designers, and that allows for a discussion of the interaction between the strategies. In this section the development of a typology for circular product design is described: a categorization of design approaches based on product and materials integrity.

The development of a typology for circular product design is important because of the following reasons[21] (1) a typology helps designers decide how to proceed when developing a product, (2) a typology helps establish a common language for the field of circular product design. No such common language exists today, but we need it if we want to begin addressing issues of critical materials. (3) A typology provides circular product design with an organizational structure which is currently missing, (4) a typology helps legitimize the nascent field of circular product design by providing a range of clearly distinct and ordered design approaches, and (5) a typology can be used in education. As materials' criticality is not currently taught in design education,[22] a typology could be a valuable tool for increasing design students' awareness of critical materials issues.

The first ordering principle of the typology is the hierarchy imposed by the inertia principle and the recycling priority order, as described above. This is not enough, however, because this hierarchy exists in a linear economy as well. We need a second ordering principle in order to create design

approaches relevant for a circular economy. One of the fundamental principles of a circular economy is that 'waste' no longer exists. A circular
economy is, in principle, a closed loop system. From a material flow perspective, resources that have entered the circular economy have to remain
accounted for at all times: before, during, and after their lifetime as useful
products.[23] It follows that product lifetime is a key concept in a circular
economy, and will be used as a second ordering principle.

9.4.1 *Defining product lifetime*

A product lifetime can end for many reasons. Often, definitions focus on the
functional lifespan of a product.[24] It is however well-know that products
stop being used for non-functional reasons as well. Ashby,[4] for instance,
distinguishes the end of a product's physical life, functional life, and technical life, as well as economical life, legal life and desirability life. In order
to include both objective and subjective reasons for the end of a product's
useful life, we propose to define product lifetime in terms of obsolescence.
A product becomes obsolete if it is no longer considered useful or significant
by its user.[25] Obsolescence does not have to be permanent, it can often be
reversed. This leads to the following definitions[23]:

> "**Product lifetime** is the duration of the period that starts at the
> moment a product is released for use after manufacture and ends at the
> moment a product becomes obsolete beyond recovery."[23]

> "**Recovery** is a term for any operation with the primary aim of reversing
> obsolescence."[23]

These definitions can be applied to materials as well, by substituting 'product' with 'material'. The difference between a material and a product is not
as clear-cut as it may seem at first sight, but for the clarity of the typology presented here, we will assume that 'product' mainly refers to finished
end-products and components.

9.4.2 *Resisting, postponing and reversing obsolescence*

We can now build on the concept of product and materials integrity by
arguing that designers in a circular economy should firstly aim to prevent a
product or material from becoming obsolete (i.e. by resisting and postpon-
ing obsolescence) and secondly, make sure that resources can be recovered
with the highest level of integrity (i.e. reversing obsolescence). These goals
can be pursued at the level of products and components, referred to as

Table 9.1:   A Typology of Circular Product Design.

Circular product design	
Design for Product integrity	Design for recycling (materials integrity)
Inherently **long** product use (resisting obsolescence)	Inherently **long** materials use (resisting obsolescence)
Physical durability: designing a product resistant to degradation over time	Physical durability: choosing materials resistant to (or stabilized against) degradation during reprocessing and subsequent use
Emotional durability: designing a product that stimulates feelings of attachment	Emotional durability: creating pleasing aesthetics with reprocessed materials
Extended product use (postponing obsolescence)	**Extended** materials use (postponing obsolescence)
Maintain: designing a product that can, with regular servicing, easily retain its functional capabilities and/or cosmetic condition	Upgrade: using additives to enhance the functional capabilities or cosmetic condition of reprocessed material, relative to the original material properties
Upgrade: enhancing a product's functional capabilities and/or cosmetic condition, relative to the original design specification	
Product **recovery** (reversing obsolescence)	Materials **recovery** (reversing obsolescence)
Recontextualize: designing a product to be re-usable in a different context than it was originally designed for, without any remedial action	Repair, refurbish and remanufacture: ensuring it is easy to separate a product's materials from potential sources of contamination during the recycling process. In the case of primary recycling, the reprocessed materials have equivalent properties compared to the original materials (equivalent to remanufacture). Secondary recycling results in lesser properties (equivalent to repair or refurbishing)
Repair, refurbish and remanufacture: designing a product to be easily brought back to working condition. In the case of remanufacture, the product is brought back to at least OEM original specification. In the case of repair and refurbish, the condition of the repaired or refurbished product may be inferior to the original specification	

design for product integrity, or at the level of materials, referred to as design for recycling (or materials integrity).

Table 9.1 gives a typology of Circular Product Design, with different design approaches to resist, postpone and reverse obsolescence. For instance, products with a high physical and emotional durability that are intended to be used for a long time, 'resist' obsolescence and operate at a high level of product integrity. Similarly, choosing materials with a high physical durability (i.e. that are resistant to, or stabilized against, degradation during reprocessing and subsequent use) helps 'resist' obsolescence in the recycling process.[20] And, creating a new aesthetic with reprocessed materials could contribute to the acceptance of imperfection as a unique material experience[26] and strengthen a product's emotional durability.[27]

In order to extend product use, or 'postpone' obsolescence, designers can create products that are easy to maintain and/or upgrade. Where maintenance is done to *retain* a product's functional capabilities and/or cosmetic condition, 'upgrading' is usually done to *enhance* its functionality or cosmetics, and is instigated by a change in the product's context of use. From a recycling perspective, 'upgrading' refers to adding virgin material or other 'additives' to a reprocessed material, in order to enhance its functional and/or cosmetic capabilities.

In order to recover products and materials, in other words, to 'reverse' obsolescence, designers can create products that are easy to re-use in a different context, and that can be repaired, refurbished or remanufactured. Repair, refurbishment and remanufacture are differentiated according to the quality of the recovered product relative to the original.[28] In the case of remanufacturing, the product is brought back to at least Original Equipment Manufacturer (OEM) specification. In the case of repair and refurbishing, the condition of the repaired or refurbished product may be inferior to the original specification.[28] This same line of reasoning is valid for recycling, with primary recycling resulting in reprocessed materials with equivalent properties as the original materials, and secondary recycling leading to a lower quality result.

9.5 Discussion

In this section we discuss the value of the typology for addressing critical materials. The typology of circular product design has the following advantages and limitations:

(1) It gives guidance. When for instance developing an electronic product, designers will first consider its physical and emotional durability. Without this, designing for ease of maintenance makes little sense, and a product that isn't durable may also not have enough residual value to be considered for repair or refurbishment. The typology has its limitations,

however. It is not suitable for materials that are supposed to biodegrade (product and material integrity is less relevant here). Also, the typology doesn't work very well for fast moving consumer goods such as toiletries and detergents. Critical materials however, occur mostly in durable goods, so this is not a major limitation.

(2) The typology establishes a common language. A first step has been made by redefining product lifetime in terms of obsolescence and bringing together the relevant design approaches for circular product design. The typology can now be used as a basis for a detailed development of design approaches and methods for circular product design.

(3) The typology provides an organizational structure. The hierarchies for product and materials integrity create a clear organizational structure. The typology brings out the parallels between product and materials integrity, stimulating designers to take both into account. However, the typology doesn't (yet) give guidance on how to address potential trade-offs and developments over time. For instance, given the current low recovery rates of critical materials, a long product life with robust and upgradeable products is possibly preferable to losing critical materials through inefficient recycling. Over time, with innovations in recycling technology, this could change and we may need to start looking at different product 'speeds' in a circular economy. Also, the typology doesn't address the potential environmental impacts of the different design approaches. The underlying assumption is that a longer product lifetime leads to an overall decrease in environmental impact, but there are possible exceptions that need to be acknowledged.[29]

(4) The typology establishes the basis for the field of circular product design. It brings together diverse and to this date unconnected design approaches under the umbrella of circular product design. One of the next steps should be the exploration of suitable circular business models. For example, in order to make a product that was designed for remanufacturing really work, obsolete products need to be consistently returned to the OEM to be remanufactured. This requires arrangements for reverse logistics and a transactional model that allows the (re)manufacturers to retain economic control of their product over time.[23] Some other interesting next steps have been indicated above, like temporal aspects and environmental impact assessments.

(5) Educational tool. The typology can be used for educational purposes, in line with the remarks made in the previous points.

9.6 Conclusion

In 2010, John Voeller[30] coined the term 'insufficient plenty' to discuss a world with "plenty of resources out there, but no guarantee that we...will have access to them." Preparing product designers for such a possible world will require them to develop skills and competencies beyond what their current design education offers. Three knowledge gaps were identified in this article. Firstly, product designers often lack even a basic understanding of what 'critical materials' are, and the instrumental role these materials have in creating high-performance products. Addressing this knowledge gap will require close collaboration with materials scientists and process engineers.

Secondly, designers need practical skills and up-to-date methods to enable them to design for maintenance, upgrading, repair, refurbishment, remanufacture and recycling; in other words: to do circular product design. The typology of circular product design presented in this chapter is a first step. Thirdly, designers need to learn how to adopt a systemic perspective towards critical materials. They need to understand the interdependencies between the different design strategies for addressing critical materials presented in this chapter, and need to be able to imagine trade-offs and possible negative consequences of their design interventions. In a circular economy, this would require designers (and companies) to monitor and manage their products and materials much closer, over time, in order to understand where these interdependencies and trade-offs might occur in practice.

These three knowledge gaps spell out a critical materials agenda for design education and design research on a methodological and a practical level. It includes bringing back a material culture in design education, invoking a deep understanding and love of the materials and processes involved in product design.

References

1. Graedel, T. (2009). Defining critical materials. In R. Bleischwitz, P.J. Welfens, and Z. Zhang (Eds.), *Sustainable growth and resource productivity — economic and global policy issues.* Sheffield, UK: Greenleaf Publishing.
2. Graedel, T.E., Harper, E.M., Nassar, N.T., and Reck, B.K. (2013). On the materials basis of modern society. *Proceedings of the National Academy of Sciences of the USA, 112*(20), 6295–6300.
3. Peck, D. (2016). *Prometheus missing: Critical materials and product design* (Doctoral thesis). Delft University of Technology, the Netherlands. Retrieved from https://repository.tudelft.nl/islandora/object/uuid%3Aa6a69144-c78d-4feb-8df7-51d1c20434ea

4. Ashby, M.F. (2009). *Materials and the environment: Eco-informed material choice* (1st ed.). Oxford, UK: Butterworth-Heinemann.

5. Karana, E., Hekkert, P., and Kandachar, P. (2008). Material considerations in product design: A survey on crucial material aspects used by product designers. *Materials & Design*, 29(6), 1081–1089.

6. Bakker, C. and Kuijer, L. (2014). More disposable than ever? Consequences of non-removable batteries in mobile devices. *Proceedings of the Going Green — CARE Innovation 2014 conference.* Vienna, Austria: CARE.

7. Greenfield, A. and Graedel, T.E. (2013). The omnivorous diet of modern technology. *Resources, Conservation and Recycling*, 74, 1–7.

8. Duclos, S.J., Otto, J.P., and Konitzer, D.G. (2010). Design in an era of constrained resources. *Mechanical Engineering*, 132(9), 36–40.

9. European Commission. (2009). *Waste Framework Directive (2008/98/EC) or Directive 2008/98/EC of the European Parliament and of the Council of 19 November 2008 on waste and repealing certain Directives.* Brussels, Belgium: European Commission.

10. Zimmermann, T. (2017). Uncovering the fate of critical metals: Tracking dissipative losses along the product life cycle. *Journal of Industrial Ecology*, 21(5), 1198–1211.

11. Habib, K. and Wenzel, H. (2016). Reviewing resource criticality assessment from a dynamic and technology specific perspective — using the case of direct-drive wind turbines. *Journal of Cleaner Production*, 112(5), 3852–3863.

12. Lim, S.R., Kang, D., Ogunseitan, O.A., and Schoenung, J.M. (2013). Potential environmental impacts from the metals in incandescent Compact Fluorescent Lamp (CFL), and Light-Emitting Diode (LED) bulbs. *Environmental Science & Technology*, 47(2), 1040–1047.

13. Stahel, W.R. (2010). *The Performance Economy* (2nd ed.). London, UK: Palgrave Macmillan.

14. Hampus, A., Söderman, M.L., & Tillman, A.M. (2016). Circular economy as a means to efficient use of scarce metals? *Proceedings of Electronics Goes Green 2016.* Berlin, Germany: Fraunhofer IZM and Technische Universität Berlin.

15. Beitz, W. (2007). Designing for ease of recycling. *Journal of Engineering Design*, 4(1), 11–23.

16. Reuter M.A., et al. (2013). *Metal recycling: Opportunities, limits, infrastructure.* Nairobi, Kenya: UNEP.

17. Ylä-Mella, J. and Pongrácz, E. (2016) Drivers and constraints of critical materials recycling: The case of indium. *Resources*, 5(4), 34.

18. Ellen MacArthur Foundation. (2013). *Towards the circular economy; economic and business rationale for an accelerated transition.* Gowes, UK: Ellen MacArthur Foundation.

19. Balkenende, R., Bocken, N., and Bakker, C. (2017). Design for the circular economy. In R.B. Egenhoefer (Ed.), *Routledge Handbook of Sustainable Design.* Oxford, UK: Earthscan.

20. Hopewell, J., Dvorak, R., and Kosior, E. (2009). Plastics recycling: challenges and opportunities. *Philosophical Transactions of the Royal Society of London B*, 364(1526), 2115–2126.

21. Teddlie, C. and Tashakkori A. (2006). A general typology of research designs featuring mixed methods. *Research in the Schools*, 13(1), 12–28.

22. Köhler, A.R., Bakker, C., and Peck, D. (2013). Critical materials: a reason for sustainable education of industrial designers and engineers. *European Journal of Engineering Education*, 38(4), 441–451.

23. Den Hollander, M.C., Bakker, C.A., and Hultink, H.J. (2017). Product design in a circular economy: development of a typology of key concepts and terms, *Journal of Industrial Ecology*, 21(3), 517–525.

24. Murakami, S., Oguchi, M., Tasaki, T., Daigo, I., and Hashimoto, S. (2010). Lifespan of commodities, part 1. The creation of a database and its review. *Journal of Industrial Ecology*, 14(4), 598–612.

25. Burns, B. (2010). Re-evaluating obsolescence and planning for it. In T. Cooper (Ed.), *Longer lasting products — alternatives to the throwaway society*. Farnham, Surrey, UK: Gower Publishing Limited.

26. Rognoli, V. and Karana, E. (2014). Toward a new materials aesthetic based on imperfection and graceful aging. In E. Karana, O. Pedgley and V. Rognoli (Eds.), *Materials experience; fundamentals of materials and design*. Oxford, UK: Elsevier.

27. Chapman, J. (2005). Emotionally durable design: Objects, experiences and empathy. London, UK: Earthscan.

28. Ijomah, W.L., McMahon, C.A., Hammond, G.P., and Newman, S.T. (2007). Development of design for remanufacturing guidelines to support sustainable manufacturing. *Robotics and Computer-Integrated Manufacturing*, 23(6), 712–719.

29. Bakker, C.A., Feng W., Huisman, J., and den Hollander, M. (2014) Products that go round: exploring product life extension through design. *Journal of Cleaner Production*, 69, 10–16.

30. Voeller, J.G. (2010). The era of insufficient plenty. *Mechanical Engineering*, 132(6), 35–39.

Chapter 10

Substitution Case Study: Replacing Niobium by Vanadium in Nano-Steels

Zaloa Arechabaleta Guenechea and S. Erik Offerman
Department of Materials Science & Engineering,
Delft University of Technology, Mekelweg 2,
2628 CD Delft, The Netherlands

The substitution of critical alloying elements in metals is a strategy to reduce the criticality of materials. Nano-steels are a novel grade of advanced high-strength steels that are suited for application in the chassis and suspension of cars and as fire-resistant steel in high-rise buildings. The high strength and ductility per unit mass make the nano-steels resource-efficient and reduce vehicle weight while maintaining crash worthiness. The excellent mechanical properties of certain nano-steels rely on the addition of small amounts (up to 0.1 wt.%) of Niobium as alloying element to the steel. Niobium is considered to be a critical raw material by the European Union due to its high economic importance as an alloying element in advanced, high-strength steel grades and due to the high supply risk related to the high degree of monopolistic production within the supply chain. This chapter describes the fundamental materials science that is needed for the substitution of the critical alloying element Niobium by Vanadium as an alloying element in nano-steels.

10.1 Introduction

The world population is expected to grow by about 30% by 2050 to approximately 9 billion people[1] and with people in emerging countries aspiring to the same lifestyle as in developed countries, the supply of resources is no longer guaranteed in the long term. Europe promulgates "resource-efficiency" as a key factor to address this challenge in the coming years and has established new enduring strategies in areas such as energy, climate change, research and innovation, industry, and transport, among others.[1] Promoting and assuring a secured, sustainable and sufficient supply of raw materials is of primary importance for the European Union (EU), as these materials play an essential role in the EU's economy, growth, and competitiveness: key economic sectors including automotive, aerospace, and construction, strongly connected to the steel industry, are directly linked to their supply.

After China, the EU is the second largest producer of steel in the world, producing 11% of the global output.[2] "Resource efficiency" in the steel industry implies, among other challenges, the reduction of energy consumption and CO_2 emissions, both directly, via modification and/or improvement in the steelmaking process, and indirectly, by providing the best suitable steel solution for a specific use. An example of the latter is the use of Advanced High Strength Steels (AHSS) for lightweight automotive applications. These types of steels, exhibiting superior strength and formability, are extremely useful for improving fuel economy and reducing greenhouse gas emissions without compromising crash worthiness. In general, it can be claimed that using steel with higher strength helps to save steel: even partial global switching to higher strength steel could save 105 million metric tons of steel a year and 20% of the costs of steel.[3]

In view of the above, the steel industry worldwide is developing AHSS to satisfy the most exigent requirements in diverse applications. Nonetheless, conventional AHSS are not always adequate for high-tech applications despite their good strength-ductility balance. In the automotive industry, for instance, new chassis and suspension designs require, in addition, stretch-flangeability and bendability.

This enhanced combination of mechanical properties cannot be obtained with conventional AHSS grades that have a multiphase microstructure. The reason for this is that voids and cracks are formed in the multiphase steels during blanking and stretch-flanging, due to stress localization at the interface between the hard phases (*e.g.* martensite, bainite and retained austenite) and the soft ductile ferrite. To overcome this hurdle, the steel industry has recently developed the so-called nano-steels, *i.e.* a new generation of AHSS. Interestingly, this new type of steel can successfully be used for advanced applications, such as specific automobile parts, particularly those parts that are important for safety, and high-rise buildings for fire resistance. In contrast to conventional AHSS, their microstructure consists of a single soft ferritic matrix, providing ductility, strengthened by nanometer-sized precipitates. The very small size of the precipitates (the hard phase in this type of steels) dispersed in the ferritic matrix is critical to achieve the excellent mechanical behavior. Figure 10.1 schematically shows the microstructures of (a) a nano-steel with interphase

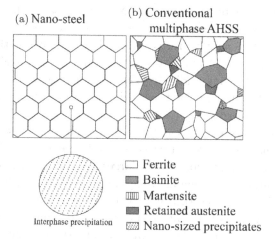

(a) Nano-steel   (b) Conventional multiphase AHSS

☐ Ferrite
■ Bainite
▥ Martensite
■ Retained austenite
▨ Nano-sized precipitates

Interphase precipitation

Fig. 10.1.   Microstructures of (a) nano-steel and (b) conventional AHSS.

precipitation and (b) a conventional multiphase AHSS containing a small fraction of bainite, martensite and retained austenite embedded in ferrite. Besides their outstanding mechanical behavior, an additional advantage of nano-steels is that they can be produced by direct cooling from austenite at high temperatures, avoiding additional and expensive heat treatments and saving energy. This is not the case for the multiphase AHSS, which are usually produced by other thermal treatments.

Nano-steels use considerable microalloying additions, including Niobium (Nb, ≤0.1 wt.%), Titanium (Ti, ≤0.35 wt.%), Vanadium (V, ≤0.45 wt.%) and Molybdenum (Mo, ≤0.7 wt.%), to form the essential precipitating species,[4–10] that is, the corresponding carbides and/or carbonitrides (MC and/or M(C,N), where M=Nb, Ti, V and Mo). Albeit these contents do not seem very high at first, they imply significant amounts of microallying additions considering the large volume of steel produced worldwide. The microalloying elements are usually expensive, meaning that the commercial and industrial implementation of nano-steels is subjected to the efficient use of these elements, and hence, to the better understanding of the precipitation reactions involved. The first nano-steels that have been developed for automotive applications are based on additions of Ti and Mo[11]: the Ti precipitates in the form of (Ti,Mo)C particles and the Mo suppresses TiC-precipitate coarsening (Ostwald ripening), keeping the precipitates small during processing. Other studies have also demonstrated a similar effect of Mo on NbC-precipitates.[12,13] Yet, quantitative data about the Mo effect on the TiC- and NbC-precipitation kinetics remains insufficient. Moreover, the EU has recently identified Nb as a critical raw material because its supply risk is high.[14] The challenge is thus to design nano-steels that are: (1) more "resource-efficient", by reducing and optimizing the microalloying additions, (2) containing less or no critical raw materials, and (3) at the same time maintaining or improving their excellent mechanical properties.

Vanadium is a promising candidate to (partially) replace Nb and Ti in nano-steels. The advantages of this element will be enumerated through this chapter. In contrast to Nb and Ti, the larger solubility of V can be exploited for different purposes, but mainly to optimize ferrite strengthening via precipitation of V(C,N). Literature on V-precipitation has provided compelling evidence that by using high Nitrogen (N) levels precipitate coarsening can be suppressed due to the reduced solubility of V(C,N) compared with VC in ferrite, which further enhances the precipitate hardening contribution. By contrast, little or nothing has been reported so far about the interaction between V and Mo, and the question to answer in the near future is if V, N, and Mo is the perfect combination for nano-steels.

From the abovementioned it can be concluded that dispersion of V(C,N) nanometer-sized precipitates is the most promising strategy to maximize precipitation strengthening and at the same time minimize the addition of alloying elements. This, however, is not straightforward and requires a deep understanding of how precipitation and the austenite to ferrite ($\gamma \rightarrow \alpha$) phase transformation interact, since both phenomena occur within the same range of temperatures during thermomechanical processing of nano-steels. In some cases, the precipitates will form at the austenite/ferrite (γ/α) interface as phase transformation proceeds. This is known as interphase precipitation (IP) and usually recognized by rows of aligned precipitates. In other cases, the precipitates will nucleate randomly from supersaturated ferrite. Different microalloying elements not only will generate different precipitation reactions with distinct kinetics, but also will influence the phase transformation (kinetics) in a different manner, which will affect the precipitation indirectly as well. Until now, it has proven to be impossible to disentangle the kinetics of both reactions, which indicates the great importance of in-situ and simultaneous use of advance characterization techniques, such as High-Resolution Transmission Electron Microscopy (HR-TEM), 3D X-ray Diffraction (3DXRD), Small Angle X-ray Scattering (SAXS), Small Angle Neutron Scattering (SANS), Neutron Diffraction (ND) and Electron Back Scattered

Diffraction (EBSD), among others, to enable precipitation optimization in nano-steels.

10.2 Nano-steels in High-tech Applications

In the following two sections, the main applications so far of nano-steels are elaborated, viz. structural fire-resistant steels and chassis and suspension systems. Note, however, that these steels are also suitable to substitute conventional High Strength Low-Alloy (HSLA) steels in other applications, for example. in oil and gas pipelines, farm machinery, industrial equipment, bridges and so on, provided that application requirements, if necessary, are satisfied, e.g. corrosion resistance, weldability, abrasion resistance, low-temperature impact toughness for line pipe and lifting and excavating applications, and fatigue properties for chassis.

10.2.1 *Fire-resistant steels*

Fire-resistant steels are structural steels, *i.e.* steels used in building constructions, which must guarantee sufficient strength at elevated temperatures to ensure that all internal and external loads applied to the structure can be carried. The exact criterion defining sufficient strength changes between different countries depending on their specific fire safety regulations. In Japan, for instance, the standards require that the yield strength (0.2% proof stress) of the steel at 600°C is at least two thirds of that specified at room temperature.[15] Less stringent regulations in Australia, United States and Europe, demand in such conditions at least one half of the room temperature yield strength.[16–18] For this type of steels, the yield strength can theoretically be calculated using the following equation[19]:

$$\sigma_y = \sigma_0 + \sigma_{ss} + \sigma_{gb} + \sqrt{\sigma_{dis}^2 + \sigma_{ppt}^2} \qquad (1)$$

where σ_0 is a friction stress and σ_{ss}, σ_{gb}, σ_{dis} and σ_{ppt} are respectively contributions from solid solution, grain boundary, dislocation and precipitation strengthening. In other words, the strength of a steel can be tailored by modifying and optimizing its microstructure, *e.g.* varying the elements in solid solution and solute contents, the microstructure grain size and density of grain boundaries, the dislocation density and the volume fraction and size of precipitates. The main reasons why a steel loses its strength at high temperature are indeed related to how the different terms in equation (1) evolve with temperature. These can be summarized as follows:

- *Enhanced mobility of dislocations.* At high temperatures, dislocations are activated by thermal energy and can easily move: dislocation glide, climb and cross slip are facilitated. When two dislocations of opposite sign encounter each other on the same slip plane, dislocation annihilation occurs. This mechanism can significantly reduce the dislocation density at elevated temperatures, and hence, lower the dislocation strengthening contribution. Note that $\sigma_{dis} \propto \rho^{1/2}$, where ρ is the dislocation density.
- *Precipitate coarsening.* The high interfacial energy of small precipitates provides the driving force for precipitate coarsening, which occurs by diffusion of solute atoms. Basically, the smallest precipitates shrink and dissolve in the steel matrix and the solute redistributes to the largest stable precipitates (Ostwald ripening), which increases the average particle size. Particle sizes larger than a critical value r*, will cause the strengthening contribution $\sigma_{ppt} \propto r^{-1}$, where r is the average particle radius, to decrease (see Fig. 10.12). Moreover, as the temperature increases, diffusion is enhanced and the rate of precipitate coarsening is accelerated.
- *Microstructure coarsening.* The mobility of grain boundaries also increases with temperature. Grain coarsening will occur in case the solute content is not high enough and/or the precipitates are not sufficiently small to pin the grain boundaries. According to the well-known Hall-Petch relation, $\sigma_{gb} \propto d^{-1/2}$, in which d is the average grain diameter, a coarser microstructure will provide a smaller grain boundary strengthening contribution. The reason is that grain boundaries are less effective in hindering the migration of dislocations as the grain size increases. Some studies have reported that subgrain boundaries (*i.e.* boundaries smaller than 15°) may also contribute to Hall-Petch strengthening.[20]
- *Less efficient solid solution strengthening.* Solute atoms also act as obstacles for the motion of dislocations and grain boundaries. Depending on the atom size, the solute concentration and the type of distortion that they produce, solute atoms may offer weak or strong resistance. At high temperatures, however, both dislocations and grain boundaries can overcome these obstacles more easily.

Fire-resistant steels make use of different approaches, based on their chemical composition, to keep the strength level, σ_y in equation (1), sufficiently high at elevated temperatures. Some steels, for instance, will simply resist the effect of a fire by being thermally stable, which is attained by adding specific elements in solid solution and increasing σ_{ss}. Others, by contrast, will provide additional strength if precipitation occurs under such extreme conditions, by instantaneously increasing σ_{ppt}. Generally, in fire-resistant steels:

- the dislocation density is kept low to avoid abrupt changes in strength at high temperatures,
- the addition of ferrite stabilizers is used to raise the temperature at which ferrite starts to transform to (soft, coarse-grained) austenite,
- elements in solid solution that produce large atomic misfit strains (strong obstacles) are preferentially employed and,
- nanometer-sized precipitates, which are slow to coarsen, are desirable.

Table 10.1 lists the chemical compositions of two commercial fire-resistant steels, containing Mo and Nb-Mo additions, respectively, manufactured by Nippon steel. The Mo steel in Table 10.1 is designed to simply withstand a fire using the effect of Mo in solid solution, whereas the Nb-Mo steel increases its strength at high temperature by combining the Mo effect with Nb nano-precipitation. It has been observed that with the latter combination, it is possible to reach failure temperatures up to 650°C for the case in which half of the room temperature yield stress is required.[21] A more recent study has investigated the fire-resistance properties of Fe-C-Mn-Nb steels with no Mo additions.[22] In this work, the C-Mn-Nb specimens are soaked at 1100°C for 45 min, rapidly cooled to 850°C and at three different cooling rates, 0.17, 1 and 100°C/s, in the temperature range from 850°C to 600°C, to be finally quenched to room temperature. Afterwards, the specimens are subjected to a standard fire test ISO 834. A reference C-Mn steel is also used for comparison purposes. The obtained results are summarized in Fig. 10.2, which clearly illustrates the contribution of Nb to the failure temperature increase. By addition of 0.1 wt.% Nb to a plain C-Mn steel with no precipitation ($n_{NbC} = 0$), the failure temperature measured during the fire test increases by 92°C. This temperature increment is related to the increase in number density of NbC precipitates, from 0 in the C-Mn steel to 10^{23} m^{-3} in the Nb-microalloyed steel after the fire test. Figure 10.2 also shows that this temperature can be further raised, by 45°C, if besides NbC-precipitation, the density of grain boundaries is increased (*i.e.*, σ_{gb} in equation (1)). It should be noted that the 45°C increment is reached for the fastest cooling rate of 100°C/s due to an increase in the density of both Low (LAGBs) and High (HAGBs) Angle Grain boundaries. The grain boundary density in the Nb-microalloyed steel changes from 0.06 to 0.64 μm^{-1}

Table 10.1:   Chemical Composition in wt.% of Two Commercial Fire-resistant Steels Produced by Nippon Steel.[23]

Steel	C	Mn	Si	Mo	Nb	S	P
Mo	0.1	0.64	0.1	0.51	—	0.05	0.009
Nb-Mo	0.11	1.14	0.24	0.52	0.03	0.02	0.009

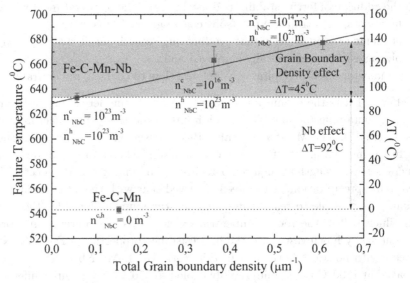

Fig. 10.2.   Failure temperature and difference in failure temperature with respect to a reference material (Fe-C-Mn) as measured during a standard fire test ISO 834 as a function of the total grain boundary density for a C-Mn-Nb alloy. The number density of NbC precipitates calculated before (n^c_{NbC}) and after (n^h_{NbC}) the fire test are also indicated.[22] Reprinted from 'Scripta Materialia', vol. 68, E. Gözde Dere, Hemant Sharma, Roumen H. Petrov, Jilt Sietsma and S. Erik Offerman, 'Effect of niobium and grain boundary density on the fire resistance of Fe–C–Mn steel', Pages 651–654, 2013, with permission from Elsevier.

by increasing the cooling rate during the $\gamma \rightarrow \alpha$ phase transformation from 0.17°C/s to 100°C/s.

10.2.2   *Chassis and suspension systems*

The chassis and suspension are the most important components of a car in order to maintain driving stability, and thus, their high reliability is indispensable. Similar to other parts of the car, weight targets also extend to the automotive chassis system and promote the use of stronger materials

(a)   (b)

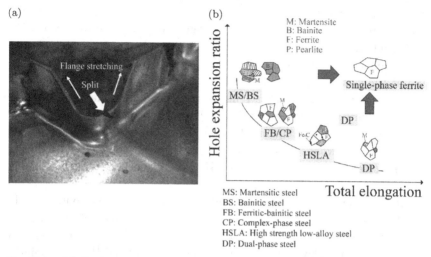

Fig. 10.3. (a) Crack initiation at a sheared edge of a component made from DP780 after flanging.[25] Reprinted with permission from Springer, in Advances in Manufacturing, 'Reverse metallurgical engineering towards sustainable manufacturing of vehicles using Nb and Mo alloyed high performance steels, Hardy Mohrbacher, 2013. (b) Microstructure effect on the balance between elongation and hole expansion ratio.

for these applications. Yet, chassis designs require very complex shapes that are only obtained after several forming processes are applied, including blanking, punching, stretch-flanging, and hole expansion operations. These processes are often limited by the strength of the steel and the presence of hard phases, particularly martensite, that are inhomogeneously distributed within the steel microstructure. Steels with several phases are not suitable for these applications,[24] as usually voids and cracks are generated during forming by decohesion of hard phase-soft phase interfaces.

    Currently, and despite their limitations, AHSS such as ferrite-bainite (FB) and/or dual-phase (DP) steels are adopted for these high-tech applications. The shortcomings are clearly illustrated in Fig. 10.3. Figure 10.3(a) shows a common problem of edge splitting during flanging operations when using a DP steel,[25] whilst Figure 10.3(b) schematically displays the balance between the hole expansion ratio and the total elongation for different AHSS microstructures. The latter demonstrates that single-phase AHSS, that is, martensitic (MS) and bainitic (BS) steels, exhibit high hole expansion ratio, but poor elongation. DP steels, conversely, display high elongation, though the capacity for hole expansion ratio is very limited. Somewhere in between lie the FB, complex-phase (CP) and HSLA steels. Figure 10.3 shows the necessity for the development of steels that, besides having a good strength-ductility balance, are also suitable for demanding forming operations. As shown in Fig. 10.3(b), the appropriate steel must adopt a single phase ferrite structure to reach the hole expansion ratio of bainitic/martensitic steels, and at the same time, keep the elongation high. Other desired properties are also high yield and tensile strengths, good bendability, fatigue strength, and weldability. The strength of the steel is maximized by grain refinement, precipitation strengthening, and by using thermally stable nanometer-sized precipitates (to avoid coarsening and the subsequent loss in strength during steel processing). This new type of AHSS described here is the promising nano-steel.

It has been observed, for instance, that in nano-steels containing Ti and Mo additions, tensile strengths of 780 MPa can be reached when the Ti/Mo atomic ratio is 1. In that case, very fine carbides (3 nm) aligned in rows (*i.e.*, interphase precipitation) have been noticed.[11] The strengthening contribution from these extremely fine precipitates exceeds by far the amount of conventional precipitation strengthening achieved to date in HSLA steels. Figure 10.4 shows the decrease in strength experienced during isothermal annealing at 650°C by two different steels, a conventional HSLA

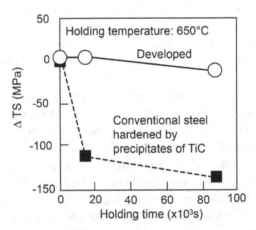

Fig. 10.4. Loss in strength undergone by a conventional HSLA steel (black squares) and a nano-steel based on Ti additions (white circles) during isothermal annealing at 650°C.[26] Reprinted by permission of Taylor & Francis Ltd (http://www.tandfonline.com) on behalf of Institute of Materials, Minerals and Mining, article title: 'Application of nanoengineering to research and development and production of high strength steel sheets', by K. Seto & H. Matsuda, Materials Science and Technology, copyright © (2013) Institute of Materials, Minerals and Mining.

steel microalloyed with Ti and a nano-steel containing Ti and Mo additions. The former is hardened with TiC-precipitation and the latter with (Ti,Mo)C precipitates that have a smaller size distribution.[26] It can be seen in Fig. 10.4 that the tensile strength TS of the nano-steel remains practically the same after annealing for very long times, whereas it abruptly decreases at 15000 s for the conventional HSLA steel, caused by faster TiC coarsening. The precipitates in the HSLA steel are bigger and thus more prone to grow. The results in Fig. 10.4 confirm the remarkably higher stability of (Ti,Mo)C-precipitates in the nano-steel. It has been reported that Mo accelerates the nucleation stage and reduces the precipitate size by reducing the interfacial energy between the precipitate and the matrix (i.e., the lattice misfit).[27,28] Nevertheless, Mo is not thermodynamically favored within the precipitate and subsequent stages of growth and coarsening require that it partitions to the matrix, so that these latter stages are consequently delayed.

Combinations of Nb and Mo additions, either alone or together with Ti, are also being investigated.[25,27,29] One of the advantages of Nb microalloying is the austenite pancaking during hot rolling that promotes finer ferritic microstructures. Fast cooling from the finish rolling temperature to the temperature range of ferrite formation and subsequent (Ti,Nb,Mo)C interphase precipitation during phase transformation, delayed by solute Nb, is the main processing strategy behind this type of steel. As observed in Fig. 10.5, this new alloy concept, using a combination of 0.04wt.%C-1.4wt.%Mn-0.09wt.%Nb, exhibits excellent bendability and high quality cutting edges, but also high hole expansion ratio and elongation.[25]

10.3 Mechanisms of Precipitation Strengthening

In order to better understand the large precipitation strengthening contribution σ_{ppt} generally obtained in nano-steels, essential for fire-resistance and chassis applications, the current section focuses on the main mechanisms by which precipitates might strengthen the steel. In general, precipitation strengthening in micro-alloyed steels arises primarily from the interaction of the moving dislocations and the precipitates, namely from:

- stress fields generated around precipitates, due to a slight misfit between the precipitates and the matrix, which hinders the motion of dislocations as depicted in Fig. 10.6, known as *coherency strengthening* (σ_{coh}),
- dislocations cutting coherent precipitates and creating both an additional precipitate/matrix interface, designated as *chemical strengthening* (σ_s),

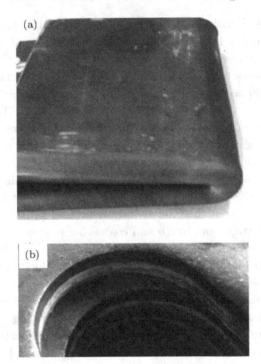

Fig. 10.5. Bendability and hole punching properties of a 0.04wt.%C-1.4wt.%Mn-0.09wt.%Nb steel.[25] Reprinted by permission from Springer, in Advances in Manufacturing, 'Reverse metallurgical engineering towards sustainable manufacturing of vehicles using Nb and Mo alloyed high performance steels, Hardy Mohrbacher, 2013.

and an anti-phase boundary, referred to as *order strengthening* (σ_{APB}) illustrated in Fig. 10.8, and

- dislocations looping around incoherent precipitates, see Fig. 10.9, which is called *dispersion strengthening* (σ_D), shown in Fig. 10.10.

Note that a precipitate is coherent if it matches perfectly with the atomic structure of the matrix at the interface (see Fig. 10.7). Thus, coherent precipitates provide continuity of slip planes between the matrix and the precipitates and are shearable. For small coherent precipitates, the coherency strengthening is given by[30]:

$$\sigma_{coh} = 4.1 M G \varepsilon^{3/2} f^{1/2} \left(\frac{r}{b}\right)^{1/2} \tag{2}$$

where M is the Taylor factor, G is the shear modulus, ε is the misfit of the precipitate, f is the volume fraction of precipitates, b is

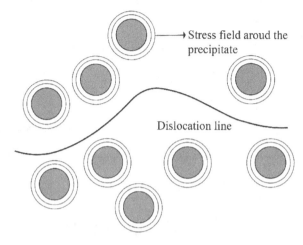

Fig. 10.6.   Interaction between a dislocation and the stress field of the precipitates.

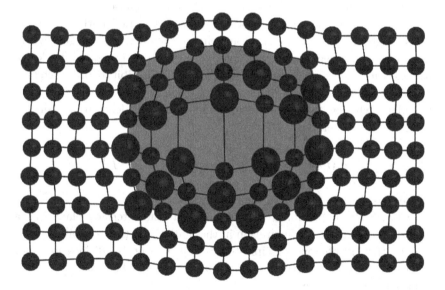

Fig. 10.7.   Schematic representation of a coherent precipitate with large strain fields.

the magnitude of the Burgers vector and r is the average precipitate radius. For larger coherent precipitates, the following equation is used instead[30]:

$$\sigma_{coh} = 0.7MG\varepsilon^{1/4}f^{1/2}\left(\frac{b}{r}\right)^{3/4} \tag{3}$$

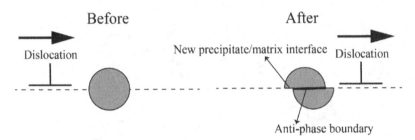

Fig. 10.8. A dislocation cutting through a precipitate coherent with the matrix.

The chemical strengthening can be estimated as[31]:

$$\sigma_s = 2MG \left(\frac{3}{\pi}\right)^{1/2} \left(\frac{\gamma_S}{Gb}\right)^{3/2} \left(\frac{b}{r}\right) f^{1/2} \tag{4}$$

in which γ_s is the energy of the precipitate-matrix interface. Usually γ_s is smaller than the energy necessary to create an antiphase boundary γ_{APB} and the chemical strengthening does not contribute significantly to increase the strength of aged alloys. If an antiphase boundary, *i.e.* a region where the order of the precipitate is disrupted (see Fig. 10.8) is created inside the particle with ordered structure, restoration of the precipitate order requires the dislocations to glide in pairs: the first dislocation creates an antiphase boundary, whereas the second returns the order. This leads to a strengthening increase given by[32]:

$$\sigma_{APB} = M \frac{\gamma_{APB}}{2b^2} \left(\left(\frac{3\pi^2 \gamma_{APB} f r}{32T} \right)^{1/2} - f \right) \tag{5}$$

where T is the line tension of the dislocation.

Conversely, a precipitate is incoherent if the interface plane has a very different atomic configuration in the matrix and the precipitate. This implies that incoherent precipitates cannot be cut by dislocations nor deformed with the matrix. Figure 10.9 displays a schematic illustration of an incoherent precipitate.

There are three different possibilities by which the dislocations can then overcome incoherent precipitates: (1) looping around the precipitate, also known as the Orowan mechanism, leaving a dislocation loop behind, (2) climbing and (3) cross-slipping. Although dislocation release at higher stresses can also occur by cross-slip, the latter two mechanisms acquire more relevance at high temperature. In precipitation strengthening by the Orowan mechanism, it is assumed that the precipitates are hard, do not

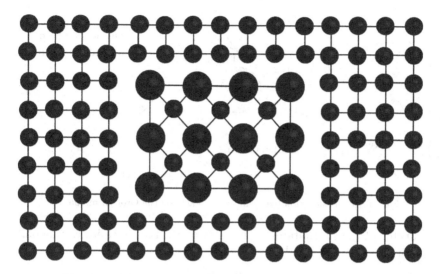

Fig. 10.9.   Schematic representation of an incoherent precipitate.

deform with the matrix and act as pinning points for dislocations. Considering the line tension of a dislocation, *i.e.* the force in the direction of the line vector that tries to shorten the dislocation, is $\sim \frac{Gb^2}{2}$, the stress necessary to bypass the precipitate can be estimated from the balance of forces between the particle resistance to dislocation motion and the line tension of dislocation as (see Fig. 10.10):

$$\sigma_D = \frac{MGb}{L} \tag{6}$$

where L is the average precipitate spacing. It should be pointed out, however, that this equation overestimates the real stress required to bypass a particle and usually represents an upper bound.[33] More accurate is the equation given by Ashby-Orowan that considers the effects of statistically distributed particles.[33] The latter is suitable for the cases in which the particle size is negligible with respect to the particle spacing. Taking into account the effect of self-interaction between dislocation lines on each side of the particle, the Ashby-Orowan equation can be expressed as[19]:

$$\sigma_D = \frac{kMGb}{2\pi\sqrt{1-\nu}L} \ln\left(\frac{x}{2b}\right) \tag{7}$$

in which k is a factor of 0.8, accounting for heterogeneity in particle distribution, ν is the Poisson's ratio, $\sqrt{1-\nu}$ is introduced as an average of the energy difference between screw and edge characters and x is the average

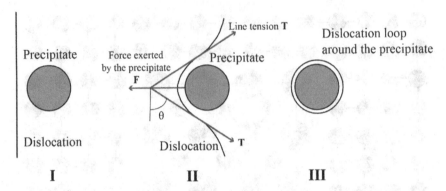

Fig. 10.10.   Dislocation bypassing an obstacle by the Orowan mechanism.

diameter of the precipitates on the slip plane. For precipitates randomly distributed in the matrix[19,34]:

$$L = \sqrt{\frac{2}{3}} \left(\sqrt{\frac{\pi}{f}} - 2 \right) \cdot r \tag{8}$$

and

$$x = 2\sqrt{\frac{2}{3}} \, r \tag{9}$$

Equations (8) and (9) are valid if interphase precipitation does not occur that is, for random precipitation, or in case it does, if the sheet spacing is larger and/or comparable to the particle spacing. If, by contrast, the sheet spacing is smaller, the interphase-precipitated particles cannot be considered randomly distributed and equations (8) and (9) are no longer applicable. In this particular case, L can be calculated as[35]:

$$L = \sqrt{r_1 r_2} \tag{10}$$

where r_1 is the mean linear inter-particle spacing along the intersection between the slip plane and the sheet plane of interphase precipitation and r_2 is the mean projected value of the perpendicular sheet spacing. r_1 and r_2 can be determined through the measured particle spacing ω, the sheet spacing, the particle aspect ratio p and the particle radius r as[35]:

$$r_1 = \frac{\omega^2}{2r} + \frac{\pi r}{2} - \frac{2r}{p \sin \theta} \tag{11}$$

$$r_2 = \frac{\lambda - 2r \sin \theta}{\sin \varphi} \tag{12}$$

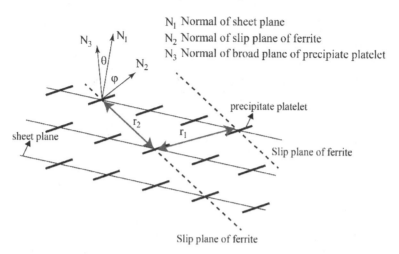

N$_1$ Normal of sheet plane
N$_2$ Normal of slip plane of ferrite
N$_3$ Normal of broad plane of precipiate platelet

Fig. 10.11. Schematic representation of the geometric orientation between the sheet plane, the carbide platelets and the ferrite slip plane as in[35]. Reprinted from Acta Materialia, Vol. 64, M.-Y. Chen, M. Gouné, M. Verdier, Y. Bréchet, J.-R. Yang, Interphase precipitation in vanadium-alloyed steels: Strengthening contribution and morphological variability with austenite to ferrite transformation, Pages 78–92, Copyright 2014, with permission from Elsevier.

in which θ is the angle between the precipitate broad plane normal and the sheet plane normal, and φ is the angle between the sheet plane normal and the slip plane normal. The geometric orientation between the sheet plane, the precipitates and slip plane are depicted in Fig. 10.6.

Precipitation strengthening involves a complex series of physical transformations that induce different strengthening mechanisms over time. In many cases, these transformations occur at such small scale that it is very difficult and tricky to identify the underlying mechanisms, giving rise to a controversy among researchers. A good example of this is the debate generated around the early stages of precipitation in microalloyed steels, for which no theory has been generally accepted so far. Several precipitate nucleation processes have been observed in microalloyed steels and these include homogeneous (uniformly and non-preferentially), interphase and heterogeneous (preferentially at specific sites such as dislocations, grain boundaries and/or vacancies) precipitation. It has been proposed that homogeneous precipitation in microalloyed steels follows the sequence Guinier-Preston (GP) zones → intermediate phase → equilibrium phase seen in non-ferrous alloys.[36] This sequence is schematically illustrated in Fig. 10.12 and can be explained as follows. During a first stage, solute-rich

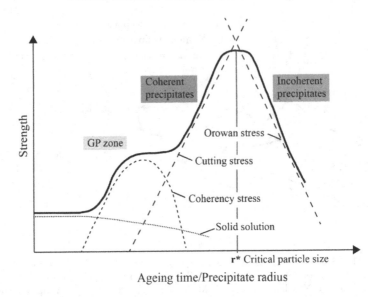

Fig. 10.12.   Schematic illustration of the precipitate strengthening contributions.

clusters (also called GP zones) are formed within the lattice driven by the high concentration of alloying elements in solid solution, which exceeds the equilibrium solubility limit. Atomic diffusion and clustering of the alloying elements is assisted by vacancies in the lattice. These solute-rich GP zones are coherent with the steel matrix and the atomic structure is continuous across the interface. In the next step, the GP zones transform into small coherent precipitates which distort the surrounding matrix due to their misfit and produce stress fields that impede dislocation motion (*coherency strengthening*). Once the dislocation has overcome these stress fields and reached the precipitate, it can pass through the particle, provided it surmounts the particle resistance (*chemical and order strengthening*), since the slip planes are coherent with the matrix. Eventually, coherent precipitates transform into incoherent ones as they grow, changing their crystal structure so that the atomic structure between the precipitate and the matrix is no longer continuous. Incoherent precipitates are distinct second-phase particles, with their own crystal structure and separated from the matrix by well-defined interfaces. Incoherent precipitates are more effective obstacles against dislocation motion (*dispersion strengthening*) and the only manner to overcome these obstacles is the dislocation bowing-out and looping around the precipitates. The maximum strengthening is attained at the moment coherent precipitates transform into incoherent ones. As soon as the latter grow slightly, the strengthening effect decreases. It should be emphasized, though, that even in that case the Orowan mechanism contributes to strength increase if compared to the condition of non-precipitation.

Regarding the nucleation stage, there is practically no reason for coherent precipitates to form in the stress field of a dislocation, essentially because the precipitate and the matrix lattices already match. Similarly, it is very unlikely that these precipitates nucleate at the grain boundaries. Only in the case of coherent precipitates with large elastic strains, for example when the precipitate nucleus is significantly larger than that of the matrix, is nucleation on vacancies favored. Semi-coherent precipitates, containing interfacial dislocations to relax the coherency strains, will nucleate much easier on dislocations. Incoherent precipitates, by contrast, with little or no atomic matching across the precipitate-matrix interface, will preferentially nucleate at the grain boundaries. As the size of coherent precipitates increases, coherent precipitates will transform into incoherent ones, which produce less lattice distortion, but have higher energy and mobility, and hence, grow faster. This means that at high temperatures incoherent precipitates will coarsen faster than coherent precipitates. Yet, coherent precipitates provide less resistance to dislocation motion. A balance between these two features can be accomplished by coherent precipitates with large strain fields.

10.4 Potential for Alternative Microalloying Elements: Replacing Nb by V.

10.4.1 *Precipitation strengthening in V-microalloyed steels*

Vanadium (V) has been used in steels as a microalloying element during the last decades, and the reason for this is that it has a great potential for ferrite strengthening. It forms nitrides (VN), carbonitrides (V(C,N)) and/or carbides (VC), depending on the temperature range and the steel's chemical composition. The solubility of these species increases in the abovementioned order, which means that VC and VN are the most and the least soluble particles, respectively. Moreover, this solubility is significantly smaller in ferrite than in austenite. Comparing to other microalloying elements that also form nitrides and carbides, like Ti, Nb and Al (Al only forms nitrides), V is the most soluble element and does not readily precipitate in austenite. Thus, although VN and V(C,N) can additionally be used for ferrite grain refinement, either by delaying austenite recrystallisation kinetics or by providing intragranular ferrite nucleation sites,[37–40] the primary objective of V addition in steel is V(C,N) or VC-precipitation strengthening. The latter usually occurs in ferrite during or after the $\gamma \rightarrow \alpha$ phase transformation upon cooling and coiling. Furthermore, the V-precipitation strengthening contribution can be enhanced by N addition, that is, at a given vanadium content, vanadium provides more strengthening when combined with N, up to the stoichiometric V:N.[41] Increasing the N content increases the driving force for V-precipitation, resulting in a smaller particle size, smaller inter-particle spacing and a greater resistance to coarsening.[42] Earlier work shows that additions of V and N significantly increase the yield stress.[43]

At relatively high temperatures in the austenitic regime, when the driving force for precipitation is small, V(C,N)-precipitates nucleate heterogeneously on dislocations and/or grain boundaries. These precipitates will lose coherency and be incoherent with the ferrite matrix afterwards. At lower temperatures, in the intercritical $(\gamma + \alpha)$ regime, V(C,N) interphase precipitation occurs during the $\gamma \rightarrow \alpha$ phase transformation. In this case, the precipitates nucleate at the migrating γ/α interface and form periodic arrays of sheets parallel to the transformation front. Note that because interphase precipitates are randomly arranged within the sheets, depending on the sheet orientation with respect to the observation plane, precipitates may appear as either well-defined lines or randomly distributed particles. Besides this, random interphase precipitation may also take place. Whether interphase precipitation occurs in sheet-like form or randomly depends on the chrystallographic characteristics of the interface between the parent austenite and the ferrite.[44] For instance, it has been observed that V(C,N) interphase precipitation is suppressed when the γ/α orientation relationship is close to the Kurdjumov-Sachs orientation relationship (K-S, $(111)_{\gamma\text{-Fe}} \| (011)_{\alpha\text{-Fe}}$ $[\bar{1}01]_{\gamma\text{-Fe}} \| [\bar{1}\bar{1}1]_{\alpha\text{-Fe}}$ [44]. It has been claimed that interphase V(C,N)-precipitates in commercial V-microalloyed steels, despite being incoherent with the ferrite matrix, follow a single variant of the Baker-Nutting (B-N) orientation relationship $(100)_{\alpha\text{-Fe}} \| (100)_{VC}$ $[011]_{\alpha\text{-Fe}} \| [010]_{VC}$).[45] Some authors have observed using Atom Probe Tomography (APT) that these precipitates consist of C, Mn and V,[46] whilst others have reported that Mn is homogenously distributed within the matrix.[47] Additionally, it has been shown that finer interphase V(C,N)-precipitates can be obtained by reducing the transformation temperature and increasing the V content, whereas the bulk C content hardly affects the particle size [47]. At these or even lower temperatures, homogeneous precipitation from supersaturated ferrite can also occur. It has been proposed that these

V(C,N)-precipitates have a coherent particle-ferrite matrix interface at the early stages of precipitation, which becomes semi-coherent and subsequently incoherent during growth.[48] Although, in principle, this agrees well with the lattice mismatch calculated for different carbonitrides in ferrite, according to which V(C,N) particles have the smallest misfit, and thus need larger sizes to lose coherency,[36,45] this hypothesis is still debatable at the present time. The controversy arises mainly from the difficulties in TEM observation and the distinction of strain field contrasts exclusively related to V(C,N) and not to any other artefact. According to ref. [46], homogeneously precipitated V(C,N) particles might have a different chemical composition from interphase precipitates. In any case, they also follow an orientation relationship of the Baker-Nutting (B-N) type, being in this case the three feasible variants possible.[36] It is clear from the above that further work is still necessary in this regard, as the distinction between coherent and incoherent precipitates is fundamental in order to estimate the real strengthening contribution of V(C,N) particles (see previous section).

10.4.2 *Is Vanadium addition the best approach?*

Vanadium is a promising candidate to (partially) replace Nb and Ti in nano-steels. It is a ferrite stabilizer element and as explained before, its higher solubility is a great advantage: while most of the V remains in solid solution in austenite and only precipitates during or after $\gamma \to \alpha$ phase transformation, other microalloying elements, such as Ti and/or Nb, precipitate at higher temperatures and are less effective at providing precipitation strengthening in ferrite. Small additions of Ti and Nb are beneficial to avoid grain coarsening during soaking in the reheating furnace and prevent austenite recrystallisation during thermomechanical processing. However, by the time the austenite starts to transform into ferrite in Ti,Nb-microalloyed steels, a significant quantity of TiC and/or NbC particles have already formed. These particles will be incoherent with the ferrite matrix and relatively large, providing limited dispersion hardening according to equations (6) and (7) and coarsening fast. It should be pointed out that V can be combined with N to further enhance precipitation hardening: for a given transformation temperature, the sheet spacing as well as the particle size of VN and V(C,N)-precipitates is smaller than that of the VC. This is not the case for Ti-,Nb-microalloyed steels, since the corresponding carbonitrides will form in austenite at even higher temperatures. Also, it should be borne in mind that when Ti is combined with V and N, to ensure a relatively fine austenite grain size, Ti is a strong nitride former and will react first with N to form TiN. This implies that less N will be available for later precipitation with V. Although high N V-microalloyed steels have sometimes been considered inadequate for welding, recent studies have shown that these steels are compatible with weldability by an appropriate choice of the microalloying additions and welding parameters.[36]

The precipitates in nano-steels should be thermally stable, which is particularly important for fire-resistant steels, and at the same time, effective in hindering the motion of dislocations. It has been seen in section 3 that coherent precipitates with large strain fields are the best approach to attain these two properties simultaneously. A very important point to consider, though, is that coherence strain fields associated with nitrides, carbides and carbonitrides have rarely been reported in the literature for microalloyed steels, other than V(C,N)-precipitates.[48] Hence, the latter seem preferable for these new types of steels. Ti- and Nb-based nano-steels usually make use of Mo additions to keep the precipitates small and increase their thermal stability. So far, there is no information available in the literature regarding the interaction of Mo and V in (Mo,V)(C,N)-precipitates. The question to answer now is whether or not the combination of V, N and Mo results in even better nano-steels, which, besides being "resource efficient", are satisfactory for high standard applications.

It has been seen that VC nanometer-sized carbides can significantly improve the mechanical properties of a conventional ferritic steel. Their contribution to the steel strength can be estimated by the Ashby-Orowan model (equations (7)–(12)). Kamikawa *et al.* have reported, for instance, yield and ultimate tensile strengths of 640 MPa and 830 MPa, respectively, in a low carbon steel with 0.1wt.%C, 0.22wt.%Si, 0.83wt.%Mn and 0.288wt.%V, after being austenitised at 1200°C for 10 min and isothermally transformed at 690°C for 300 s.[19] Moreover, the measured uniform and total elongation are relatively high, viz. 10% and 20%, respectively. VC-precipitates of 4.5 nm diameter are the reason behind this substantial improvement. The precipitation strengthening contribution (σ_{ppt} in equation (1) calculated by equations (7), (8), and (9)) yields for this particular case 385 MPa. However, it should be pointed out that this value decreases to 200 MPa as the isothermal holding time at 690°C is increased from 300 s to 48 h, caused by VC-precipitate coarsening. Chen *et al.* have observed that a given V(C,N) volume fraction, the precipitation strengthening associated with interphase precipitation, calculated by means of equations (6), (10), (11) and (12), may vary between 100 MPa and 300 MPa, depending on the particle size and arrangement, *i.e.* sheet spacing and inter-particle spacing.[35] Similarly, other authors have reported precipitation strengthening contributions of both interphase and random (V,Ti)C precipitates of 300 MPa.[49] Due to their beneficial effect, it has also been proposed that V(C,N)-precipitates can be used to develop "nano-precipitated DP steels",[50] containing ferrite, martensite and a dispersion of very fine V(C,N) precipitates in ferrite. Mechanical experiments conducted on these microstructures have shown that V(C,N) nano-precipitation significantly increases the strength of DP steels, with almost no loss in ductility when the ferrite fraction is $\leq 50\%$. Besides, precipitation strengthening notoriously improves the strength-ductility balance. Strain partitioning between the ferrite and the martensite is suppressed by the fine dispersion of precipitates, which also act as sources for dislocation multiplication.

10.4.3 *Industrial implications*

The desirable properties of nano-steels are attained basically through ferrite grain refinement and nanometer-sized precipitates of the corresponding carbonitrides. Industrial and commercial implementation of these steels requires the maximization of these two effects at the lowest cost. For Nb-microalloyed steels, usually controlled rolling (CR) is used as thermomechanical processing, including two well-differentiated stages: deformation at high temperatures in the recrystallisation region and deformation below the stop recrystallisation temperature. The latter produces a deformed austenite microstructure prior to phase transformation and enhances ferrite nucleation at deformation bands in the austenite grain interior, providing additional ferrite grain refinement. However, powerful mills are necessary under these conditions because of the steel's high resistance to deformation. Moreover, deformed austenite is not the only approach to obtain a fine ferrite microstructure in the later processing stages.[51] In the case of V-microalloyed steels, most of the V remains in solid solution in this temperature regime, which allows the use of recrystallisation controlled rolling (RCR) as thermomechanical processing. The latter is, conversely, not suitable for Nb-microalloyed steels. As an advantage to CR, higher finish rolling temperatures can be employed in RCR to reduce the roll force requirements. Accelerated cooling combined with lower cooling temperatures after RCR and V(C,N) nano-precipitation might be an appropriate strategy to reach the high-standard requirements.[51,52] The accelerated cooling contributes to ferrite grain refinement (increases σ_{gb} in equation (1)) and can be used to compensate for the larger austenite grain sizes obtained after soaking, in the absence of other microalloying elements, and when the finish rolling temperature is increased. In addition, the use of V as a microalloying element has other benefits if compared to Nb: excellent castability with minimal cracking, reduced reheating/soaking requirements, and predictable strengthening over a large alloy range. Low ductility in microalloyed steels is attributed to strain-induced precipitation of carbonitrides at the austenite grain boundaries. The limited strain-induced precipitation of V(C,N) makes V less detrimental in this regard.[53] Also, it should be pointed out that high N levels from electric arc furnaces are no longer a problem in V-microalloyed steels.

The kinetics of the austenite grain growth during soaking can be explained by equation (13) when temperature and time are the main controlling variables[54]:

$$D^2 - D_0^2 = K \exp\left(-\frac{Q}{RT}\right) \tau \tag{13}$$

in which D_0 and D are the mean austenite grain sizes at times τ_0 and τ, respectively, K is a constant, Q is the activation energy for grain growth, R is the gas constant, and T is the temperature. Austenite grain size is usually controlled during soaking by small additions of Ti and/or Nb, which are also strong nitrides and carbides formers and have smaller solubility than V in austenite. Higher temperatures than that of soaking are necessary in CR and RCR to dissolve the latter precipitates, which thus prevent grain coarsening. On the other hand, non-dissolved VN and/or V(C,N) precipitates can also be effective in controlling the austenite grain size at lower temperatures, in warm working conditions such as those applied in warm forging, where the soaking temperature is usually low (e.g., below 900°C).[54] By studying the kinetics of austenite grain growth in three different steels, with chemical compositions (in wt.%) 0.4C-1Mn-0.004N, 0.4C-1Mn-0.04N and 0.4C-1Mn-0.04N-0.078V, Staśko *et al.* have shown that undissolved V(C,N) particles inhibit austenite grain growth effectively, keeping the grain size small ($\sim$5.45–8.1 μm) in the range of temperatures between 840–1000°C. In the temperature range of 1050–1200°C, by contrast, austenite grain growth occurs due to the complete dissolution of these precipitates. Indeed, in the latter case, the N dissolved in the austenitic matrix promotes grain growth by reducing the activation energy for grain boundary migration. It is suggested in that work that interstitial elements, such as N and C, dissolved in austenite reduce the binding energy between Fe atoms in the matrix favoring the austenite grain growth in the absence of precipitates. Similar results have been found by Adamczyk *et al.* for two steels with chemical compositions (in wt.%) 0.27C-1.4Mn-0.35Si-0.2V-0.025Al-0.016N and 0.25C-1.4Mn-0.33Si-0.15V-0.011Al-0.018N. They have seen in their work that V(C,N) and AlN particles hinders the austenite grain growth in the two steels up to 1000°C. Above this temperature, the density of undissolved precipitates is small and precipitate coarsening occurs, causing rapid austenite grain growth.

The effect of V in hot working conditions has largely been studied.[55,56] It is well-known that the effect of V solute drag on the static recrystallisation kinetics, caused by V in solid solution, is small compared to that of Nb. This effect has been quantified by the so-called solute retardation parameter (SRP), which quantifies the delay observed in recrystallisation time by the addition of 0.1 wt.% of a microalloying element (e.g. Nb, Ti, Mo or V) to a C-Mn base steel. It has been shown that SRP(Nb)=222, whereas SRP(V)=13.[57] By contrast, although less effective than Nb(C,N), precipitation of V(C,N) particles on dislocations (strain-induced precipitation) has also been shown to affect the static recrystallisation kinetics of austenite, mainly in the low temperature regime.[37] Moreover, depending on the soaking temperature, which will define the amount of undissolved precipitates, V can also delay the austenite recrystallisation kinetics by non-dissolved V(C,N). In contrast to strain-induced V(C,N) precipitation, undissolved V(C,N) particles are already present before deformation and cannot stop recrystallisation completely. Yet, they effectively delay static recrystallisation at low temperatures corresponding to warm working conditions[58]]. Both strain-induced and undissolved V(C,N) can act as nucleation sites for the formation of intragranular ferrite in later stages, contributing to a significant ferrite grain refinement.[38,39]

References

1. European Commission. (2011). *A resource efficient Europe — flagship initiative under the Europe 2020 strategy.* Brussels, Belgium: European Commission.
2. European Comission. (2013). *Action plan for a competitive and sustainable steel industry in Europe.* Brussels, Belgium: European Commission.
3. Allwood, J. and Cullen, J. (2012). Sustainable materials — with both eyes open. Cambridge, UK: Cambridge Institute for Sustainability Leadership.
4. Rijkenberg, R.A. and Hanlon, D.N. (2013). *WO2013167572A1.* Automotive chassis part made from high strength formable hot rolled steel sheet. Tata Steel Europe. Geneva, Switzerland: WIPO.

5. Rijkenberg, R.A. (2014). *WO2014122215A1.* A high-strength hot-rolled steel strip or sheet with excellent formability and fatigue performance and a method of manufacturing said steel strip or sheet. Tata Steel Europe. Geneva, Switzerland: WIPO.

6. Funakawa, Y., Shiozaki, T., Tomita, K., Saito, T., Nakata, H., Sato, K., Suwa, M., Yamamoto, T., Murao, Y., and Maeda, E. (2003). *EP1338665A1.* High tensile hot rolled steel sheet and method for production thereof. JFE Steel Corporation. Munich, Germany: EPO.

7. Ariga, T., Nakajima, K., and Funakawa, Y. (2013). *EP2554705A1.* Hot-dip galvanized steel sheet with high tensile strength and superior processability and method for producing same. JFE Steel Corporation. Munich, Germany: EPO.

8. Kariya, N., Takagi, S., Shimizu, T., Mega, T., Sakata, K., and Takahashi, H. (2006). *EP1616970A1.* High strength hot-rolled steel plate. JFE Steel Corporation. Munich, Germany: EPO.

9. Yokota, T., Kobayashi, A., Seto, K., Hosoya, Y., Heller, T., Hammer, B., Bode, R., and Stich, G. (2007). *EP1790737A1.* A high strength steel excellent in uniform elongation properties and method of manufacturing the same. JFE Steel Corporation and ThyssenKrupp Steel. Munich, Germany: EPO.

10. Kömi, J., Keltamäki, K., Intonen, T., Kinnunen, H., Outinen, J., Porter, D., and Rasmus, T. (2008). *EP2235226A1.* Method for selecting composition of steels and its use. Rautarukki Oyj. Munich, Germany: EPO.

11. Funakawa, Y., Shiozaki, T., Tomita, K., Yamamoto, T., and Maeda, E. (2004). Development of high strength hot-rolled sheet steel consisting of ferrite and nanometer-sized carbides. *ISIJ International*, 44(11), 1945–1951.

12. Lee, W.B., Hong, S.G., Park, C.G., Kim, K.H., and Park, S.H. (2000). Influence of Mo on precipitation hardening in hot rolled HSLA steels containing Nb. *Scripta Materialia*, 43(4), 319–324.

13. Zhang, Z., Yong, Q., Sun, X., Li, Z., Wang, Z., Zhou, S., and Wang, G. (2015). Effect of Mo addition on the precipitation behavior of carbide in Nb-bearing HSLA steel. *HSLA Steels 2015, Microalloying 2015 and Offshore Engineering Steels 2015.* Hangzhou, Zhejiang Province, China: The Chinese Society for Metals and Chinese Academy of Engineering.

14. European Comission. (2014). *On the review of the list of critical raw materials for the EU and the implementation of the Raw Materials Initiative.* Brussels, Belgium: European Commission.

15. Fushimi, M., Chikaraishi, H., and Keira, K. (1995). *Development of fire-resistant steel frame building structures.* Tokyo, Japan: Nippon Steel.

16. Standards Australia. (1985). AS 1530.4. *Fire resistance tests of elements of building construction.* Sydney, Australia: Standards Australia.

17. American Society of Testing Materials. (1996). ASTM E119. *Standard test method for fire tests of building construction and materials.* Philadelphia, PA, USA: American Society of Testing Materials.

18. ECCS Technical Committee 3: Fire safety of steel structures. (1983). *European recommendations for the fire safety of steel structures.* Amsterdam, the Netherlands: Elsevier.

19. Kamikawa, N., Sato, K., Miyamoto, G., Muruyama, M., Sekido, N., Tsuzaki, K., and Furuhara, T. (2015). Stress-strain behaviour of ferrite and bainite with nano-precipitation in low carbon steels. *Acta Materialia*, 83, 383–396.

20. Liu, Q., Huang, X., Lloyd, D.J., and Hansen, N. (2002). Microstructure and strength of commercial purity aluminium (AA 1200) cold rolled to large strains. *Acta Materialia*, 50, 3789–3802.

21. Sha, W. and Kelly, F.S. (2004). Atom probe field ion microscopy study of commercial and experimental structural steels with fire resistant microstructures. *Materials Science and Technology*, 20(4), 449–457.

22. Dere, E.G., Sharma, H., Petrov, R.H., Sietsma J., and Offerman, S.E. (2013). Effect of Niobium and grain boundary density on fire-resistance of Fe-C-Mn steel. *Scripta Materialia*, 68(8), 651–654.

23. Sha, W., Kelly, F.S., and Guo, Z.X. (1999). Microstructure and properties of Nippon fire-resistant steels. *Journal of Materials Engineering and Performance*, 8(5), 606–612.

24. Lacroix, G., Pardoen, T., and Jacques, P.J. (2008). The fracture toughness of TRIP-assisted multiphase steels. *Acta Materialia*, 56, 3900–3913.

25. Morhbacher, H. (2013). Reverse metallurgical engineering towards sustainable manufacturing of vehicles using Nb and Mo alloyed high performance steels. *Advances in manufacturing*, 1(1), 28–41.

26. Seto, K. and Matsuda, H. (2013). Application of nanoengineering to research and development and production of high strength steel sheets. *Materials Science and Technology*, 29(10), 1158–1165.

27. Jang, J.H., Heo, Y.U., Lee, C.H., Bhadeshia, H.K.D.H., and Suh, D.W. (2013). Interphase precipitation in Ti-Nb and Ti-Nb-Mo bearing steel. *Materials Science and Technology*, 29(3), 309–313.

28. Jang, J.H., Lee, C.H., Heo, Y.U., and Suh, D.W. (2012). Stabiliy of (Ti,M)C (M=Nb, V, Mo and W) carbide in steels using first-principles calculations. *Acta Materialia*, 60, 208–217.

29. Chen, C.Y., Chen, C.C., and Yang, J.R. (2014). Microstructure characterization of nanometer carbides heterogeneous precipitation in Ti–Nb and Ti–Nb–Mo steel. *Materials Characterisation*, 88, 69–79.

30. Gladman, T. (1999). Precipitation hardening in metals. *Materials Science and Technology*, 15(1), 30–36.

31. Martin J.W. (1998). *Precipitation hardening* (2nd ed.). Oxford, UK: Butterworth-Heinemann.

32. Martin J.W. (1998). *Precipitation hardening* (2nd ed.). Oxford, UK: Butterworth-Heinemann.

33. Gradman, T. (2002). *The physical metallurgy of microalloyed steels*. London, UK: Maney Publishing.

34. Kamikawa, N., Abe, Y., Miyamoto, G., Funakawa, Y., and Furuhara, T. (2014). Tensile behaviour of Ti,Mo-added low carbon steels with interphase precipitation. *ISIJ International*, 54(1), 212–221.

35. Chen, M.Y., Gouné, M., Verdier, M., Bréchet, Y., and Yang, J.R. (2014). Interphase precipitation in vanadium-alloyed steels: strengthening

contribution and morphological variability with austenite to ferrite phase transformation. *Acta Materialia*, 64, 78–92.

36. Baker, T.N. (2009). Processes, microstructure and properties of vanadium microalloyed steels. *Materials Science and Technology*, 25(9), 1083–1107.

37. Medina, S.F., Quispe, A., and Gómez M. (2003). Strain induced precipitation effect on austenite static recrystallisation in microalloyed steels. *Materials Science and Technology*, 19(1), 99–108.

38. Ishikawa F., Takahashi, T., and Ochi, T. (1994). Intragranular ferrite nucleation in medium-carbon vanadium steels. *Metallurgical and Materials Transactions A*, 25(5), 929–936.

39. Medina, S.F., Gómez, M., and Rancel L. (2008). Grain refinement by intragranular nucleation of ferrite in a high nitrogen content vanadium microalloyed steel. *Scripta Materialia*, 58(12), 1110–1113.

40. Furuhara, T., Yamaguchi, J., Sugita, N., Miyamoto, G., and Maki, T. (2003). Nucleation of proeutectoid ferrite on complex precipitates in austenite. *ISIJ International*, 43(10), 1630–1639.

41. Glodowski, R.J. (2002). *Experience in producing vanadium-microalloyed steels*. Paper presented at the International Symposium on Thin-Slab Casting and Rolling, Guangzhou, China.

42. Karmakar, A., Mandal A., Mukherjee, S., Kundu, S., Srvastava, D., Mitra, R., and Chakrabarti, D. (2016). Effect of isothermal holding temperature on the precipitation hardening in vanadium-microalloyed steels with varying carbon and nitrogen levels. ArXiv:1607.02721.

43. Korchynsky M. and Stuart H. (1970)The role of strong carbide and sulfide forming elements in the manufacture of formable high strength low alloy steels. *Proceedings of the Symposium on low alloy high strength steels*. 17–27. Nuremberg, Germany: Metallurgical Companies.

44. Zhang, Y.J., Miyamoto, G., Shinbo, K., and Furuhara, T. (2013). Effects of α/γ orientation relationship on VC interphase precipitation in low-carbon steels. *Scripta Materialia*, 69(1), 17–20.

45. Morales E.V., Gallego J., and Kestenbach H.J. (2003). On coherent carbonitride precipitation in commercial microalloyed steels. *Philosophical Magazine Letters*, 83(2), 79–87.

46. Nöhrer, M., Zamberger, S., Primiga, S., and Leitner H. (2013). Atom probe study of vanadium interphase precipitates and randomly distributed precipitates in ferrite. *Micron*, 54–55, 57–64.

47. Zhang, Y.J., Miyamoto, G., Shinbo, K., and Furuhara, T. (2017). Quantitative measurements of phase equilibria at migrating α/γ interface and dispersion of VC interphase precipitates: Evaluation of driving force for interphase precipitation. *Acta Materialia*, 128, 166–175.

48. Baker, T.N. (2016). Microalloyed steels. *Ironmaking and steelmaking*, 43(4), 264–307.

49. Chen, J., Lu, M., Tang, S., Liu, Z., and Wang, G. (2014). Influence of cooling paths on microstructural characteristics and precipitation behaviours in a low carbon V-Ti microalloyed steel. *Materials Science and Engineering A*, 594, 389–393.

50. Kamikawa, N., Hirohashi, M., Sato, Y., Chandiran, E., Miyamoto, G., and Furuhara, T. (2015). Tensile behaviour of ferrite-martensite dual phase steels with nano-precipitation of vanadium carbides. *ISIJ International*, 55(8), 17811–1790.
51. Zajac, S., Siwecki, T., Hutchinson, B., and Atlegård, M. (1991). Recrystallisation controlled rolling and accelerated cooling for high strength and toughness in V-Ti-N steels. *Metallurgical and Material Transactions A*, 22(11), 2681–2694.
52. Zhang, J., Wang, F.M., Yang, Z.B., and Li, C.R. (2016). Microstructure, precipitation and mechanical properties of V-N-alloyed steel after different cooling processes. *Metallurgical and Materials Transactions A*, 47(12), 6621–6631.
53. Li, Y. and Milbourn, D. (2015). The influence of vanadium microalloying on the production of thin slab casting and direct rolled steel strip. *HSLA Steels 2015, Microalloying 2015 and Offshore Engineering Steels 2015*. Hangzhou, Zhejiang Province, China: The Chinese Society for Metals and Chinese Academy of Engineering.
54. Staśko, R., Adrian, H., and Adrian, A. (2006). Effect of nitrogen and vanadium on austenite gain growth kinetics of a low alloy steel. *Materials Characterisation*, 56(4–5), 340–347.
55. White, M.J. and Owen, W.S. (1980). Effects of vanadium and nitrogen on recovery and recrystallisation during and after hot-working some HSLA steels. *Metallurgical and Material Transactions A*, 11(4), 597–604.
56. Andrade, H.L., Akben, M.G., and Jonas, J. J. (1983). Effect of Molybdenum, Niobium and Vanadium on static recovery and recrystallisation and on solute strengthening in microalloyed steels. *Metallurgical and Material Transactions A*, 14(10), 1967–1977.
57. Jonas, J.J. (1984). Mechanical testing for the study of austenite recrystallization and carbonitride precipitation. In D.P. Dunne and T. Chandra (Eds.), *Conference on High Strength Low Alloy Steels*. Wollongong, Australia: University of Wollongong.
58. García-Mateo, C., López, B., and Rodriguez-Ibabe, J.M. (2001). Static recrystallisation kinetics in warm worked vanadium microalloyed steels. *Materials Science and Engineering A*, 303(1–2), 216–225.

Chapter 11

Strategies towards Carbon Nanomaterials-Based Transparent Electrodes

Amal Kasry* and Ahmed A. Maarouf†

*Faculty of Engineering, and Mostafa El-Sayed Nanotechnology Research Center,
The British University in Egypt (BUE), El Sherouk City,
Suez Desert Road, Cairo 11837, Egypt

†Department of Physics, Institute for Research & Medical Consultations,
Imam Abdulrahman Bin Faisal University, P.O. Box 1982,
Dammam 31441, Saudi Arabia

In a resource-constrained world, the demand for new renewable resources becomes a must. Two of the rapidly developed technologies are photovoltaic applications to act as efficient sources of solar energy, and touch screen-dependent communication technologies. Both require the use of transparent conducting layers that achieve the main requirements of very high conductivity and very high transparency. In this chapter, we describe our efforts to use carbon nanomaterials, specifically graphene, as transparent conducting electrodes. Graphene sheet resistance was reduced, while keeping its transparency high enough to be suitable for the two mentioned applications. The lowest value achieved was 22 Ω/square at 90% transparency which is close enough to the values of the conventional currently used transparent electrodes.

11.1 Introduction

The huge increase in the world's population has led to constraining the available resources which are not renewable. This generated the demand for finding new resources or dealing with the available ones in order to satisfy the continuous increase in the world's demands. One of the most important requirements for life is energy. The major current sources of energy — coal, oil and natural gas — are gradually being depleted and their use also has environmental problems. Renewable energy is an expression that is increasingly being used to describe energy sources like the sun or wind which will not be depleted with use. Solar cells that are used as a source of energy in homes, streets, and some industrial locations, are becoming an important technology in modern life, to such an extent that some countries started announcing that by the next century, solar cells would be the main, if not the only, source of energy. Solar cells consist of a number of main elements, one of which is the transparent electrodes. These electrodes must have very high conductivity as well as very high transparency.

In conventional thin film solar cells, the transparent conducting electrodes (TCE) consist of a relatively thick film (50–100 nm) typically composed of Indium Tin Oxide (ITO)[1] or Al-doped ZnO.[2] The latter avoids using the rare element Indium. Although ITO and Al-doped ZnO films have the desired optical and electrical properties, they require expensive deposition techniques and they are brittle, which make them difficult to use on flexible substrates. Hence, the development of alternative TCE materials is desirable to achieve the performance metrics of low cost and compatibility with flexible substrates, while maintaining acceptable engineering performance characteristics of a sheet resistance in the range of 10 Ω/square, at an optical transparency of above 90%. Graphene, in either single or multi-layer form,[3–8] and carbon nanotube thin films [9–12] are strong competitors to meet this challenge. However, while very high transparencies are achieved with these carbon-based films, their electrical resistivity is still very high (>100 Ω/square at >90% transmittance) to replace conventional TCEs; grain boundaries, defects, and impurities are factors limiting the conductivity of the carbon film.[13,14]

Within the last decade, and since its isolation in 2004, graphene has been rapidly considered as a leading candidate to replace conventional materials currently used in many applications, including being used as a transparent conducting electrode. This is because of its high transparency and exceptional transport properties. Graphene, a single layer of graphite, possesses remarkable optical and electrical properties that have stimulated a vast amount of research in the fields of condensed matter physics and materials science.[15–18] At room temperature, charge carriers in graphene can travel thousands of interatomic distances without scattering, resulting in very high carrier mobilities.[1] In addition to its remarkable electronic properties, graphene is optically transparent, flexible and has high chemical and mechanical stability. And it has a relatively low manufacturing cost. All this makes graphene an ideal candidate for a transparent conducting electrode (TCE). Several methods exist for the preparation of single- to few-layer graphene films. Solution-based methods include chemical exfoliation with organic solvents[6,19] and chemically reducing graphite oxides.[20] Graphene is also grown on metal substrates (i.e., nickel or copper) via chemical vapor deposition (CVD).[8,21–25]

In this chapter we will show our efforts to develop using graphene as a TCE for several applications. Graphene preparation, characterization, and treatment to achieve the required properties will be discussed.

In this work, graphene was mainly prepared by a chemical vapor deposition method (CVD) as this growth method yields high-quality large-area films.[8,22] Large-area graphene films were grown on Cu foils as first reported by the Ruoff group. Despite the high-quality graphene growth, to date the sheet resistance (RS) of a single graphene layer grown with this method (1200 Ω/square at 97% transmittance) remains too high for the sheet to be used as a TCE.[8]

We have used several strategies in order to reduce the sheet resistance of graphene while keeping its high transparency. The first strategy was to combine stacking of multi-graphene layers with chemical doping in order to increase the number of charge carriers and consequently increase the conductivity.[26] The second was by designing a metal busbar structure in contact with the graphene layers to decrease the distance that the electrons have to travel and increase the conductivity, while decreasing the transparency by just 4%.[27] The third was by creating a regular lattice of pores in a graphene sheet, then subsequently doping these structures (which we call a doped graphene nanomesh).[28] And finally the fourth was by graphene/carbon nanotube hybrid, where a monolayer of carbon nanotubes was used to connect the grains in the graphene layers,[29] this led to reducing the electron scattering, which consequently led to increasing the sheet resistance while not losing any transparency.

11.2 Graphene Preparation and Characterization

As previously mentioned, the graphene used in this work was prepared by the CVD method. A piece of Cu foil (25 nm thick, Sigma-Aldrich) was placed in a 1 in. diameter quartz furnace tube at low pressure (60 mTorr). Prior to processing, the system was flushed with 6 sccm of forming gas (5% H2 in Ar) for 2 h at a pressure of 500 mTorr to remove any residual oxygen and water present in the system. The concentrations of oxygen and water in the chamber were monitored with a residual gas analyzer (Ametek, Dycor Dymaxion). The Cu foil was then heated to 875°C in forming gas (6 sccm, 500 mTorr) and kept at this temperature for 30 min to reduce native CuO and increase the Cu grain size. After reduction, the Cu foil was exposed to ethylene (6 sccm, 500 mTorr) at 875°C for 30 min. The sample was cooled in forming gas (6 sccm, 500 mTorr). Poly methyl methacrylate (PMMA) was spin-coated on top of the graphene layer formed on the Cu foil, and the Cu foil was then dissolved in 1 M iron (III) chloride. The remaining graphene/PMMA layer was thoroughly washed with deionized water and transferred to a quartz substrate. Subsequently, the PMMA was dissolved in hot acetone (80°C) for 1 h. The substrate with graphene was rinsed in methanol and dried in a stream of nitrogen. Multilayers of graphene were prepared by stacking individually grown graphene layers on top of each other until the desired number of layers was obtained. After each graphene layer addition, a transmission spectrum (Perkin-Elmer Lambda 950 UV-vis spectrometer) was obtained using a blank quartz sample as a reference for subtraction. The sheet resistance was also measured using a manual four-point probe apparatus (Signatone, probe distance 1.5 mm).[26]

Several techniques have been used to characterize the graphene sheets. Fig. 11.1(a) is an atomic force microscopy (AFM) image of a graphene film after it was transferred from copper to a quartz substrate. The image shows a single layer of graphene divided by several-nanometer-high folds in the

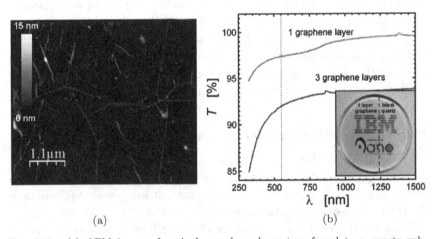

(a)   (b)

Fig. 11.1. (a) AFM image of a single graphene layer transferred to a quartz substrate. (b) Plots of transmittance vs wavelength for monolayer (green) and trilayer (blue) graphene on a quartz substrate. Inset: optical image of a quartz slide with half of the slide covered with a single graphene sheet. [Copied with Permission; Copyright © 2010, American Chemical Society].

film that were formed during cooling of the substrate. Once the film was transferred to a transparent substrate, a UV-vis-NIR absorption spectrum was obtained. Fig. 11.1(b) shows UV-vis-NIR spectra of monolayer and trilayer graphene that were transferred individually to a quartz substrate. As expected from the band structure, the absorption spectrum is flat and rather featureless. The single layer transmits 97% of the light at 550 nm, as expected for a single layer of graphene. The transmittance is high over a broad spectral range, making graphene advantageous for photovoltaic applications. Raman spectroscopy was also performed to confirm the growth and the transfer of one layer graphene. A Four-Probe method was used to monitor the sheet resistance after the transfer and after the treatment. The sheet resistance of one layer of graphene directly after the transfer was 1000–1200 Ω/square.[26]

11.2.1 *Stacking and doping of graphene layers*

Two approaches were pursued to reduce the sheet resistance of graphene[26]: stacking of graphene layers on top of each other and/or chemical doping. Stacking of graphene layers essentially adds channels for charge transport; however, this approach simultaneously reduces the transparency of the system.[8] In addition, since the sheet resistances of the individual layers remain unchanged, alternative approaches such as chemical doping must be considered.[6,29,30] Graphene is classified as a zero-band-gap semiconductor, where the density of states vanishes at the Dirac point, as a result, undoped graphene has a low carrier density and a high sheet resistance. When graphene is exposed to air, there would be unintentional dopants, which will lead to the Fermi level being not at the Dirac point. Chemical doping further increases the carrier concentration and thus further reduces the resistance of the film.

The key results of this work were that stacks of graphene films of up to eight layers could be effectively p-doped with nitric acid. The films were doped either after each layer was stacked (interlayer-doped) or after the last layer was stacked (last-layer-doped) as shown in Scheme 11.1(a). The interlayer doping method yields better optoelectronic properties. The sheet resistance is reduced by a factor of 3 upon exposure to nitric acid, yielding films with sheet resistance of 90 Ω/square at a transmittance of 80% (at 550 nm). The nitric acid p-dopes the films, thus increasing the carrier concentration and reducing the sheet resistance (Scheme 11.1(b)). A network resistor model was developed to describe the transport in the

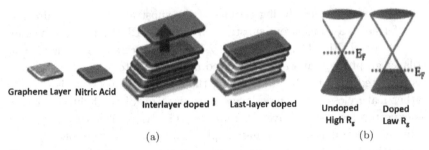

Graphene Layer Nitric Acid

Interlayer doped Ⅰ Last-layer doped

Undoped
High R$_s$

Doped
Law R$_s$

(a)   (b)

Scheme 11.1.   (a) Schematic illustrating the two different doping methods pursued here. In the interlayer-doped case, the sample is exposed to nitric acid after each layer is stacked, whereas in the last-layer-doped case, the film is exposed to nitric acid after the final layer is stacked. (b) Illustration of the graphene band structure, showing the change in the Fermi level due to chemical p-type doping. Graphene/Carbon nanotubes-Metal Busbar Hybrid. [Copied with Permission; Copyright © 2010, American Chemical Society].

stacked graphene films. The model describes a characteristic channel length where all of the graphene layers are active in transport and indicates that each additional layer represents an additional transport channel which leads to an increase in the conductivity. The experimental data shows a linear increase in conductivity as a function of the number of layers, which is in excellent agreement with the model.[26]

Figure 11.2(a) shows plots of transmittance (measured at 550 nm) as a function of the number of stacked graphene layers, for both the interlayer-doped and last-layer doped stacked graphene films. The measured values clearly follow the Beer-Lambert law[31] shown as the fitted black line. A higher transmittance is obtained for the interlayer-doped films than the last-layer doped films, as shown in Fig. 11.2(a). This was attributed to the enhancement in the transmission of the interlayer-doped films due to the removal of amorphous carbon species or other impurities from each separate graphene layer by nitric acid. These impurities either can form during the graphene synthesis or are residues from the PMMA resist stripping process. In the case of last-layer-doped stacked graphene films, the graphene system is exposed to nitric acid after the stacking is complete, leaving possible impurities between the layers, which lowers the total transmittance of the stack. In the interlayer-doped case, the transmittance decreases by 2.5% for each added layer, indicating that the film behaves as a set of individual graphene layers.

The sheet resistance as a function of the number of graphene layers for both the interlayer-doped and last layer-doped films is shown in

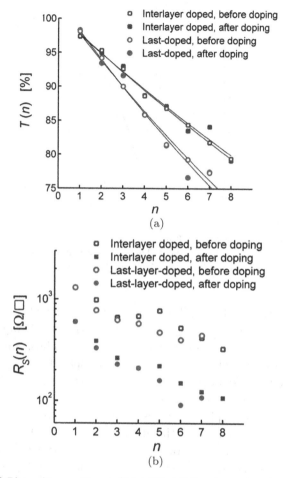

Fig. 11.2. (a) Plots of transmittance {T(n) [%] at 550 nm} as a function of the number of graphene layers n for both the interlayer-doped (blue data) and the last-layer-doped (red data) cases. The black lines are fits using the Beer-Lambert law. Interlayer doping produces films with higher transmittance values. (b) Plots of the sheet resistance (RS) as a function of the number of graphene layers for both the interlayer-doped (blue) and last-layer-doped (red) cases before and after doping. The sheet resistance is reduced by a factor of 3 with nitric acid doping. [Copied with Permission; Copyright © 2010, American Chemical Society].

Fig. 11.2(b). In both cases, the sheet resistance is reduced by a factor of 3, indicating efficient p-type doping with nitric acid. The resistance was comparable in the interlayer-doped and last-layer-doped cases, which is due to one of two possibilities; the first is that the nitric acid could intercalate

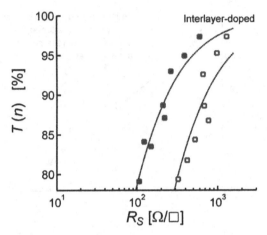

Fig. 11.3. Plot of transmittance {T(n) [%] at 550 nm} vs sheet resistance RS before (O) and after (•) doping with nitric acid. The curves are guidelines for visual purposes only. [Copied with Permission; Copyright © 2010, American Chemical Society].

into the graphene stacks, resulting in equivalently doped films, while the second is that, since nitric acid is volatile, the dopant could evaporate before another layer is added, yielding equivalently doped films. Since the sheet resistance of the two films is similar and the optical data (Fig. 11.2(a)) show that interlayer doping yields a higher transmittance, only the interlayer doping method was considered in our data analysis.

Figure 11.3 shows plots of transmittance versus sheet resistance for stacked graphene films in which the films were interlayer doped. Each point represents the addition of a graphene layer before (O) or after (square) doping. The resistance decreases with transparency as additional transport channels are added. The sheet resistance decreases by a factor of 3 upon doping with nitric acid, reaching a minimum value of 90 Ω/square at a transmittance of 80%. This result highlighted the potential of graphene as a transparent electrode, since films with a transmittance of 93% have a resistance of 250-90 Ω/square, a value already suitable for many display applications.

11.3 Graphene-Metal Busbar Hybrid

Another potential way to enhance the performance of the carbon-based transparent electrodes by decreasing their sheet resistances is to deposit a microscale metallic busbar and finger pattern to be in contact with the film itself to increase the photo current collection. According to the design, the metallic pattern is to cover only a small fraction of the surface area, reducing the overall film sheet resistance without significantly reducing the overall optical transparency of the electrode. In this work, a metal busbar microstructure was introduced to decrease the effective sheet resistance of both graphene and carbon nanotube films to a value such that they are suitable for use as transparent conducting electrodes.[27]

A suitable metallic busbar pattern was first designed to be deposited on the carbon-based films and we estimated the theoretical efficiency of the pattern. We then experimentally verified the enhanced performance of the composite film when two different metals (Pd and Cu), which have different contact resistances to the carbon layer, are used. Both graphene and nanotube carbon-based films in mono- and multi-layer forms are considered in this configuration. The resulting composite 2-component overlayer microstructure results in a TCE with a high optical transparency and a low electrical resistance, which satisfies the engineering specifications for TCEs and could potentially replace conventional oxide-based materials.

An example of a pattern of busbars and fingers is illustrated in Fig. 11.4(a). This metalized pattern forms an overlayer on the film and performs the task of collecting current from the film, which itself collects current from the underlying device. The role of the metal grid was thought of as reducing the distance the charge has to travel in the pseudometallic carbon-based film (distance "x" in Fig. 11.4(a)), reducing the effective sheet resistance of the composite layer. Fig. 11.4(a) shows two busbars plus two sets of fingers dividing up a square space on the surface for the purpose of current collection. The busbars are relatively thick, while the much thinner fingers form an array extending transversely to the busbars. A significant voltage is applied between the two busbars in Fig. 11.4(a) (Left). This voltage appeared across the narrow spacing between the finger end and busbar in Fig. 11.4(a) (Right). To minimize the possibility of electrical breakdown between the finger end and busbar, the finger ends were terminated in a rounded cross-section in order to minimize the electric field crossing the busbar-finger end junction. It was verified using a 2-probe method that the majority of current flows from finger to finger in the design since the resistance between two fingers is overall 73% lower

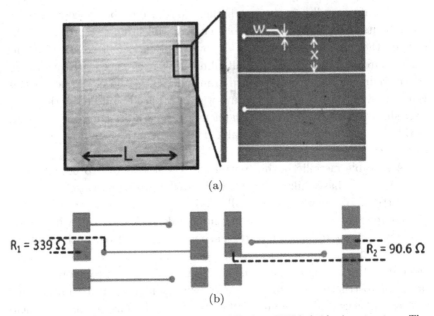

(a)

(b)

Fig. 11.4.   (a) Optical photograph of the metal busbar-CNT hybrid microstructure. The thick vertical metal lines (left figure) are busbars, the thin horizontal metal lines (right figure) are fingers. The high optical transmittance of the film after metal deposition is seen from the relatively small metal-covered area. The critical dimension "x" between metal fingers (right figure) determines the maximum distance that charge has to travel in the film. L = 1 cm, x = 500 μm, w = 20 μm. (b) A scheme representing the resistance measured by the 2-probe method showing the resistance between the fingers to be 25% of that between the gap between the finger and the electrode. [Copied with Permission; Copyright © 2012, ELSEVIER; License no. 3915870627594].

than that between the end of the finger and the closer electrode as indicated in Fig. 11.4(b).

The proposed busbar configuration, implemented with Cu, can theoretically reduce the sheet resistance by a factor of 1000, while limiting the optical absorption to only 4%. Experimental sheet resistance and optical transparency data are presented for both Pd and Cu, and for mono- and multi-layer graphene as well as nanotube films. It was found that the metal busbar microstructure decreases the sheet resistance by a factor of 8 and 70 on graphene and nanotube films respectively, a sufficient resistance reduction to enable utilization as a TCE. The contact resistance between the metal grid and carbon film is believed to limit the ultimate performance (Fig. 11.5).

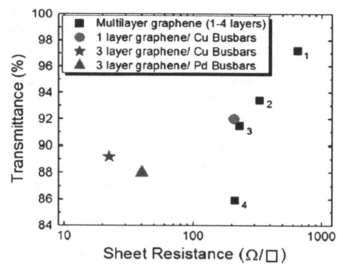

Fig. 11.5. Transparency versus sheet resistance for single and multilayer graphene without and with Cu and Pd busbar microstructures. Without busbars multilayer graphene has a lower sheet resistance but at heavy cost in reduced optical transparency. Addition of busbar/finger arrays adds a constant offset of only 4% to the transparency while dramatically reducing the sheet resistance of the graphene layers. [Copied with Permission; Copyright © 2012, ELSEVIER; License no. 3915870627594].

11.4 Doped Graphene Nanomesh as Transparent Electrodes

Graphene has been used as a transparent conductor. To lower the sheet resistance of graphene, it has to be doped. Due to the sp2 nature of the bonding in graphene, dopants physisorbed on graphene are usually unstable, resulting in the subsequent increase of the sheet resistance.

Therefore, developing a stable controlled way of doping graphene is of great interest.

Graphene nanomeshes (GNMs) are graphene-based structures formed by creating a lattice of nano-sized pores in a graphene layer. The pore lattice details control the electronic properties of the resulting structures, which could be semimetallic or semiconducting with a fractional eV gap,[33-44] which makes them potential candidates in the transistor world. In addition, GNMs offer a solution to the relative chemical inertness of graphene. Pore edges can be utilized for controlled functionalization (chemical modification), an advantage that graphene lacks. Combined with the other properties of graphene, such as high electronic mobility and high optical transparency, these two properties make GNMs excellent candidates for many applications.

Recently, GNMs have been suggested as superior support templates for metal catalytic nanoparticles,[45] thereby offering a novel solution to the nanoparticle agglomeration problem. The pore edges of GNMs can be controllably functionalized to bind the desired nanoparticles of various sizes, and hence decrease their undesired mobility.

Controlled, stable, chemical doping of pore-edge passivated GNMs has been recently proposed.[28] A neutral dopant brought to the vicinity of a GNM pore may undergo a charge transfer reaction with the GNM, and therefore gets ionized. The resulting ion is electrostatically trapped in the pore by the local dipole moments of the pore edge functional groups. The GNM is doped, moving the Fermi level of the system into the valence or the conduction bands (see Fig. 11.6). First principles calculations show that the GNM band structure is not affected by the doping process. Furthermore the dopant state involved in the charge transfer is far from the Fermi level, and hence is not expected to affect electronic transport in a GNM-based device. This doping approach, termed chelation doping, offers a stable, controlled, and rigid band way of doping GNMs.[28]

GNMs have been fabricated by several groups using a block copolymer lithography approach, with pore-size distributions in the 20–40 nm range.[45–48] Doped GNM-based transistors exhibit an ON-OFF ratio which is an order of magnitude larger than that of pristine graphene, but with lower electrical conductivity.[46] Sub-nanometer non periodic pores were recently fabricated.[47]

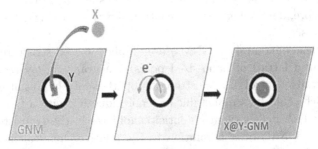

Fig. 11.6. A scheme showing the possibility of the charge transfer by introducing a neutral dopant to the graphene nano-pore. [Copied with Permission; Copyright © 2013, American Chemical Society].

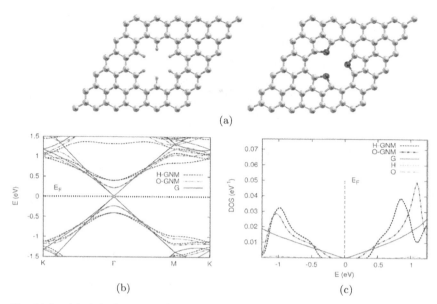

Fig. 11.7. (a) A hydrogen- and an oxygen-passivated GNM (H-GNM and O-GNM. (b) The band structures of the two passivated systems and pristine graphene. (c) The density of states (DOS) of H-GNM, O-GNM and graphene. [Copied with Permission; Copyright © 2013, American Chemical Society].

Passivation of pore edges with chemical moieties different from carbon leads to the creation of edge dipoles. These dipoles provide the electrostatic trap that hosts the dopant species. Figure 11.7(a) shows a hydrogen- and an oxygen-passivated GNM (H-GNM and O-GNM, respectively). As we see, all chemical bonds are satisfied, therefore, the passivated GNM is chemically stable. The band structures of the two passivated systems and pristine graphene are shown in Fig. 11.7(b). First principles calculations using LDA predict a gap of 0.7 eV for the H-GNM and 0.4 eV for the O-GNM. Away from the gap region, the bands are linear, with a group velocity half that of pristine graphene. The density of states (DOS) of H-GNM, O-GNM and graphene are shown in Fig. 11.7(c). The projected DOS (PDOS) show that the passivating species do not have any significant contribution in the linear region or at the band edges, and are therefore expected not to cause any resonant scattering.[49]

We now turn to the n-doping of GNM's. Passivating the pore edge with oxygen leads to the creation of pore edge dipoles due to the electronegativity mismatch between carbon and oxygen. This makes it possible to host

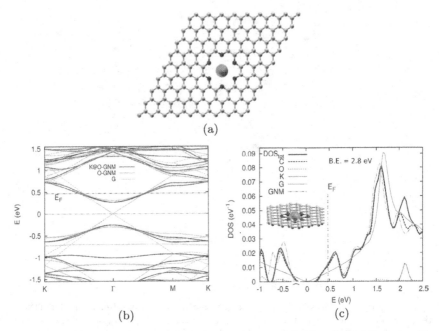

(a)

(b)   (c)

Fig. 11.8.   (a) Doping the O-GNM with a potassium atom by donating its 4s electron to the O-GNM Skelton. (b) The band structures of the doped and undoped systems indicating that the potassium chelation preserves the O-GNM band structure. (c) The density of states of the potassium-doped O-GNM, showing contributions of different species. The undoped O-GNM and pristine graphene are included for comparison. [Copied with Permission; Copyright © 2013, American Chemical Society].

positive ions in the pore region. Electron-donating elements (e.g. potassium, sodium, and lithium) can be used as dopants. A potassium atom brought to the vicinity of the pore donates its 4s electron to the GNM skeleton, and thus dopes the GNM (Fig. 11.8(a)). The doping is ultra-stable, with an LDA-predicted binding energy of 2.8 eV.

A comparison of the band structures of the doped and undoped systems indicates that the potassium chelation preserves the GNM band structure (Fig. 11.8(b)) (rigid band doping) in the region of interest, merely shifting the Fermi level into the conduction band. The DOS and PDOS of the potassium-doped GNM confirms this doping picture (Fig. 11.8(c)), and the location of the potassium empty 4s state is far from the Fermi energy, and hence is not expected to affect the transport properties of the doped GNM.

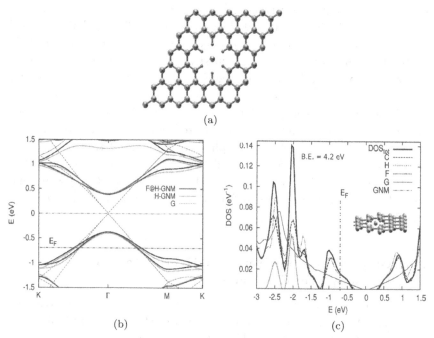

Fig. 11.9.   (a) H-GNM doped with fluorine atom. (b) The band structures of the fluorine-doped and undoped systems. (c) The total density of the fluorine-doped H-GNM, showing the contributions of different species. The undoped H-GNM and pristine graphene are included for comparison. [Copied with Permission; Copyright © 2013, American Chemical Society].

A p-doped GNM system is formed by chelating an electron accepting moiety in a hydrogen passivated pore. A fluorine atom brought close to the pore of the GNM results in the system lowering its total energy by transferring an electron from the graphene skeleton to the fluorine atom (Fig. 11.9(a)). The fluorine ionizes and docks in the pore, electrostatically binding to the pore edge dipoles. The GNM is now p-doped. The band structures of the doped and undoped systems are shown in Fig. 11.9(b). The doping causes no significant change in the band curvatures, and the Fermi level of the chelated system is in the linear region of the valence band Therefore, chelation p-doping of GNMs occurs in a rigid band way.

The rigid-band doping picture is confirmed by inspecting the DOS of the fluorine-doped H-GNM system. Fig. 11.9(c) shows the total DOS, with the carbon, hydrogen, and fluorine contributions. The total DOS of the H-GNM and that of pristine graphene are shown for comparison. The Fermi level of the fluorine-doped H-GNM system indicates that the GNM is p-doped. The chelated and unchelated systems have very similar DOS, except that the fluorine occupied 2p states on the left side of the gap. These states are 1.3 eV away from the Fermi level and are therefore too far to obstruct low energy transport.

Therefore, ion chelation doping offers a novel way to stably dope graphene structures, making it possible to harvest the exceptional electronic and optical properties of graphene in the field of transparent electrodes. The doping occurs in a rigid band way, which leaves the GNM spectrum unchanged. Dopants are tightly bound to the GNM pore, and their electronic signatures are far from the Fermi level where the electrode would operate.

11.5 Graphene/Carbon Nanotube Hybrid Material as Transparent Electrodes

Reducing the graphene sheet resistance is the primary goal of research attempting to use graphene as a transparent electrode. In addition to the aforementioned methods of stacking of graphene layers and patterning of metal busbars, a novel approach has been recently proposed to lower the sheet resistance of graphene.

Grain boundaries in CVD-grown graphene are connected with a thin bridge-like conducting material, conducting carbon nanotubes (CCNT). The CCNT bridges lower the electronic scattering, which decreases the sheet resistance. The CCNT density is small enough not to reduce graphene's transparency. Using this approach a three-fold reduction of the sheet resistance has been achieved without any loss of transparency.[29]

Three resistances control the electronic properties of the hybrid system: (1) the graphene sheet resistance as imposed by its average grain size,[50,51] vacancies, adatoms, and other defects which act as scattering centers for electrons, increasing the sheet resistance; (2) the contact resistance between CCNTs, which depends on the details of the two tubes, including their chiralities, their relative orientation, the crossing angle, and Fermi level of the junction,[52,53] and (3) the contact resistance between the CCNT and the graphene sheet, which is expected to vary with the nanotube chirality and orientation with respect to the underlying graphene lattice.

The graphene-CCNT hybrid film is fabricated by depositing CCNTs from a dilute solution on a quartz substrate,[54,55] then transferring a CVD prepared graphene sheet[26] onto the top of the CCNTs. A monolayer, and a bilayer, prepared under the same conditions and transferred to quartz substrates without the CCNTs, act as a control. The optical transmission of the three systems is measured using a UV spectrometer, and the sheet resistance is obtained from a 4-probe measurement.

Various adsorbents existing between the CCNTs and the graphene sheet create an electrically insulating layer that prevents the coupling between the CCNTs and the graphene, which makes the pristine graphene, the bilayer graphene, and the graphene-CCNT hybrid samples have almost the same resistance.

Annealing the three samples at 600°C for 10 minutes in vacuum removes any impurities and improves CCNT/graphene contact.

The resistance of the annealed graphene-CCNT hybrid system is about half that of the annealed graphene, with nearly the same transmission. Annealing at this high temperature in vacuum removes various adsorbents and enhances the coupling of the nanotubes to the graphene. SEM images of the graphene-CCNT system before and after annealing confirm this picture (Fig. 11.10). It is clear that CCNTs can be distinguished in the image due to charging. After annealing the CCNTs cannot be distinguished as they are now coupled to the graphene. The graphene grain boundaries are clear in both images and are typically of size ~300 nm.

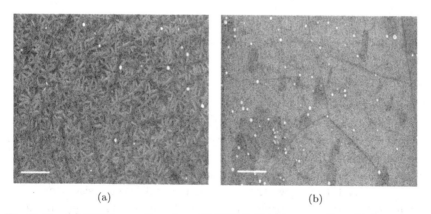

(a)   (b)

Fig. 11.10.   (a) SEM image of graphene-CCNT hybrid before annealing. (b) SEM image of graphene-CCNT hybrid after annealing where the nanotubes in the monolayer are not visible any more due to the coupling with the graphene. The scale bar is 600 nm. [Copied with Permission; Copyright © 2016, ELSEVIER; License no. 3915871103315].

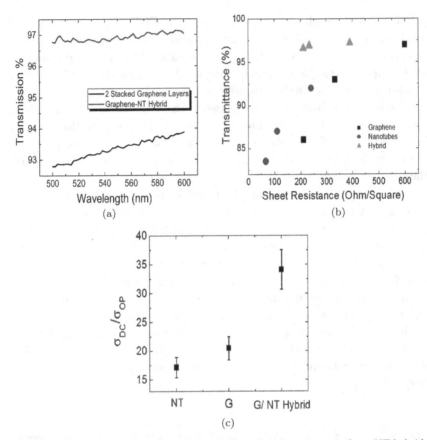

Fig. 11.11.   (a) The transmission of two stacked graphene layers vs. graphene-NT hybrid. (b) Transmission vs sheet resistance. (c) The optical density of the different films showing the increase of the sheet resistance of the graphene-CCNT hybrid film at a very high transparency. [Copied with Permission; Copyright © 2016, ELSEVIER; License no. 3915871103315].

Figure 11.11(a) shows the optical transparency of a sample of the hybrid system as a function of wavelength over the visible range, in comparison to the transparency of a bilayer graphene sample. At 97%, the transparency of the hybrid system is higher than that of the graphene bilayer.

Figure 11.11(b) shows the transmission and sheet resistance results for hybrid samples with different CCNT densities, multilayer graphene, and thick carbon CCNT films. The samples were annealed then doped by nitric acid as previously described. The transparency of the hybrid samples is equal to that of single layer graphene, which illustrates the monolayer nature of the hybrid material. The lowest-density CCNT hybrid sample has a resistance half that of single layer graphene. Tripling the CCNT monolayer density decreases the resistance to about one third that of single layer graphene. Further increase in the CCNT monolayer density does not significantly reduce the hybrid resistance. The same resistance could be achieved using a thick CCNT film with an optical transparency of 90%, and few layers of graphene with a transparency of 85%.

The performance of transparent electrodes can be evaluated through the ratio of DC to optical conductivity (σ_{DC}/σ_{Op}). This ratio can be extracted by measuring the optical transparency, $T(\lambda)$:

$$T(\lambda) = \left(1 + \frac{188.5\,\Omega}{R_S}\frac{\sigma_{Op}(\lambda)}{\sigma_{DC}}\right)^{-2},$$

where σ_{DC} and σ_{Op} are the electrical and optical conductivities respectively, R_S is the sheet resistance, and λ is the wavelength (550 nm) at which the transmission is measured. Figure 11.11(c) shows this ratio for the three layers where the hybrid has $\sigma_{Op}/\sigma_{DC} = 34$, which is very close to the ITO value of Ref. [56].

Figure 11.12 shows the Raman scattering data graphene, CCNT, and hybrid samples. The graphene spectrum exhibits the typical features of a

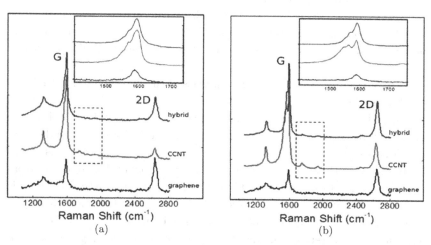

Fig. 11.12.   Raman scattering spectra of graphene monolayer, CCNT monolayer, and the hybrid. (a) Sample with lowest density CCNT monolayer. (b) Sample with highest density CCNT monolayer. Insets show the spectra at the G-peak region. [Copied with Permission; Copyright © 2016, ELSEVIER; License no. 3915871103315].

single graphene layer. We first consider the hybrid sample made with the lowest density CCNT monolayer, which we show in Fig. 11.12(a). The sample spectrum shows the characteristic signature of the nanotubes, including the splitting of the G-peak at $1600\,cm^{-1}$, in addition to the M and iTOLA modes at ~ 1750 and $\sim 1930\,cm^{-1}$.[57,58] The hybrid sample spectrum shows that the fine structure at the G-peak, observed in the CCNT sample, is now flattened (see inset), and that the M and iTOLA modes disappear. This is because some of the Raman modes of the CCNTs are suppressed by their coupling to the graphene layer. Furthermore, the low density hybrid 2D-peak slightly shifts up by $\sim 5\,cm^{-1}$ compared to the 2D-peaks of the graphene and the CCNT samples. By comparison, bilayer graphene shows a 2D-peak shift of $19\,cm^{-1}$.[59] This shows that the CCNT monolayer in the hybrid sample is effectively less than that of a second graphene layer.

Figure 11.12(b) shows the Raman data for highest density CCNT monolayer, and its hybrid sample. The CCNT monolayer data shows more resemblance to the typical nanotube Raman spectrum, with a clear splitting of the G-peak, and stronger M and iTOLA modes. The G-peak of the hybrid sample is broadened (see inset), and the M and iTOLA modes are clearly suppressed. This can be explained through the coupling between the CCNTs and the graphene, which significantly affects the CCNT Raman spectrum. This demonstrates the monolayer nature of the CCNT film. This is further confirmed by the $\sim 7\,cm^{-1}$ shift happening in the 2D-peak of the hybrid, which is still less than that of an effective second graphene layer.[59] This agrees with our transmission results previously discussed.

To understand the electronic properties of the hybrid system, we developed a simple mean-field type model for the graphene-CCNT system. This is justified by the fact that the nanotubes are randomly dispersed on the graphene sheet. We will further assume that all nanotubes are conducting, which is justified by the fact that the graphene-nanotube system is chemically doped, although purification methods are being developed to separate conducting from insulating tubes.[60] In the low nanotube-density limit where the nanotube-network percolation is weak, contributions from the nanotubes to the graphene conductance will arise from the nanotubes crossing graphene scattering boundaries (e.g. a nanotube crossing over a grain boundary). That is, a CCNT establishes another conductance path between neighboring regions, and thus one can model the nanotube through a local modification of the resistance of the graphene sheet. We further assume that the contact resistance between crossing CCNTs is much higher than that between a CCNT and graphene.[52,53] This is justified by the observed large decrease of the hybrid resistance for CCNT film densities that are well below the CCNT-network percolation limit.

We will treat the graphene sheet as a continuum resistive medium, which we discretize into a square grid of tiles. The tile edge, aG, is taken to be smaller than the smallest length in the system (the nanotube length LT). Neighboring tiles are coupled to each other through a resistance rG (assuming that the tile itself has negligible internal resistance), determined by the experimentally measured neat graphene sheet resistance. The graphene-CCNT hybrid system is constructed by adding the nanotubes to the graphene layer at random positions and orientations. This is modeled by adding a local parallel resistance rGT when a CCNT crosses two graphene tiles. CCNTs are considered to be of uniform size, LT ~350 nm, as inferred from SEM images of the physical system. Figures 11.13(a) and 11.13(b) present a schematic of a graphene sheet with grains and the CCNTs acting as conducting bridges connecting the grains.

The sheet resistance of the modeled hybrid system is determined using a 4-probe resistance calculation. A sample size of side Ls = 40 microns was simulated. The tile size is taken to be aG = 0.25 microns to match the typical graphene grain size determined from SEM images.[50,51] The tile size is smaller than the nanotube length LT, so that each CCNT is

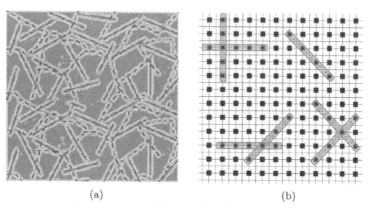

(a)   (b)

Fig. 11.13.   (a) Schematic showing graphene sheet with a layer of nanotubes. The dotted lines mark the graphene grains and the CCNTs are shown as red lines. (b) A 2D resistance network model of the graphene sheet. The CCNTs increase the sheet conductance by locally changing the model resistance. [Copied with Permission; Copyright © 2016, ELSEVIER; License no. 3915871103315].

guaranteed to bridge at least two tiles, thereby simulating the coupled graphene CCNT system. Reassuringly, further decrease in the tile size aG does not lead to a significant change in the calculated sheet resistance. The resistance parameter rG is set by fitting the calculated neat graphene sheet resistance to the experimentally measured value. The CCNT density, $\rho 2D$, and length, $LT = 0.35$ microns, are inferred from the SEM images of Fig. 11.10(a). The inter-probe distance, d4p, of the 4-probe calculation is taken to be 4 microns, so that it spans many graphene-CCNT grains. This way, we have $aG < LT \ll d4p \ll Ls$. This order secures a quantitatively correct description of the system as experimentally realized and measured.

The conductance gain of the hybrid system resulting from the graphene/CCNT coupling can be quantitatively described through the parameter η, defined by:

$$\eta = \frac{R_G - R_H}{R_G}$$

where R_G is the sheet resistance of graphene, and R_H is the sheet resistance of the hybrid system.

In Fig. 11.14, we plot the experimental values of η for the annealed and doped hybrid system with its calculated values of η of our simulated system.

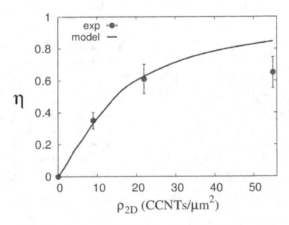

Fig. 11.14. A plot of h vs the CCNT 2D density r2D. The model predicts a linear dependence on the CCNT density at lower densities below or at weak percolation of the CCNT network. For higher densities, not all CCNTs contribute to lowering the graphene resistance, and $\square$ saturates. The deviation between the model and the experiment at high density is because 3D effects are not considered in the model. [Copied with Permission; Copyright © 2016, ELSEVIER; License no. 3915871103315].

The graphene-CCNT coupling modeled by the resistance parameter r_{GT} is obtained by fitting the model to the experiment as described above. The model predicts a linear dependence of η on the density. The experimental η saturates at high CCNT densities. This is explained as follows: as the density increases above the percolation threshold, not all CCNTs are in contact with the graphene, and since the CCNT-CCNT resistance is much higher than that of the CCNT-graphene, the fractional decrease in the resistance, η, will saturate. Such 3D effects are not included in our simple model, which predicts a further increase in η at high CCNT densities, where some CCNTs lie on each other rather than on the graphene sheet.

11.6 Conclusions

Transparent electrodes are currently considered one of the most important elements in several industrial applications like solar cells manufacturing and touch screen technologies. The huge demand for these two applications, which are just two examples out of many, besides the fact that the current technologies use considerably rare, limited, and costly manufactured materials, makes it essential to use new materials that cost less, are easier to prepare, and perform similarly, if not better. Carbon nanomaterials like graphene and carbon nanotubes satisfy these specifications. In this chapter, we tried to introduce our efforts in this field of enhancing the properties of carbon nanomaterials to replace the currently used ones. The sheet resistance of graphene was first successfully reduced to 90 Ω/square at 80% transparency, which is good enough for touch screen technologies but not for photovoltaic applications, by following the strategy of combining stacking of multi-graphene layers with chemical doping in order to increase the number of charge carriers. This was later improved to 22 Ω/square at 90% transparency using a carefully designed metal busbar and fingers pattern in contact with the carbon layers, which led to reducing the distance that the electron has to travel and consequently reducing the sheet resistance.

In the same context, other strategies, like graphene-metal busbar hybrid and graphene nanomesh, were followed in order to enhance the performance of the transparent electrodes by reducing the sheet resistance while keeping the transparency relatively high.

The hybrid material made from CCNTs atop a graphene sheet significantly reduced the sheet resistance compared to pristine graphene, with a negligible decrease in transparency. The CCNT film reduces the system sheet resistance by providing alternative current paths which bridge over different scattering regions. The differential reduction in resistance is maximal at low CCNT densities, where the CCNT network has weak or no percolation, indicating that the alternative current paths created are due to bypassing the graphene domain edges. A simple resistance network model successfully explains the experimental behavior. Furthermore, the CCNT network increases the dopant binding to the graphene sheet, thereby enhancing the doping stability. This hybrid graphene-CCNT system has potential applications in various TCE applications. Combining the CCNT-graphene hybrid with the micro-busbar array (described in 27) might produce a very effective TCE with many technological applications where we postulate the CCNT's will improve busbar adhesion and possibly contact resistance.

References

1. Carlson, D.E. and Wronski, C.R. (1976). Amorphous silicon solar cell. *Applied Physics Letters*, 28(11), 671–673.
2. Breivik, T.H. *et al.* (2007). Nano-structural properties of ZnO films for Si based heterojunction solar cells. *Thin Solid Films*, 515(24), 8479–8483.
3. Wang, X. *et al.* (2008). Transparent, Conductive Graphene Electrodes for Dye-Sensitized Solar Cells. *Nano Letters*, 8(1), 323–327.
4. Contreras, M.A. *et al.* (2007). Replacement of transparent conductive oxides by single-wall carbon nanotubes in Cu (In, Ga) Se_2-based solar cells. *The Journal of Physical Chemistry C*, 111(38), 14045–14048.
5. Watcharotone, S. *et al.* (2007). Graphene-silica composite thin films as transparent conductors. *Nano Letters*, 7(7), 1888–1892.
6. Blake, P. *et al.* (2008). Graphene-based liquid crystal device. *Nano Letters*, 8(6), 1704–1708.
7. Kim, K.S. *et al.* (2009). Large-scale pattern growth of graphene films for stretchable transparent electrodes. *Nature*, 457, 706–710.
8. Li, X. *et al.* (2009). Transfer of large-area graphene films for high-performance transparent conductive electrodes. *Nano Letters*, 9(12), 4359–4363.
9. Gruner, G.J. (2006). Carbon nanotube films for transparent and plastic electronics. *Journal of Materials Chemistry*, 16, 3533–3539.
10. Hu, L. *et al.* (2004). Percolation in transparent and conducting carbon nanotube networks. *Nano Letters*, 4(12), 2513–2517.
11. Wu, Z. *et al.* (2004). Transparent, conductive carbon nanotube films. *Science*, 305(5688), 1273–1276.
12. Blackburn, J.L. *et al.* (2008). Transparent conductive single-walled carbon nanotube networks with precisely tunable ratios of semiconducting and metallic nanotubes. *ACS Nano*, 2(6), 1266–1274.
13. Neophytou, N. and Kienle, D. (2006). Influence of defects on nanotube transistor performance. *Applied Physics Letters*, 88(24), 242106.

14. Yazyev, O.V. and Louie, S.G. (2010). Electronic transport in polycrystalline graphene. *Nature Materials*, 9, 806–809.
15. Geim, A.K. and Novoselov, K.S. (2007). The rise of graphene. *Nature Materials*, 6, 183–191.
16. Geim, A.K. (2009). Graphene: Status and prospects. *Science*, 324(5934), 1530–1534.
17. Zhang, Y., *et al.* (2005). Experimental observation of the Quantum Hall Effect and Berry's Phase in graphene, *Nature*, 438, 201–204.
18. Novoselov, K.S., *et al.* (2004). Electric field effect in atomically thin carbon films. *Science*, 306(5696), 666–669.
19. Hernandez, Y., *et al.* (2008). Production of graphene by liquid-phase exfoliation of graphite. *Nature Nanotechnology*, 3, 563–568.
20. Park, S. and Ruoff, R.S. (2009). Chemical methods for the production of graphenes. *Nature Nanotechnology*, 4, 217–224.
21. Li, X., *et al.* (2009). Transfer of large-area graphene films for high-performance transparent conductive electrodes. *Nano Letters*, 9(12), 4359–4363.
22. Sutter, P., *et al.* (2008). Epitaxial graphene on ruthenium. *Nature Materials*, 8, 406–411.
23. Coraux, J., *et al.* (2008). Structural coherency of graphene on Ir(111), *Nano Letters*, 8(2), 565–570.
24. Yu, Q., *et al.* (2008). Graphene segregated on nickel surfaces and transferred to insulators. *Applied Physics Letters*, 93(11), 113103.
25. Reina, A., *et al.* (2009). Large area, few-layer graphene films on arbitrary substrates by chemical vapor deposition. *Nano Letters*, 9(1), 30–35.
26. Kasry, A., *et al.* (2010). Chemical doping of large-area stacked graphene films for use as transparent, conducting electrodes. *ACS Nano*, 4(7), 3839–3844.
27. Kasry, A., *et al.* (2012). High performance metal microstructure for carbon-based transparent conducting electrodes. *Thin Solid Films*, 520(15), 4827–4830.
28. Maarouf, A.A., *et al.* (2013). Crown graphene nanomeshes: Highly stable chelation-doped semiconducting materials. *Journal of Chemical Theory and Computation*, 9(5), 2398–2403.
29. Maarouf, A.A., *et al.* (2016). A graphene–carbon nanotube hybrid material for photovoltaic applications. *Carbon*, 102, 74–80.
30. Jung, N., *et al.* (2009). Charge transfer chemical doping of few layer graphenes: Charge distribution and band gap formation. *Nano Letters*, 9(12), 4133–4137.
31. Farmer, B., *et al.* (2009). Chemical doping and electron-hole conduction asymmetry in graphene devices. *Nano Letters*, 9(1), 388–392.
32. Skoog, D.A. (1988). *Transparencies to accompany fundamentals of analytical chemistry*. Philadelphia, USA: Saunders College Publishing.
33. Skoog, D.A., *et al.* (1988). *Accompany Fundamentals of Analytical Chemistry*, Philadelphia, USA: Saunders College Publishing.
34. Fürst, J.A., *et al.* (2009). Electronic properties of graphene antidot lattices. *New Journal of Physics*, 11, 095020–095039.

35. Pedersen, T., *et al.* (2008). Optical properties of graphene antidot lattices. *Physical Review B*, 77(24), 245431.
36. Liu, W., *et al.* (2009). Band-gap scaling of graphene nanohole superlattices. *Physical Review B*, 80(23), 233405.
37. Petersen, R. and Pedersen, G.P. (2009). Quasiparticle properties of graphene antidot lattices. *Physical Review B*, 80(11), 113404.
38. Vanevic, M., *et al.* (2009). Character of electronic states in graphene antidot lattices: Flat bands and spatial localization. *Physical Review B*, 80(4), 045410.
39. Martinazzo, R., *et al.* (2010). Symmetry-induced band-gap opening in graphene superlattices. *Physical Review B*, 81(24), 245420.
40. Baskin, A. and Král, P. (2011). Electronic structures of porous nanocarbons. *Scientific Reports*, 1, 36.
41. Afzali-Ardakani, A., *et al.* (2014). *US8835686B2*. Controlled assembly of charged nanoparticles using functionalized graphene nanomesh. International Business Machines Corporation. Alexandra, Virginia, USA: USPTO.
42. Nistor, R.A., *et al.* (2011). The role of chemistry in graphene doping for carbon-based electronics. *ACS Nano*, 5(4), 3096–3103.
43. Newns, D.M. (1969). Self-consistent model of hydrogen chemisorption. *Physical Review*, 178(3), pp. 1123–1135.
44. Hewson, A.C. and Newns, D.M. (1974). Effect of the image force in chemisorption. *Japanese Journal of Applied Physics*, 13(2–2), 121.
45. Liang, X., *et al.* (2010). Formation of bandgap and subbands in graphene nanomeshes with sub-10 nm ribbon width fabricated via nanoimprint lithography. *Nano Letters*, 10(7), 2454–2460.
46. Safron, N.S. *et al.* (2011). Semiconducting two-dimensional graphene nanoconstriction arrays fabricated using nanosphere lithography. *Small*, 7(4), 492–498.
47. Bai, J. *et al.* (2010). Graphene nanomesh. *Nature Nanotechnology*, 5, 190–194.
48. Guo, J. *et al.* (2014). Crown ethers in graphene. *Nature Communications*, 5, 5389–5394.
49. Wehling, T.O., *et al.* (2010). Resonant scattering by realistic impurities in graphene. *Physical Review Letters*, 105(5), 056802–056805.
50. Huang, P.Y. (2011). Grains and grain boundaries in single-layer graphene atomic patchwork quilts. *Nature*, 469, 389–392.
51. Yu, Q. (2011). Control and characterization of individual grains and grain boundaries in graphene grown by chemical vapour deposition. *Nature Materials*, 10, 443–449.
52. Fuhrer, M.S. (2000). Crossed nanotube junctions. *Science*, 288(5465), 494–497.
53. Maarouf, A.A. and Mele, E.J. (2011). Low-energy coherent transport in metallic carbon nanotube junctions. *Physical Review B*, 83(4), 045402.
54. Tulevski, G.S., *et al.* (2007). Chemically assisted directed assembly of carbon nanotubes for the fabrication of large-scale device arrays. *Journal of the American Chemical Society*, 129(39), 11964–11968.

55. Engel, M. (2008). Thin film nanotube transistors based on self-assembled, aligned, semiconducting carbon nanotube arrays. *ACS Nano*, 2(12), 2445–2452.

56. Choi, K.H. *et al.* (1999). ITO/Ag/ITO multilayer films for the application of a very low resistance transparent electrode. *Thin Solid Films*, 341(1–2), 152–155.

57. Dresselhaus, M.S. *et al.* (2005). Raman spectroscopy of carbon nanotubes. *Physics Reports*, 409(2), 47–99.

58. Maultzsch, J. *et al.* (2005). Radial breathing mode of single-walled carbon nanotubes: Optical transition energies and chiral-index assignment. *Physical Review B*, 72(20), 205438.

59. Graf, D. *et al.* (2007). Spatially resolved raman spectroscopy of single- and few-layer graphene. *Nano Letters*, 7(2), 238–242.

60. Tulevski, G.S. *et al.* (2013). High purity isolation and quantification of semiconducting carbon nanotubes via column chromatography. *ACS Nano*, 7(4), 2971–2976.

Chapter 12

Sustainability in Mining

J.H.L. Voncken and M.W.N. Buxton

Delft University of Technology,
Faculty of Civil Engineering and Geosciences,
Department of Geosciences and Engineering,
Section Resource Engineering,
Stevinweg 1, 2628 CN Delft, The Netherlands

Sustainability is often defined as: *the ability to continue a defined behavior indefinitely.* However, considering the nature of mining operations, this cannot be meant with the phrase "Sustainability in Mining". Sustainability in the mining industry should be understood in the same way as sustainability in environmental science: *meeting the resources and services needs of current and future generations without compromising the health of the ecosystems that provide them.* A number of aspects of this are addressed in this chapter: use of energy, use of water, land disruption, reducing waste (involving solid waste, liquid waste, and gaseous waste), acid rock drainage when dealing with sulfide minerals, and restoring environmental functions at mine sites after mining has been completed. To do everything in an environmentally sound way is costly, but in the end necessary. Regarding this, it is concluded that governmental regulations concerning emission of waste, storage of waste and re-use of the land after mining are essential to provide a sustainable form of mining and mineral processing.

12.1 Introduction

Sustainability is often defined as: *the ability to continue a defined behavior indefinitely.* Clearly, this cannot be meant with the phrase "Sustainability in Mining". Once you have taken the ore, or coal, or industrial rocks, or industrial minerals out of the Earth, they are gone. The material will be processed and used. You cannot then put them back again. Therefore, sustainability in the mining industry should be understood in the same way as sustainability in environmental science: *meeting the resources and services*

*needs of current and future generations without compromising the health of
the ecosystems that provide them.*[1]

For the mining industry this means: to reduce the environmental impact
of mining and minimize the footprint of its activities throughout the mining
cycle, including work to restore ecosystems when mining has been terminated. In 1972, the Club of Rome predicted that many critical resources
would soon be exhausted.[2] Although their prediction was proven to be false,
nowadays again many people think that many critical resources may soon
be completely depleted.

There is, however, no limit in sight for the supply of raw materials.[3]
Advances in technology, product substitution, and reduction in pollution
ensure a "sustainable" supply. The concept of a usable resource is an
economic definition. Therefore improvements in extraction and processing
technology allow the redefinition and transformation of previously uneco-
nomic "waste" into economically viable ore. (e.g.[4]). This is clearly shown
for copper. Calculation of available resources and reserves has steadily
increased over time and increased demand has been met by increased supply (e.g.[5]). Available supplies are currently estimated to be in excess of
40 years. This figure has remained relatively constant between 30 years and
60 years of supply from the 1980's up to the present day.[6]

There are a number of aspects that should be addressed[7]:

- *Use of energy*
- *Use of water*
- *Land disruption*
- *Reducing waste*
- *Acid rock drainage when dealing with sulfide minerals.*
- *Restoring environmental functions at mine sites after mining has been
  completed.*

12.2   Use of Energy

Mining and metal processing often are very energy-intensive processes.
Trucks and excavators use diesel fuel, and for grinding ore, a lot of electricity is used. Also to refine copper, aluminum and zinc a lot of energy
is required. Coal (usually in the form of coke) is needed to smelt iron ore,
and subsequently make steel. Other environmental impacts arise from the
extraction of fossil fuels (coal, oil, and gas) and from the infrastructure
required to produce energy. Also these processes produce greenhouse gases,
and there is an increased risk of environmental contamination.

If the energy consumption in mining can be reduced, this can lead to reduction in the emission of greenhouse gases. Sources of fossil fuels will last longer, and additionally, operating costs are reduced. This again leads to a decrease in the cost of the commodity which is being mined.

12.3 Use of Water

In mining, water is used for a number of activities.[8,9]

- Lubrication
- Cooling
- Agglomeration
- As a medium enabling particles to be acted on (grinding and separating)
- Dust control
- Meeting the needs of the workers on site
- Mineral processing and metal recovery
- Preventing the mine from flooding

The amount of water used by a mine depends on the size of the mine and the activities. The latter are:

- the mineral being extracted
- the extraction process used

A mine that produces a metal may use the process of flotation[1,a] to separate ore minerals from waste. On the other hand, coal mines, salt mines and gravel mines use much less water.

Water needed for lubrication is usually applied in drilling. Water for cooling is used in many types of machinery in the mine. In agglomeration, water is used to obtain a suspension from which particles may settle through the process of flocculation.[b]

---

[a] *Flotation is a separation process, where finely ground ore is mixed with water to which additives have been added that alter surface properties of ore minerals and waste minerals. The particles become either hydrophobic (water shunning) or hydrophyllic (water loving). By blowing air bubbles through the suspension while stirring, the hydrophobic particles will adhere to the air bubbles, and rise to the surface, forming a foam layer. The hydrophyllic particles stay in the suspension. The foam is skimmed off at the surface, while the remaining ore pulp is removed as waste.*

[b] *Flocculation in the field of chemistry, is a process wherein colloids come out of suspension in the form of floc or flake, either spontaneously or due to the addition of a clarifying agent. The action differs from precipitation in that, prior to flocculation, colloids are merely suspended in a liquid and not actually dissolved in a solution.[10]*

Dust control is necessary. Dust must be prevented by spraying water, for a number of reasons. Dust is harmful to people, and harmful for equipment. Dust in the atmosphere may also lead to a dangerous situation where combustible dust particles become statically charged, and may discharge leading to a so-called dust explosion. Other sources of ignition in such a case may be sparks from machinery, friction or electric arcing, hot surfaces, or simply fire. Dust explosions have frequently happened in coal mines, often cause a large amount of damage, and may cause loss of lives.[11]

The workers on site need water to drink, and water to clean themselves after completion of their work (taking a bath or shower).

In mineral processing and metal recovery, water is used in a number of separation techniques, such as density separation techniques (for instance the *shaking table*,[c] see Fig. 12.1; and *the jig*,[d] see Fig. 12.2, and by surface chemistry methods[12] (*froth flotation, Fig. 12.3*).

Even in coal mining (in a coal washery) water is used to separate small coal particles (which float) from fine rocks particles (which do not).

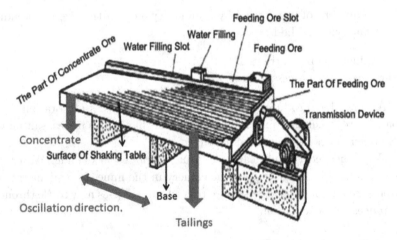

Fig. 12.1.   Shaking table. Original image from.[13] Adapted.

---

[c] *The shaking table is a piece of equipment where a thin film of water loaded with ore particles is floated over a vibrating, inclined surface, leading to a separation of particles due to differences in specific gravity, size, and shape.*

[d] *A jig is a piece of equipment, in which ore is brought into a tank with water, of which the floor can go upward and downward in a slow pulsating movement, as a result of which the particles are thrown upwards and consequently sink according to specific gravity, which in the end results in a bed of particles that is stratified on basis of density of the particles.*

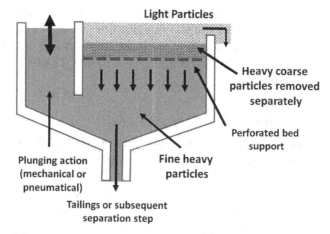

Fig. 12.2. Schematic picture of the operation of a jig. Original image from.[14] Redrawn.

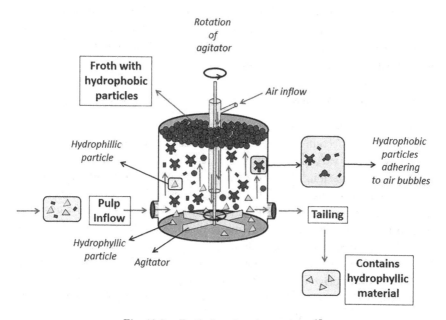

Fig. 12.3. Froth flotation, image from.[15]

To prevent a mine from flooding, usually water has to be pumped away (except in desert areas). This may have in itself a harmful effect. The large scale pumping of ground water in the German open pit lignite mines near Cologne also cause a significant lowering of the water table in adjacent Dutch areas, including the Dutch natural reserve "De Meinweg", near Roermond. Here, because of the pumping of water in, for instance, the German open pit lignite mine at Hambach, the land is drying out and many species of plants and animals are becoming endangered. The Hambach mine is currently some 370 m deep.[16]

12.4 Land Disruption

Mining activities use land at every stage of the mining cycle (exploration, construction, operation, closure, and post closure). However, the use of land can be diminished. The overall footprint of the mining area can be reduced by reducing the amount of waste produced and stored, and by planning mining sites around existing infrastructure where possible. Although at present, production of metals requires excavation of the ore, in the future, bio-mining, using bacteria to break down the ore minerals, may lead to an even larger reduction in the area of land used. In theory, only wells would need to be drilled, the solution with the bacteria pumped down, and, after a certain predetermined amount of time, pumped up again.[7,17] However, land disruption can never be completely avoided.

12.5 Reducing Waste

Mine waste includes solid waste, mine water, dust from mining, and waste from mineral processing (solid waste, process water, dust particles). Their composition can vary significantly, and so does their potential for environmental contamination.[18] Waste management plans are required to design and build appropriate storage facilities for the large amounts of waste produced at most mining sites. This includes rock waste dumps for rocks that had to be excavated, but are not mineralized, and tailing ponds for waste water from mineral processing plants.[7]

Using the proverb "prevention is better than cure", it is more economical and more effective to prevent pollution than to clean it up later. Therefore dedicated plans should be made for handling the anticipated waste before the mining operation starts.

What can be done to minimize waste?[7]

(1) Use of cleaner production techniques.
(2) Environmental control technologies.
(3) Use of waste as a raw material.
(4) Process re-engineering to reduce the amount of waste.

Water management strategies should be used to minimize the amount of waste water, and if possible, to clean it to an acceptable quality before releasing it. Over the past decades, in many countries, formal legislation has been passed in order to set acceptable standards for human impact on air, water and land.[7]

12.5.1 *Solid waste*

Solid waste from mining include overburden and waste rock, tailings remaining after mineral processing, and residues from any further treatment, such as leaching residues, water treatment residues, and slags.[8]

Mine wastes can be used on-site or off-site, stored in waste heaps, or used in leaching operations in order to recover additional valuable elements. Similarly, tailings may be used on- or off-site, stored in tailing ponds, or used in leaching operations.

However, the presence of waste disposal and storage facilities results in loss of land and can be harmful for the environment (loss of natural ecosystems, erosion, and leaching of materials from such facilities by wind and rain). Tailing facilities and waste heaps can collapse, leading to disasters involving the environment and the humans living there.[8]

12.5.2 *Liquid waste*

Water produced on a mine site can contain a wide variety of contaminants. Some of the more common contaminants are listed in Table 12.1. Not all of these will always be found in a particular mining operation.

These, and other impurities in water, have to be reduced in concentration to levels acceptable for the purpose for which the water is intended to be used.

Several techniques are available to clean the water. The technique to be applied depends on the required water quality, and also on the quantity.

These techniques are[8]:

- *Thickening, clarification and filtration.*
  Well known, and widely used.
- *Precipitation.*
  A common pre-treatment method.
- *Membrane technologies.*
  Techniques employed are micro-filtration, ultra-filtration, nano-filtration, reverse osmosis, and electro-dialysis.

Table 12.1: Common Contaminants in Mine Water and their Sources. (after[8]). Not Every Contaminant is Present at Every Mine Site.

Contaminant	Typical Source
Metals, Fe, Mn, Cr, Zn, Cu, etc.	AMD (Acid Mine Drainage) from oxidation of pyrite in mines, waste rocks piles, and tailings dams
Sulfate ions	AMD from oxidation of pyrite in mines, waste rocks piles, and tailings dams
Acidity (H^+ ions)	AMD from oxidation of pyrite in mines, waste rocks piles, and tailings dams
Cyanide	Spilling from gold leaching operations, and seepings from gold tailings dams
Suspended solids	Inadequate underground settling, run-off from surface, tailings dams, rock piles etc.
Sodium ions	From groundwater, artesian water, and land-locked inland lakes, addition of sodium-based reagents
Chloride ions	From groundwater, see above.
Nitrogen compounds	Wastes from explosives, sewage and other domestic wastes
Phosphate ions	Sewage and domestic wastes
Radionuclides	AMD attack on radionuclide bearing rocks
Microbes	Human fecal contamination, run-off from livestock grazing.

- *Ion exchange.*
  Used to demineralize water, and to recover specific elements. Used mainly as a last step in water cleaning.
- *Biological processes.* Can be useful to remove organic materials, nutrients, and sulfate, and to neutralize Acid Mine Drainage (AMD).

12.5.3 Gaseous waste

The nature of gaseous waste depends on the process. When fuel (oil, natural gas or coal) is burned to produce heat, the major gaseous products are CO_2 and H_2O. When carbon (or coke) is applied as a reductant to reduce oxides, large quantities of CO_2 (and sometimes CO) are produced. During roasting or smelting of sulfide concentrates, SO_2 is a major product. Also small quantities of SO_3 may be produced.[8]

When air is used as input, Nitrogen is also invariably a part of the off-gas, but is not considered a waste product, since it is an input which generally passes through the process without any chemical reactions.[8]

Gas streams, however, almost always contain particulate matter: entrained fine solid or molten particles arising from the material in the reactor, or formed from reactions in the reactor. This off-gas usually also contains small quantities of other, possibly harmful, gases in addition to N_2, CO_2, H_2O, and sometimes SO_2.[8]

The basic functions of a gas handling system include some or all of the following[8]:

- Containment and capture of process gas at the reactor exit.
- Cooling of the gas to a suitable temperature for subsequent handling, cleaning, or both.
- Separation of particulate matters (fumes and dusts)
- Transfer of the gas to gas cleaning, sulfur fixation, or the stack.

Several types of devices are used for separating solids from gas streams: settling chambers, cyclones and other centrifugal separators, bag filters, scrubbers, and electrostatic precipitators.[8] Gas cleaning systems may also contain activated carbon filters, to remove harmful chemicals such as dioxins, which may form in some reactors, e.g. as happened in the sinter plant of former Dutch phosphorus producer Thermphos.[19]

12.6 Acid Rock (Mine) Drainage

Acid Rock Drainage (ARD), also known as Acid Mine Drainage (AMD) is produced when sulfide-bearing material is exposed to oxygen and water. The production of AMD usually, but not exclusively, occurs in iron sulfide-aggregated rocks. The process occurs naturally, but mining can promote AMD simply through increasing the amount of sulfides exposed to water and air. Naturally occurring bacteria can enhance AMD production by assisting in the breakdown of sulfide minerals.[20]

AMD is often the major source of contaminated water at a mine site. The major mineral involved is usually pyrite, which oxidizes when exposed to water and air. The process is actually complex as it involves chemical, biological, and electrochemical reactions. Also it varies with environmental conditions. Factors of influence are, for instance, pH, pO_2, the specific surface and morphology of pyrite, the presence of bacteria and/or clay minerals, as well as hydrological factors.[21] Pyrite in mining waste or coal overburden is initially oxidized by the atmospheric O_2 producing H^+, SO_4^{2-}, and Fe^{2+}. The divalent iron can be further oxidized by O_2 into Fe^{3+}. This in turn hydrolyses, and precipitates as amorphous iron hydroxide releasing additional amounts of acid.[21] In the initial stage, pyrite oxidation is rather slow, but as acid production increases, and the pH in the vicinity of pyrite decreases to below 3.5, formation of ferric hydroxide is hampered, and the activity of Fe^{3+} in solution increases.[21] Then oxidation of pyrite by Fe^{3+} becomes the main mechanism for acid production in mining waste. At low pH, an acidophyllic, iron oxidizing bacterium (Thiobacillus ferrooxidans) catalyzes and accelerates the oxidation of Fe^{2+} to Fe^{3+}.[12]

The following reaction equation summarizes the process[22]:

$$4FeS_2 + 15O_2 + 14H_2O = 4Fe(OH)_3 + 8SO_4^{2-} + 16H^+$$

Prevention of acid mine drainage can be opted for ("source control"), but is not always realistic. Another possibility is "migration control". As both oxygen and water are needed to continue the formation of AMD, excluding either one (or both) should lead to prevention of AMD. However, completely sealing off abandoned mines from an influx of air and/or water is not easily done, and therefore usually, other approaches are better. One of these approaches is to minimize the production of AMD by blending acid-generating and acid-consuming materials (Johnson and Hallberg, 2005). This may be, for instance, adding apatite to pyritic mine waste in order to precipitate Fe^{3+} as ferric phosphate. In that way, its potential to act as an oxidant of sulfide minerals is reduced. Due to formation of coatings on the added phosphate, this technique often is only a temporary solution. The only alternative is to minimize the impact that polluted mine water has on the environment: migration control measures. These can be the continuous application of alkaline materials to neutralize acidic mine waters and precipitate metals, or the use of natural or constructed wetland ecosystems.[22]

12.7 Restoring Environmental Functions at Mine Sites after Mining has been Completed

Mining is a temporary activity. The life time of a mine ranges from a few years, to several decades. A mine is closed once the mineral resource is exhausted, or the mining operation is no longer profitable.[23] Usually, before a mining permit is granted, plans for closure of the operation are required. Also it must be demonstrated that the mine site will not be a threat to the health of the environment or society in the future. Depending on the site, the mine may be intended for other human uses, or restored to its pre-mining use. Such matters should increasingly be included in the original mine plan, and financial assurances are often required, in the event that the responsible company is unable to complete the closure as planned.[9]

After termination of mining activities, land restoration should be carried out. As a matter of fact, three different terms should be considered:

- Restoration
- Reclamation
- Rehabilitation

Restoration is defined as "the replication of site conditions prior to disturbance".[24]

Reclamation is defined as "rendering a site habitable to indigenous organisms".[24]

Rehabilitation is defined as "disturbed land will be returned to a form and productivity in conformity with a prior land use plan including a stable ecological state that does not contribute substantially to environmental deterioration and is consistent with surrounding aesthetic values".[24]

In some cases, restoration in the strict sense may be impossible.

Mining is also a destructive activity. To reduce environmental harm, careful management and regulation are necessary. In developing countries, this may, however, become second to economic growth. In brief, the environmental impacts caused by mining, are[24]:

- Ecosystem disturbance and degradation
- Habitat destruction
- Adverse chemical impacts (from improperly treated wastes)
- Loss of soil-found carbon (to the atmosphere).

12.8 Concluding Remarks

As can be learned from the previous discussion, sustainability in mining and mineral processing has many aspects, ranging, among others, from use of land for waste rock dumps, use of water, use of energy, up to the use of poisonous chemicals, and emission of waste gasses, waste fluids, and gaseous waste. To do everything in an environmentally sound way is costly, but in the end necessary, although especially in developing countries the economic aspects will probably be most important. Regarding the latter, governmental regulations concerning emission of waste, storage of waste and re-use of the land after mining are essential to provide a sustainable form of mining and mineral processing.

References

1. Morelli, J. (2011). Environmental sustainability: a definition for environmental professionals. *Journal of Environmental Sustainability*, 1(1), 2.
2. Meadows, D.H., Meadows, D.L., Randers, J., and Behrens, W.W. III. (1972). *The limits to growth: A report for The Club of Rome's project on the predicament of mankind.* Washington D.C., USA: Universe Books.
3. Batterham, R.J. (2017). The mine of the future — even more sustainable. *Minerals Engneering*, 107, 2–7.
4. Meinert, L.D., Robinson, G.R., and Nassar, N.T. (2016). Mineral resources: Reserves, peak production and the future. *Resources*, 5(1), 14.
5. USGS. (2016). *Mineral commodity summaries.* Virgina, USA: USGS.
6. Graedel, T.E., Gunn, G., and Espinolza, L.T. (2014). Metal resources, use and criticality. In G. Gunn (Ed.), *Critical metals handbook.* New York, USA: Wiley.
7. MiningFacts. (2017a). How can mining become more environmentally sustainable? Retrieved from https://www.fraserinstitute.org/categories/mining.
8. Rankin, W.J. (2011). *Minerals, metals and sustainability — meeting future material needs* (1st ed.). Leiden, the Netherlands: CRC Press/Balkema.
9. MiningFacts. (2017b). What are the water quality concerns at mines? Retrieved from https://www.fraserinstitute.org/categories/mining.
10. Flocculation. (n.d.) In *Wikipedia.* Retrieved on 23 October 2018, from https://en.wikipedia.org/wiki/Flocculation#Civil_engineering.2Fearth_sciences
11. Dust explosion. (n.d.) In *Wikipedia.* Retrieved on 23 October 2018, from https://en.wikipedia.org/wiki/Dust_explosion.
12. Wills, B.A. and Napier-Munn, T.J. (2006). *Wills' Mineral Processing Technology* (7th ed.). Oxford, UK: Butterworth-Heinemann.
13. Baichy. (n.d.) 6S shaking table. Henan, China: Baichy. Retrieved from https://www.alibaba.com/product-detail/6S-laboratory-shaking-table_60169029982.html.
14. MET-SOLVE Laboratories Inc. (n.d.) Mineral processing introduction. Retrieved from http://met-solvelabs.com/library/articles/mineral-processing-introduction.
15. Voncken, J.H.L. (2016). Mineral processing and extractive metallurgy of the rare earths. In J.H.L. Voncken (Ed.), *The rare earth elements — an introduction.* Switzerland: Springer.
16. RWE. (2017). The Hambach opencast mine. Retrieved from https://www.group.rwe/der-konzern/organisationsstruktur/rwe-power.
17. Moskvitch, K. (2012, March 21). Biomining: How microbes help to mine copper. *BBC News.* Retrieved from http://www.bbc.com/news/technology-17406375.
18. Mining Facts. (2017d). How are waste materials managed at mine sites? Retrieved from https://www.fraserinstitute.org/categories/mining.
19. Mennen, M., Dusseldorp, A., Mooij, M., and Schols, E. (2010). *De verspreiding van dioxinen rond Thermphos.* Bilthoven, the Netherlands: RIVM.

20. Akcil, A. and Koldas, S. (2006). Acid mine drainage (AMD): causes, treatment and case studies. *Journal of Cleaner Production*, 14(12–13), 1139–1145.
21. Evangelou, V.P. (1995). *Pyrite oxidation and its control.* Boca Raton, Florida, USA: CRC press.
22. Johnson, D.B. and Hallberg, K.B. (2005). Acid mine drainage remediation options: a review. *Science of The Total Environment*, 338(1–2), 3–14.
23. Mining Facts. (2017c). What happens to mine sites after a mine is closed? Retrieved from https://www.fraserinstitute.org/categories/mining.
24. Tripathi, N., Singh, R.S., and Hills, C.D. (2015). *Reclamation of mine-impacted land for ecosystem recovery.* Hoboken, New Jersey, USA: Wiley-Blackwell.

Part IV

Recycling as a Critical Material
Mitigation Strategy

Chapter 13

How to Get Stuff Back?

Jan-Henk Welink

Delft University of Technology, Mechanical,
Maritime and Materials Science, Mekelweg 2,
2628 CD, Delft, The Netherlands

In order to recycle or re-use products that contain Critical Raw Materials (CRMs), these products have to be collected from consumers and professional organizations (e.g. businesses) first. Waste Electrical and Electronic Equipment (WEEE) is collected for economical, environmental, and public health and safety reasons. Lessons can be learned from the collection of WEEE, such as how to influence and stimulate consumers to collect WEEE separate from other waste, and how to stimulate and train companies in separating waste.

13.1 Introduction

Critical Raw Materials (CRMs) are to be found in all kinds of products; e.g. Electrical and Electronic Equipment (EEE) and alloys. Retrieving the CRMs from disposed or End-of-Life (EoL) products starts with collecting these products in such a manner that separation techniques in combination with hydro- and pyro-metallurgy are able to operate with the highest possible yield. Products are also retrieved for partial or complete re-use of the product, such as remanufacturing activities. To comply to legislation the collection of products is organized by governments or producers. The effectiveness of the collection of disposed products depends on the behavior of consumers and professional organizations.

13.2 Organizing the Return of Disposed Products

Waste Electrical and Electronic Equipment (WEEE) has been collected for recycling purposes throughout the world. The drivers for governments, NGOs and companies to collect WEEE are[23]:

- Economical. In the case that the value of the metals is higher than the cost of retrieval: e.g. collection and recycling costs.
- Environmental. By recycling and (partial) re-use WEEE will not be land-filled or incinerated, preventing leaching of harmful substances to the environment. Recycling will reduce the global demand for metal production, reducing greenhouse gases and harmful substances formed by mining and processing ore. Recycling also saves energy in comparison to the processing of ore to retrieve virgin metals.
- Public health and safety reasons. Recovery of metals in the informal sector, as well as landfilling or incineration with poor or no environmental standards can create harmful substances of such a level that public health and safety are compromised.

Current practices worldwide of returning WEEE are[23]:

- Official take-back systems, via municipalities, retailers or commercial pick-up services.
- Disposal with mixed residual waste to landfills and incineration. If the metals of WEEE are not separated before or after (from the bottom-ash) incineration or landfilling, toxic leachate from landfills or incinerators ash or harmful air emissions from incinerators will compromise the environment and public health.
- Collection outside the official take-back systems. WEEE is collected by individual waste dealers or companies for metal and/or recycling or export. In many cases this export is not legal.
- Informal collection and recycling in developing countries, usually by self-employed people. If the collected WEEE does not have any value, it is landfilled or incinerated causing damage to the environment and public health.

In order to organize the return of WEEE, different countries have adopted legislation. The European Union adopted the WEEE directive (2002/96/EC) focused at collection targets. With the Home Appliance Recycling Law (HARL) and Small Appliance Recycling Law, Japan wants to increase the recycling rate. Australia aims to improve this with the

National Waste Policy and the National Television and Computer Recycling Scheme. India has "Guidelines for environmentally sound management of e-waste" focused on the collection and recycling of e-waste. Indonesia and the Philippines are working on e-waste legislation.[47] China started extended producer responsibility practice for WEEE recycling in 2011.[23]

The aim of the WEEE directive (2002/96/EC) of the EU is "encouraging the design and production of electrical and electronic equipment which takes into full account and facilitates their repair, possible upgrading, reuse, disassembly, and recycling". For this "each producer should be responsible for financing the management of the waste from his own products". The financial and physical responsibility of the stakeholders varies per EU country for:[36]

- Producers: the set-up of own collection points or for provision of collection containers, and the funding of collection systems, sorting, trans-shipment, treatment, and bulk transport.
- Local authorities: the obligation to WEEE take-back or to own WEEE management by direct trading.
- Retailers: the obligation to WEEE take-back.
- Recyclers: the optional direct trading of WEEE and the obligation to downstream reporting of WEEE streams.
- Coordinating body/clearing house: the registering of WEEE put on market, managing take-back of WEEE on request and joint communication on WEEE collection.

In China, the formal and in most cases state-controlled recycling of WEEE is financed via a product tax for five types of appliances: TV-sets, air conditioning, washing machines, refrigerators, and computers. The recyclers have to collect WEEE through their own collection schemes or via traders. Municipalities and retailers do not have a role in WEEE collection. The WEEE management is the responsibility of six governmental agencies. The structure of these agencies with interdependent, aligned, and sometimes overlapping responsibilities makes effective and efficient WEEE management more difficult.[36]

13.3 Stimulating Consumers

The effectiveness of collection systems of WEEE and other EoL-products that contain CRMs depends on the motivation of consumers to separate these EoL-products from the residual waste. The motivation of consumers can be explained by psychological models. These models can explain the behavior and encourage consumers to engage in waste separation. Based on the behavior of consumers, the effectiveness of collection systems and communication about waste separation can be enhanced.

13.3.1 *Psychological models*

Regularly used psychological models to explain the waste separation behavior of consumers are:

- TPB: Theory of Planned Behaviour[4,43,44]
- IMB: Information-Motivation-Behaviour Skills model[38]
- CADM: multilevel Comprehensive Action Determination Model[22]

The TPB is based on the hypothesis that the behavior of an individual follows directly from the intention of an individual. The intention of an individual is influenced directly by three factors:

- Attitude towards the behavior: the individual's evaluation (favorable or unfavorable) of performing the behavior. Example: An individual can be involved in social and societal responsible actions and therefore waste separation is important for that individual.
- Subjective norm: the individual's perception of social pressure to perform or not to perform the behavior. Example: A student stops waste separating in a student house (dorm) because fellow students have a negative attitude towards waste separation.
- Perception of control: perception of the individual's own ability to carry out the behavior. Example: An individual wants to separate waste but thinks that the recycling bins are too far away or are difficult to find.

In the model it is assumed that personality, past experiences and demographic characteristics also affect the individual's behavior, and indirectly affect the above three factors of the model, and are therefore included in the model.

CADM (multilevel Comprehensive Action Determination Model) is based on the TPB model. The model assumes that behavior in waste separation is determined by:[38]

- The intentions of an individual
- Conditions given by the circumstances of an individual
- Habits of an individual. These are influenced by the perception of control (TPB model) and the intentions of the individual, because habits develop over time.

The CADM model studies actual existing barriers and facilities that influence behavior (in this case waste separation) and the link to the subjective barriers and facilities (experienced by the individual).

The IMB-model (Information-Motivation-Behavioral skills model) assumes a connection between the three factors:[22]

- Information: Information known by an individual related to engaging in waste separation behavior. Examples are the knowledge of the locations of waste bins for waste separation, or the knowledge on how and what to bring to these bins.
- Motivation. The motivation of an individual is determined by his/her own perception of the behavior, and the need for the behavior to conform to a social norm.
- Skills: The simple skills of an individual to separate waste for recycling and to offer recyclable waste at regular collection days.

The IMB model implies that information and motivation for waste separation defines the specific skills to separate waste. The skills determine the behavior of waste separation.

13.3.2 *Influencing and stimulating behavior*

The recycling behavior of consumers can be influenced positively (encouraging) or negatively. This section discusses the behavior of consumers, the general primary consumer attitudes on waste separation, promotion opportunities (pay-per-bin, education, and regulation), and finally, the influence of external circumstances on the consumer.

13.3.2.1 *Waste separation behavior of consumers*

Seacat and Northrup[38] researched the waste separation behavior of consumers with the IMB model and concluded that more waste separation occurs when there is more information available about waste separation and when people become more motivated. These aspects determine the waste separation abilities (skills) of consumers. The researchers emphasize that collection systems are used more if these systems are simple to use according to consumers. The convenience of separate waste collection systems is an important condition for waste separation.

The TPB model is based on the assumption that the behavior of a consumer depends on the attitude towards that behavior, subjective norms (social pressure) and the perception of control (perception of the possibility and difficulty of waste separation). Tonglet *et al.*[44] conclude that the knowledge and opportunity for waste separation are important factors to encourage waste separation, but consumers should not be deterred by the problems that physical waste separation may entail, such as lack of convenience, place for bins at home, and available time. Besides this, experience with waste separation in the past and its importance for society are also important factors. Another study[25] confirmed this. This study looked at the effect of the perception of distance to facilities for waste separation and the actual distance. A significant relationship was found between the perception of the distance to waste separation facilities and waste separation behavior, rather than between the actual distance and waste separation behavior. The perception of that distance had an even larger influence on waste separation than the intention to separate waste.

The CADM-model assumes that waste separation behavior, as in the TPB-model, also depends on the intentions of an individual, the conditions created by the circumstances of an individual and the habits of a individual. The conclusions of the study of Klöckner and Oppedal[22] are in line with the two studies described above; waste separation behavior is determined by the intentions and habits of an individual and by the conditions created by that individual's circumstances. On the individual's personal level, waste separation behavior is determined by the intention to separate waste and by the waste separation habits the individual has developed, but external circumstances in the waste separation system play an important role. A study of waste separation practices shows that people who have the intention to separate waste also have the perception that the collection system makes waste separation easy.

The above mentioned psychological studies are in line with the conclusions of econometric and statistical research conducted in Sweden.[5,17] These Swedish studies show that if the infrastructure of collection systems is considered as marginally helpful by consumers, waste separation then depends on strong personal, moral and social norms. The publicity required to maintain these strong norms amongst the public becomes less important when consumers perceive the waste collection systems as more convenient. This is confirmed by a study on oral communication.[6] In this study it appears that oral communication by door-to-door visits on the separate collection of food waste has little impact on the separate collection behavior. Access to collection systems, however, has a major impact. Research on the separate collection of WEEE[37] suggests that internal factors such as social pressure and attitude are the most important variables to explain the intention of households to collect WEEE. The second most important variable is the convenience of separate waste collection. A study showed that if separate waste collection systems are considered to be sufficiently convenient and available, more waste is separated in mass (kilograms) and types of recyclable material.[12] This is also shown in a study on separate waste collection behavior in schools.[33]

Case: CFLs

A case study on recycling Compact Fluorescent Lights (CFL)[45] shows that availability and convenience of separate waste collection systems are important factors. This study was conducted in the state of Maine in the United States. In 2009, 520 people were questioned on the separate collection of discarded CFLs. That few CFLs were collected appeared to be the result of:

- Insufficient knowledge about recycling CFLs
- The lack of convenience of collection systems

The researchers recommended waste collectors to make the collection of CFLs more convenient and to make any access to information to consumers on the collection of CFLs more simple.

13.3.2.2 *General attitudes of consumers*

In the previous section researchers concluded that separate waste collection depends on the attitude of consumers and the waste collection systems. This section describes three cases about consumer attitudes, the personal and social circumstances that determines if a consumer will separately collect waste.

Case: Batteries

In a case study on the collection of batteries in Switzerland,[18] 1,000 consumers were surveyed by mail. This study shows that the factors that positively influence waste separation behavior are:

- Knowledge about recycling batteries
- "Organizing" battery collection at home
- Disagreeing with reasons why not to separately collect batteries

The attitude of the respondents towards an ecologically better waste collection (better for the environment) and confidence in the authorities who collect waste, did not seem to be reasons to collect batteries separately.

Case: WEEE collection in Cardif

In a study on the collection of WEEE at waste collection sites in the Cardiff area (United Kingdom) nearly 5,000 people were asked to complete a survey.[11] The survey showed that households who already collect other types of materials separately (e.g. paper, glass, cans and plastic) were also more likely to collect WEEE. Many respondents who collected separated waste indicated that they would visit other waste collection sites as well, if they could bring their separated WEEE. The researchers also noted that through separated waste collection consumers became more aware about this topic and asked for more opportunities for separate waste collection of other types of waste materials. The households did not want to have more information about the importance of separated waste collection, but wanted information on how to collect waste materials and EoL-products separately.

Case: Organic waste

In Wyre (UK), a survey was conducted with more than 2,500 completed surveys on collection of green waste. The response rate was 50%. The study[46] on these surveys showed that respondents aged 25 to 44 years were the least likely to separately collect organic waste, possibly due to a lack of time. The survey also showed that the public desired more information and services to facilitate the collection of waste separated by type.

13.3.2.3 *Financial incentives*

In Minnesota (USA), researchers[41] analyzed data from the collection of recyclable waste on the effect of differentiated rates (pay-per-bag) for separated waste collection, on informing the public, and on waste collection systems. The researchers concluded that differentiated rates for waste disposal significantly increase the amount of separately collected recyclable materials.

In a survey among residents of the London borough, Havering, respondents seemed to have a lower response to financial incentives than to improving the plan on separate waste collection itself, in particular the waste collection infrastructure and support.[40] The financial incentives that could possibly have an impact were, according to those surveyed:

- rewards over penalties; and,
- financial rewards at the community level (e.g. for improvements in the neighborhood) or as tax refunds or reduction in the municipal taxes, rather than individual rewards.

In 2005 and 2006, in Portsmouth (United Kingdom) providing financial incentives were compared to informing people at the door ("doorstepping") on separated waste collection, and with a feedback card in the mail-box when the separated waste collection was done incorrectly, or not at all.[42] The feedback card was encouraging and had an informative tone. The "doorstepping" was regarded as successful, but only relatively few households were reached. The feedback card was very effective and relatively cost-effective. The financial incentives consisted of rewards for households if more recyclable materials were collected separately. The study did not show that financial incentives were effective as a method to encourage recycling. Only 13% of respondents mentioned that financial incentives were the main reason to collect and separate recyclable waste.

13.3.2.4 *Effect of housing type*

Researchers studied the impact of the housing type on separate waste collection in the London borough of Havering.[39] The researchers counted the number of households that collected separated recyclable waste, and compared that to the number of households in a street. The numbers showed that streets with detached houses have more households that collect separated waste than streets with terraced houses or flats. The researchers believe that separated waste collection will be improved at terraced houses and flats when social interactions are improved. In improving these social interactions, street architecture can have an influence.

13.3.2.5 *Effect of income on separate waste collection*

The effect of income of households on separate waste collection has been researched in several studies. However, these studies do not give a definitive conclusion. A study conducted in the state of Minnesota in the United States,[41] shows a marginal decrease in income at the increase in the volume of separately collected materials: 0.2 percentage points per US $1,000 increase in income. A study conducted in Belgium[14] with data from 2003 is in line with the study of Sidique *et al.*[41]: in households with a higher income per capita less recyclable waste is collected separately.

A study conducted in Cardiff (United Kingdom) showed that households with an income of less than £10,000 collected less WEEE for recycling than households with higher incomes.[11] The researchers do note that less WEEE was found in the residual waste from households with low incomes. The researchers explain this effect at households with low incomes on:

- Less availability of suitable transport (often a car) to bring the WEEE to a recycling center, which in many cases was three to four kilometers away.
- Keeping electrical and electronic products longer and giving them away to others rather than disposing of them as waste.

13.3.2.6 *Social groups: students and British-Asians*

Researchers[33] have studied the attitude on separate waste collection of students at a high school in California in the United States. It was found that more recyclable waste was collected after it was explained which materials were to be collected and how they were to be collected. The collection systems had to be accessible and consistently at the same location. The usefulness of separated waste collection was already known to the students; among students this was already the norm. The researchers recommend that communication to students about separated waste collection should not focus on the importance of recycling. The advice was to make students ambassadors in separated waste collection campaigns toward adults (teachers, parents). The researchers noted that students have a broader perspective on the world and are more future-oriented than adults often assume.

A British study showed that the recycling behavior of the British Asian community in Burnley was no different from the other inhabitants of Burnley.[26]

13.3.3 *Effectiveness of collection and delivery systems*

The effectiveness of collection systems for recyclable waste depends on different variables. These variables are the frequency of pick-ups, the size of containers for recyclable and residual waste, and the set-up of collection points where consumers could bring their recyclable discarded products. This section describes these variables.

13.3.3.1 *Frequency of pick-ups and the size of containers*

In the UK, an econometric study was carried out on the collection or "pick up frequency of "dry recyclebles" (e.g. waste paper, cans, plastics), "compostables" (organic waste), and residual waste.[1] In the UK recyclables such as waste paper, cans, and plastic bottles are collected in one container or one plastic bag. This collected "dry recycleble" material is mechanically separated. The researchers studied data from different regions in the UK on the collection rate of the three above-mentioned types of discarded materials. A relationship was found between the frequency of the pick-up of the residual waste and the percentage of waste that has been recycled. It appeared that the less often residual waste was collected, the more "dry recyclables" and organic waste was collected. The researchers recommended against the plans of local authorities to collect more garbage. The researchers found that the type of bin for "dry recyclables" determines the amount of "dry recyclables" collected. Collection with the wheelie bin with a maximum of 120 liters gave the most "dry recyclables". A plastic bag, a wheelie bin of 180 to 240 liters and more than 240 liters also provide more collection. The researchers found an increase in collected organic waste when this material was collected only once a week. However, this was not the case for all seasons.

A study carried out in Belgium with data from 2003[14] showed that the collection of recyclable materials increased when

- residual waste was collected once every two weeks instead of once a week
- there is an extra collection system for organic waste; for example an additional "green bin".

In addition to the influence of pay-per-bag, which caused an increase of separated waste collection, a collection system for organic waste had the largest positive influence on more separated waste collection.

13.3.3.2 *Waste collection points and collection systems: A case of computer monitors and TVs*

In the State of Maine in the United States, computer monitors and televisions have been collected by municipalities since 2004. Each municipality has its own method to collect these EoL-products. The data related to these collection methods and the amount of computer monitors and televisions collected was analyzed in an econometric research project.[1] This research showed that if consumers had to pay for the disposal of their EoL-products at a collection point, less EoL-products were collected. However, the more often a collection point was opened, the more computer monitors and televisions were collected per capita. The distance to the collection point was not relevant. If recyclable waste was separately collected (without computer monitors and televisions) in a municipality, fewer computer monitors and televisions were brought to the collection points. The researchers recommended that the municipalities in Maine cancel the financial contribution for the disposal of computer monitors and televisions, that they increase the hours and days a collection point is open for the convenience of the consumer, and that they consider picking up computer monitors and televisions in homes.

13.3.3.3 *Effects of collection and bring-systems*

Researchers analyzed data from the collection of recyclable waste in the State of Minnesota in the United States.[41] The statistical study focused on the effect of differentiated rates on the collection of recyclable materials (pay-per-bag), public information, and collection and delivery systems. Also the interaction of collection and bring-systems was studied. A bring-system is a system where the consumer brings the (recyclable) waste to a collection point, instead of having it collected at home. The study found that collection and bring- systems were more effective when combined. However, when the collection and bring- systems were implemented separately, both had little impact on the promotion of separated waste collection for recycling.

13.3.3.4 *Collection systems and consumer attitudes*

In an econometric study in Sweden,[5,17] the effect on separated waste collection of social and personal norms, feeling morally responsible, knowledge of separated waste collection, and the set-up of collection systems was researched. Both studies showed that personal, moral and social norms were important factors for separated waste collection if the infrastructure of the collection systems were not very widespread or if the individual had to make more effort to separately collect a recyclable material. Hage *et al.*[17] concluded that publicity for recycling is important to keep morale high for separated waste collection, but this is becoming less important as it is easier for households to separately collect waste for recycling. These findings are in line with research on the convenience and availability of collection systems.[12] If convenience and availability are sufficient, more materials are recycled by weight and by types of recyclable waste. The research of Best and Kneipp[7] shows a similar conclusion: collection systems (opposed to bring-systems) had a positive impact on waste separation behavior; the attitude of consumers however had a small impact. No relationship was found between the type, size and color of the container and the volume of separately collected material.[24]

13.3.3.5 *Collection systems for mobile phones*

In the United Kingdom (UK) researchers[31] made an overview of the collection systems for cell phones. In the UK, organizations that collect cell phones can be divided into five groups:

1. Shops: including online shops
2. Charitable organizations and NGOs (Non Governmental Organizations)
3. Producers of cell phones
4. Network enterprises
5. "RRR companies": Reuse, Recycling and Repair. These organizations are solely concerned with collecting cell phones, with the intent of reselling repaired cell phones or recycling the cell phones. These RRR companies often give money for a collected cell phone.

Cell phones were collected through mail or courier or collected in stores. Collection via free post was most common. The second most common method was by pick-up by a courier, which is the most common practice of RRR companies. The researchers counted 102 different collection programs. The researchers recommended that collecting organizations focus on collection points in places where people gather, such as libraries, shopping malls and schools.

13.3.4 *Effectiveness of communication*

In interviews at households in the districts of Kensington and Chelsea in London,[35] it was found that more than 50% of the interviewees had a lack of information about separated waste collection, which was the reason to not separate waste. The largest factor contributing to lack of visits to waste collection points was a lack of information about the existence of these points. The effect of the implementation of Robinson and Read's research, the door-to-door personal interviews with households in combination with local and national campaigns, was that in these neighborhoods the most waste was collected and separated for recycling. In the state of Minnesota in the United States, research[41] showed that public awareness resulted in more separated waste collection. The researchers calculated that one dollar per person per year for communications would raise recycling by 2%. In order to make communication effective, communication methods must be selected that are appreciated by the audience. In a study[28] in Rushcliffe in the UK, 1,000 people were asked for their preference for the best means of communication. The panel indicated a preference for:

- brochures (79%)
- newspapers (34%) and
- personal letters (33%)

In a survey in the town of Rushcliffe, 75% of the respondents indicated that marketing and communications had influenced them to recycle more, and 70% indicated that newsletters were the most effective way of communication about separate waste collection.

In a study at Michigan State University in the United States,[21] students and staff were surveyed about separated waste collection. The survey showed that the respondents had no need for communication about the benefits of separated waste collection. They wanted to know what and how to separate waste and where the collections points were. The researchers concluded that in publicity on separated waste collection the approach should be differentiated to different types of audiences. Students had a need for the promotion of separated waste collection, and the staff had a preference for personal contact on this issue. The researchers recommended placing collection boxes at convenient places in the university, in order to communicate with the public on where and how material is collected for recycling.

13.4 Stimulating Companies

13.4.1 *Introduction*

Companies try to reduce the cost of waste and to be resource efficient for economic reasons. Waste minimization is also an important topic from the QHSE policy (Quality, Health, Safety, Environment) of a company. An example of classic ways a company can reduce waste and recycle is found in a study on the waste reduction in the hospitality industry[10]:

- Minimize waste by
  - o the reduction of one-way (single-use) products
  - o prioritizing reusable/refillable products.
  - o redesigning, re-specifying and/or customizing products and practices in order to use less material that will finally end up as waste.
- Optimize and/or demand longer service life of products.

- Recycle waste and steer waste towards a useful application, for example exchanging useful materials with another company that uses it as a raw material.
- Organize the reuse/recycling loop

Franchetti[13] found that implementing ISO14001 in companies operating in the United States reduced the amount of solid waste. The study showed that in addition to the ISO14001 certification, the total number of employees and the cost of disposing of the residual waste were significant for the reduction of residual waste.

Although companies try to reduce the cost of waste by waste minimization, they do find barriers. Owners of SMEs in the UK were concerned about the environment, but this has not translated into concrete measures.[34] A majority (59%) of the business owners who did not separate waste pointed out that this was due to external factors, while 26% indicated that internal factors were the reason. External factors are not being able to give away the materials or to recycle, that the municipality does not recycle all materials, lack of facilities, the advice on collection containers, government support, the containers and/or space for containers or notification of municipal recycling, the problem of locating a suitable collector and unreliable storage. Internal barriers are cost, lack of knowledge, feasibility, the sorting material, space, time, or the willingness of staff to comply.

The researchers find that by implementing environmental management, SMEs will obtain lasting economic benefits. The researchers therefore recommended setting up training programs for entrepreneurs on environmental management including waste minimization and separation for recycling.

13.4.2 *Training of companies*

In order to train companies in waste minimization and separated waste collection for recycling, different types of trainings were developed.

In the United Kingdom, companies were helped in the prevention and recycling of waste by the Envirowise Program.[15] This program provided practical advice based on existing practices in the industry. Envirowise also had a free help line for businesses. In addition to providing information and assistance, a considerable amount of effort was put into marketing, to help the companies overcome barriers that stopped them taking action. The program was started in 1994 and by 2000 125 million GBP was saved per year by reducing the use of raw materials and by reducing waste.

In a large waste minimization project in the UK focused on agricultural companies, various means were used to transfer knowledge.[3] Of the farms that signed up, 52% were invited through mailings and newsletters, and 26% came in contact with the program by word of mouth. Knowledge was spread by:

- training meetings
- hotline
- auditing quantities of energy and waste, and how to comply with regulations. Part of the auditing was an analysis of the utility company's bills.

The researchers found a strong relationship between the level of a company's commitment and costs saved. Of companies with a strong commitment, on average GBP 945 was saved compared to the average saving of GBP 520.

In the UK, research was conducted on the transferring of knowledge via "business clubs" in the food and beverage sector.[19] The "business clubs" were specially developed for waste minimization. In the "business clubs", training and workshops were given, followed by interactive "reporting back" sessions of the "project champions" in their business. The experience was that in the meetings the following took place:

- cross-fertilization of ideas
- stimulation of innovation, motivation and knowledge and
- encouragement of management strategies.

The disadvantages of this approach to companies were that

- many companies were in the same supply chain, resulting in no exchange of sensitive/confidential information
- the "project champions" knew more of the subject and therefore felt that they "gave more than received", while the other participants in meetings were annoyed about the compliments the project champions got.

WMCs (Waste Minimization Clubs) were set up in the years 2000 to 2005. They were publicly funded and had been a success in several areas in the UK. The WMCs' approach was appreciated by many companies[32] because the companies

- felt encouraged by the progress of other companies
- felt obliged to work towards goals, because of the WMC
- felt reassured that other companies also had the same problems

- got experience in different methods
- had a sense of community

Although WMCs were appreciated by companies, there were also a few issues. Competing companies were not always willing to share information. Meetings were often poorly planned and did not give enough new information and training. In the beginning the costs of membership were found to be high, which caused doubts about participation. Reasons that companies did join the WMCs were, next to cost reduction company image, environmental concerns, sustainable development and the pressure from legislation, from the customers in the supply chain, or from the media.

WMCs resulted in 10% waste reduction in businesses and one or more jobs created for every GBP £2000 invested after 24 months.[9] In one program, the ratio of cost reduction to the cost of joining the WMC was 10 to 1, after 24 months. In another program this was the case for 70% of the companies.

In West Sussex in the United Kingdom 308 (mainly) SMEs participated in WMCs resulting in an annual cost reduction of GBP £214,000 (in 2003) and an annual saving of 1437 tons of residual waste.[2] The companies were helped through workshops, newsletters, audits, subsidies, and a telephone helpline. The companies were approached to join a WMC by:

- presentations at business meetings and trade associations
- presentations with direct mailing followed by a phone call
- direct mailing followed by a phone call
- flyers in the newsletters of local government
- personal contact with references to local government, environmental agencies or environmental programs.
- press releases on radio, in local and national newspapers, and in local government publications.

13.4.3 *Taking products back*

In order to be less dependent on critical raw materials, companies can take products back for:

- re-use
- repair for reuse
- cleaning or "upgrading" with new parts (refurbish)
- re-assembly (re-manufacture)

- retrieving its parts for the manufacturing of new products
- recycling. In many cases the company collects products themselves, so that it can maintain control over its resources.

An example of taking products back are industrial nickel-cadmium and lithium batteries.[8] The company taking back these batteries had financial and strategic reasons, viz:

- The company acquired raw materials themselves for their production of industrial batteries.
- The expertise of recyclers in recovering metals from the batteries and other products already existed and had been proven and monitored in Europe.
- A part of the metal for the production of new batteries had already been purchased from the recycling industry.
- The company experienced price fluctuations of raw materials and therefore wanted more control over their raw materials acquisition.

There were also marketing reasons:

- In the industrial sector, any new restrictions on a product or service is a new reason for further differentiation of the identity of the product. The product distinguishes itself from other products on the market.
- The purchasers of batteries also demanded a responsible processing of discarded products. The company added a new service to their portfolio: recycling of batteries.

Matsumoto[27] listed conditions to be taken into account for the recovery of old products:

- Ownership of the product
- Ability to collect the old products
- Cost advantages of recycling, remanufacturing etc.
- Preferences of consumers
- No conflicts with the market of new products
- Non-economic motives (responsibility towards the environment, strategic control over raw materials)
- Organizational structure of the company, fitting operations on remanufacturing, recycling etc.
- Legislation (e.g. WEEE directive)

Matsumoto used these conditions for setting up a computer model which simulates the market for reuse. The model also includes modeling decision-making of major actors in the recycling market.

To take products back a reverse logistics system has to be set up. In this field, mathematical models and computer models are developed to assess the feasibility and cost implications of take-back operations. For the recycling of WEEE, a model was developed in the United States based on the modeled behavior of actors that make decisions, such as recyclers, producers and consumers. The model can calculate the material flows in a complex recycling network.[30] In another study, a mathematical algorithm has been developed in order to set up a closed cycle in the supply chain.[20] The researchers concluded in the case of battery recycling that by integrating forward and reverse logistics one-third of costs were saved. Another study, using a reverse logistics network model,[29] showed that in the whole operation the remanufacturing process had the largest cost and not the reverse logistics. Mathematical models were used to optimize a collection structure for WEEE in Portugal.[16] The results gave the collector support for strategic expansion with more collection points in the vicinity of the largest sources of WEEE in Portugal.

Conclusion

Waste Electrical and Electronic Equipment (WEEE) is collected for economical, environmental, and public health and safety reasons. This chapter describes how to organize and stimulate the collection of WEEE from consumers and professional organizations (e.g. businesses). These lessons learned could also be applied to other products containing CRMs, such as industrial alloys.

References

1. Abbott, A., Nandeibam, S., and O'Shea, L. (2011). Explaining the variation in household recycling rates across the UK. *Ecological Economics*, 70(11), 2214–2223.
2. Ackroyd, J., Coulter, B., Phillips, P.S., and Read, A.D. (2003). Business excellence through resource efficiency (betre): An evaluation of the UKs highest recruiting, facilitated self-help waste minimisation project. *Resources, Conservation and Recycling*, 38(4), 271–299.
3. Ackroyd, J., Jespersen, S., Doyle, A., and Phillips, P.S. (2008). A critical appraisal of the UK's largest rural waste minimisation project: Business excellence through resource efficiency (betre) rural in East Sussex, England. *Resources, Conservation and Recycling*, 52(6), 896–908.

4. Ajzen, I. (1991). The theory of planned behaviour. *Organizational Behavior and Human Decision Processes*, 50(2), 179–211.
5. Andersson, M. and von Borgstede, C. (2010). Differentiation of determinants of low-cost and high-cost recycling. *Journal of Environmental Psychology*, 30(4), 402–408.
6. Bernstad, A., la Cour Jansen, J., and Aspegren, A. (2013). Door-stepping as a strategy for improved food waste recycling behavior — evaluation of a full-scale experiment. *Resources, Conservation and Recycling*, 73, 94–103.
7. Best, H. and Kneip, T. (2011). The impact of attitudes and behavioral costs on environmental behavior: A natural experiment on household waste recycling. *Social Science Research*, 40(3), 917–930.
8. Broussely, M. and Pistoia, G. (2007). *Industrial applications of batteries, from cars to aerospace and energy storage*. Oxford, UK: Elsevier.
9. Clarkson, P.A., Adams, J.C., and Phillips, P.S., Third generation waste minimisation clubs: a case study of low cost clubs from Northamptonshire, UK. *Resources, Conservation and Recycling* 36(2), 107–134.
10. Cummings, L.E. (1992). Hospitality solid waste minimization: a global frame. *International Journal of Hospitality Management*, 11(3), 255–267.
11. Darby, L. and Obara, L. (2005). Household recycling behaviour and attitudes towards the disposal of small electrical and electronic equipment. *Resources, Conservation and Recycling*, 44(1), 17–35.
12. Domina, T. and Koch, K. (2002). Convenience and frequency of recycling: implications for including textiles in curbside recycling programs. *Environment and Behavior*, 34(2), 216–238.
13. Franchetti, M. (2011). ISO 14001 and solid waste generation rates in US manufacturing organizations: an analysis of relationship. *Journal of Cleaner Production*, 19(9–10), 1104–1109.
14. Gellynck, X., Jacobsen, R., and Verhelst, P. (2011). Identifying the key factors in increasing recycling and reducing residual household waste: A case study of the Flemish region of Belgium. *Journal of Environmental Management*, 92(10), 2683–2690.
15. Gibson, M. (2001). The work of Envirowise in driving forward UK industrial waste reduction. *Resources, Conservation and Recycling*, 32(3–4), 191–202.
16. Gomes, M.I., Barbosa-Povoa, A.P., and Novais, A.Q. (2011). Modelling a recovery network for WEEE: A case study in Portugal. *Waste Management*, 31(7), 1645–1660.
17. Hage, O., Söderholm, P., and Berglund, C. (2009). Norms and economic motivation in household recycling: Empirical evidence from Sweden. *Resources, Conservation and Recycling*, 53(3), 155–165.
18. Hansmann, R., Bernasconi, P., Timo Smieszek, T., Loukopoulos, P., and Scholz, R.W. (2006). Justifications and self-organization as determinants of recycling behavior: The case of used batteries. *Resources, Conservation and Recycling*, 47(2), 133–159.
19. Hyde, K., Miller, L., Smith, A., and Tolliday, J. (2003). Minimising waste in the food and drink sector: using the business club approach to facilitate training and organisational development. *Journal of Environmental Management*, 67(4), 327–338.

20. Kannan, G., Sasikumar, P., and Devika, K. (2010). A genetic algorithm approach for solving a closed loop supply chain model: A case of battery recycling. *Applied Mathematical Modelling*, 34(3), 655–670.

21. Kaplowitz, M.D., Yeboah, F.K., Thorp, L., and Wilson, A.M. (2009). Garnering input for recycling communication strategies at a Big Ten University. *Resources, Conservation and Recycling*, 53(11), 612–623.

22. Klöckner, C.A. and Oppedal, I.O. (2011). General vs. domain specific recycling behaviour — applying a multilevel comprehensive action determination model to recycling in Norwegian student homes. *Resources, Conservation and Recycling*, 55(4), 463–471.

23. Kumar, A., Holuszko, M., and Espinosa, D.C.R. (2017). E-waste: An overview on generation, collection, legislation and recycling practices. *Resources, Conservation and Recycling*, 122, 32–42.

24. Lane, G.W.S. and Wagner, T.P. (2013). Examining recycling container attributes and household recycling practices. *Resources, Conservation and Recycling*, 75, 32–40.

25. Lange, F., Brückner, C., Kröger, B., Beller, J., and Eggert, F. (2014). Wasting ways: Perceived distance to the recycling facilities predicts pro-environmental behavior. *Resources, Conservation and Recycling*, 92, 246–254.

26. Martin, M., Williams, I.D., and Clark, M. (2006). Social, cultural and structural influences on household waste recycling: A case study. *Resources, Conservation and Recycling*, 48(4), 357–395.

27. Matsumoto, M. (2010). Development of a simulation model for reuse businesses and case studies in Japan. *Journal of Cleaner Production*, 18(13), 1284–1299.

28. Mee, N., Clewes, D., Phillips, P.S., and Read, A.D. (2004). Effective implementation of a marketing communications strategy for kerbside recycling: A case study from Rushcliffe, UK. *Resources, Conservation and Recycling*, 42(1), 1–26.

29. Mutha, A. and Pokharel, S. (2009). Strategic network design for reverse logistics and remanufacturing using new and old product modules. *Computers and Industrial Engineering*, 56(1), 334–346.

30. Nagurney, A. and Toyasaki, F. (2005). Reverse supply chain management and electronic waste recycling: a multitiered network equilibrium framework for e-cycling. *Transportation Research Part E*, 41(1), 1–28.

31. Ongondo, F.O. and Williams I.D. (2011). Mobile phone collection, reuse and recycling in the UK. *Waste Management*, 31(6), 1307–1315.

32. Phillips, P.S., Pratt, R.M., and Pike, K. (2001). An analysis of UK waste minimization clubs: key requirements for future cost effective developments. *Waste Management*, 21(4), 389–404.

33. Prestin, A. and Pearce, K.E. (2010). We care a lot: Formative research for a social marketing campaign to promote school-based recycling. *Resources, Conservation and Recycling*, 54(11), 1017–1026.

34. Redmond, J., Walker, E., and Wang, C. (2008). Issues for small businesses with waste management. *Journal of Environmental Management*, 88(2), 275–285.

35. Robinson, G.M. and Read, A.D. (2005). Recycling behaviour in a London Borough: Results from large-scale household surveys. *Resources, Conservation and Recycling*, 45(1), 70–83.
36. Salhofer, S., Steuer, B., Ramusch, R., and Beigl, P. (2016). WEEE management in Europe and China — A comparison. *Waste Management*, 57, 27–35.
37. Saphoresa, J.D.M., Ogunseitan, O.A., and Shapirod, A.A. (2012). Willingness to engage in a pro-environmental behavior: An analysis of e-waste recycling based on a national survey of U.S. households. *Resources, Conservation and Recycling*, 60, 49–63.
38. Seacat, J.D. and Northrup, D. (2010). An information–motivation–behavioral skills assessment of curbside recycling behavior. *Journal of Environmental Psychology*, 30(4), 393–401.
39. Shaw, P.J. (2008). Nearest neighbour effects in kerbside household waste recycling. *Resources, Conservation and Recycling*, 52(5), 775–784.
40. Shaw, P.J. and Maynard, S.J. (2008). The potential of financial incentives to enhance householders' kerbside recycling behavior. *Waste Management*, 28(10), 1732–1741.
41. Sidique, S.F., Lupi, F., and Joshi, S.V. (2010). The effects of behavior and attitudes on drop-off recycling activities. *Resources, Conservation and Recycling*, 54(3), 163–170.
42. Timlett, R.E. and Williams, I.D. (2008). Public participation and recycling performance in England: A comparison of tools for behaviour change. *Resources, Conservation and Recycling*, 52(4), 622–634.
43. Tonglet, M., Phillips, P.S., and Bates, M.P. (2004). Determining the drivers for householder pro-environmental behaviour: waste minimisation compared to recycling. *Resources, Conservation and Recycling*, 4(1), 27–48.
44. Tonglet, M., Phillips, P.S., and Read, A.D. (2004). Using the Theory of Planned Behaviour to investigate the determinants of recycling behaviour: a case study from Brixworth, UK. *Resources, Conservation and Recycling*, 41(3), 191–214.
45. Wagner, T.P. (2011). Compact fluorescent lights and the impact of convenience and knowledge on household recycling rates. *Waste Management*, 31(6), 1300–1306.
46. Williams, I.D. and Kelly, J. (2003). Green waste collection and the public's recycling behaviour in the Borough of Wyre, England. *Resources, Conservation and Recycling*, 38(2), 139–159.
47. Yoshida, A., Terazono, A., Ballesteros, F.C., Nguyen, D.Q., Sukandar, S., Kojima, M., and Sakata, S. (2016). E-waste recycling processes in Indonesia, the Philippines, and Vietnam: a case study of cathode ray tube TVs and monitors. *Resources, Conservation and Recycling*, 106, 48–58.

Chapter 14

Challenges in Advanced Solid Waste Separation

Maarten C.M. Bakker

Delft University of Technology,
Stevinweg 1, 2628CN Delft, The Netherlands
m.c.m.bakker@tudelft.nl

Efficient mechanical and sensor-based separation of mixed solid waste into valuable secondary materials is a critical step in recycling and in the preservation of primary resources. To position the associated research field, we discuss backgrounds and links to societal and political developments within and outside the classical waste chain. An encompassing theory detailing separation processes is briefly introduced and explained using appealing examples. The roles of science and industry in the development of innovative technologies is clarified, finishing with examples of physical principles that are behind many contemporary separation technologies. Throughout, scientific challenges and technological advances are identified.

14.1 Introduction

Solid waste separation was traditionally the domain of a network of small businesses that relied mainly on manual labor and a means of transport, but this has developed into a modern waste management industry. The main drivers behind the changes were the increasing complexity and volumes of the waste, environmental legislation, and the sector's ambition to recycle more and better. Due to this impressive growth and matching societal impact, the academic world has recognized solid waste as an important field of science, technology and education. A transformation of our society into a circular economy model will widen the research scope. For example, the integration of product design, manufacturing and efficient recycling processes will help to close material loops and minimize material losses. Complementarily, the traditional business models of the recycling chain may have to be adapted or reinvented, as specialized recycling companies need to collaborate more closely to create a viable supply chain of **secondary raw materials** (SRM) as a realistic alternative to virgin raw materials.

Separation is a key step in the recycling chain where mixed solid waste is converted back to reusable materials. Most of the solid waste is generated by sources such as building and demolition, industry, municipalities, agriculture, and automobiles. These waste streams may be subdivided into either source-specific or materials-specific streams,[1] which then reveal in higher detail the origins and complex composition of the waste. As a principle, a separation technique aims to concentrate a target group of objects (i.e. pieces of waste) from the mixed waste, for which it utilizes their shared unique **attributes**. An attribute may be any physical or end-of-life (EOL) related product property of a piece of waste that can be utilized by a solids separation technique. In contrast, phase changing separation, as investigated in the chemical and metallurgical fields, requires the objects to be brought into a state of suspension or liquid melt before separation can be effected. The solids and phase changing separation methods may be complementary for end-of-life products with intrinsically mixed material compositions, such as waste from electrical and electronic equipment (WEEE), or alloys in metal scrap. In such cases solids separation can enhance the contents of the targeted materials or elements to a level where phase-changing separation can operate efficiently and attain the purity that is required by the market. Herein the focus is on solids separation, because the physics of the phase-changing techniques is quite different. This work aims to provide a view onto some of the major challenges in the field of solid waste separation. To this end, we give a brief overview of the field and typical research topics, propose a few possible scenarios in which the research connects to innovations and societal changes taking place in and outside the traditional waste chain, discuss in some detail the relation between waste composition and separation performance, discuss the mechanical and sensor based techniques, and show how seemingly simple physical principles can be utilized in effective separation techniques.

14.1.1 *Roots of separation science and technology*

In the 1970s public awareness and political mind-set converged on the idea that the protection of our environment was just as urgent as public health,[2] which used to be the main motivation behind waste management. This resulted in the national and EU legislation[3] that is aimed at protecting and preserving the air, water, land, and all biological life forms. The need for even greater protection boosted the development of technology and logistics of what is today a recognized waste management industry. In the EU, this industry drives the two main waste management options of recycling and waste-to-energy, as well as the least favored option of landfill. Especially in northern EU countries landfill is mainly reserved for hazardous types of wastes. The recycling option has gained preference due to increasing public awareness of the impact of materials production and waste generation on our society and living environment, and due to the political notion that many materials have a strategic value for the economy. Management options can also complement each other to provide a better waste solution. For example, non-recyclable residues from solid waste separation with high calorific contents may find their way to an incineration plant, while bottom ash residue from a waste incinerator contains a variety of metals that can be efficiently recovered using metal separators.

The state of the art in separation science and technology evolved naturally from fields like agriculture, chemistry and mining. But soon after waste management had evolved it was recognized that solid waste forms a class of materials on its own. After all, it consists of a wide variety of natural and man-made materials and comes in all shapes and sizes. To cope with such high complexity, the waste branch had to adapt the adopted technologies, thus sparking the evolution in separation technology that continuous to this day. In another response to the ever-increasing complexity of our waste, creative minds developed novel technologies based on previously unexplored physical principles that are unique to waste separation. For example, eddy current separators[4] concentrate non-ferrous metals and near-infrared sensor sorters[5] concentrate plastics according to material type.

14.1.2 *The circular economy*

Products in the linear economy were as a rule used only once and then discarded as waste in landfills and incinerators, while the concept of a circular economy does away with the notion of waste altogether.[6] Instead, the lifespan of an EOL product should be continued indefinitely, preferably on the basis of its former functional value as in product reuse or refurbishment. If that is not an option, then on the basis of the intrinsic materials value as in recovery as SRM, which are the products of solids separation. To realize a circular economy, the present-day waste chains must be turned into united supply chains. This requires two major steps: the development of SRM-tailored quality standards and ensuring supply certainty. But even when these two hurdles are overcome and the circle is complete, the lifespan of SRM remains finite because no material or method of separation is perfect. Materials are subject to natural degradation, pollution, and wear, and to some degree will be mixed up with other wastes, causing them to finally end up as useless residue. This implies that the number of life cycles a material can sustain is finite and, moreover, that definitive solutions to this challenge must come from outside the traditional waste chain. We discuss four scenarios with a possible impact on solid waste separation. Each scenario could apply to a circular model that can support economic growth, innovations in product design and technology, and new business models.

Scenario 1: The producing industries may invest in reducing product complexity by making products easier to separate into pure materials by using fewer material combinations and by avoiding miniaturization involving material combinations. This improves the effectiveness of separation, increases SRM recovery rates and minimizes dilution of minor materials. In addition, sustainable material resources are needed to at least compensate for the loss of non-sustainable materials in the EOL phase.

Scenario 2: This scenario is based on the concept of leasing. The consumer pays for a product like a service, but the product remains the property of the producer. When buying a new product the consumer turns in the old one. The advantage is that the producer becomes the problem owner and the party responsible for the collection of retail-returns. This reduces the amount and variability of the wastes cycling through the public waste chain, provided the responsible producer also organizes the reuse, refurbishing, or recycling of the EOL.

Scenario 3: In a deposit scheme such as for PET bottles or glass beer bottles in the Netherlands the EOL are returned to the producer through retail outlets. In principle this is a prime example of circular reuse. However, the variety in consumer products is practically infinite so this scheme cannot possibly be extended to all EOL. But in a realizable future a product could be provided with a unique tag such as a bar code that reveals for example the manufacturer and the materials composition. Such a tag could also serve for price and storage information as is already common practice in retail. The main issue is that fully automated sensors should be able to read this tag in the EOL phase, for example at the recycling plant. Thanks to the tag, valuable dedicated materials can be returned to the original producer in a way similar to the deposit scheme. If the materials hold no special value for the original manufacturer, the tag reveals materials information that can facilitate efficient recycling by another route.

Scenario 4: As we have entered the information age, it is expected that Information and Communication Technology (ICT) will also find its way into the secondary supply chains of the circular economy. The value added by information and data in the 'upper world' of production, consumption, and services offers a glimpse of what may lie in store once comparable ICT infrastructure is in place in the secondary supply chains with what used to be waste. The chains can provide information and quantitative data through sensor systems on the origins, volumes, and composition of the waste, the efficiency of collection and separation processes, and on the composition, quality, and available volumes of the produced SRM. It is expected that this information can generate a trading and commercial value independent of what is generated at the whimsical secondary materials markets.

14.1.3 *Sensor sorting and quality inspection*

During the last thirty years, sensor technology in waste separation was boosted by the rapid development of advanced sensors, in particular the types that allow moving objects to be scanned without making physical contact. Though sensors were around for much longer, it wasn't until the technology reached a mature level of performance and price that it became more widely adapted in the waste industry. In an application, the waste objects are deposited on a conveyor on which they should be in rest before reaching the sensor. The sensor scans each object individually after which software interprets the sensor data to decide in real-time on the presence of the targeted unique attributes. If detected, a signal is sent to an actuator that mechanically moves the object to a separate stream. The price of a sensor system tends not to be determined by the physical sensor, but rather by the actuator system and the electronic hardware required to process all the acquired data in real-time. A wide variety of sensors is available to detect all kinds of object attributes, be it geometrical (e.g. length, height), physical (e.g. magnetic or chemical properties), or related to the functional product origins (e.g. color, shape, material build-up).

Actuators also come in a wide variety of types, where air jets and systems for mechanical pushing or hitting the object are favorites. A major research challenge that could mean a real breakthrough in sensor sorting is the development of a new generation of compact, low-energy actuators that can be stacked along the length of a conveyor belt. This would enable the extraction of multiple products with different desirable properties in one processing step. Key in this development would be that all actuators are activated through one and the same sensor unit to limit the costs. In principle, the possible number of stacked actuators need only be limited by the length of the conveyor. This development would result in a considerable cost saving for sorting, because the present generation of sensor sorters can only make two or perhaps three products in one step.

A distinctive advantage of combining different sensors in one system is that it can identify different useful product properties in one step. For example, in the case of sorting waste plastics one could discern the polymer type, color, and presence of polymer additives at the same time to sort a tailor-made product for the end user.

Since a sensor scans each individual waste object, the separation of large objects is more efficient than small objects, because throughput capacity is typically measured as a mass rate. When feeding a sensor sorter the objects should form either a monolayer on a conveyor (2D sensors) or be put in a moving single file of interspaced objects (fixed point sensor). It is a challenge to achieve the highest feeding rate that the sensor can sustain without making too many errors, which requires the densest possible homogeneous distribution of objects without overlapping or touching.

Sensors can also be employed without an actuator to inspect the composition of a batch of SRM, which operation is called **quality inspection**. In this case, the feed rate of the sensor may be higher than its maximum capacity if the objective is to sample the waste stream. By scanning a representative number of samples per time interval, the sampling statistics return an average composition per time interval with associated uncertainty. The uncertainty may be reduced by increasing the time interval or by increasing the sampling rate, whichever is cheaper or more practical.

14.1.4 *Waste collection systems*

Waste was traditionally discarded through an efficient scheme of compacting collection trucks and disposed of in landfills or incineration plants. This strengthened the public belief that waste has no value and can easily be made to 'disappear'. This applied to all waste streams with perhaps the exception of the most hazardous for which more strict regulations were in place. Here, we focus on household waste because we all experience this type of waste first hand. As recycling gained momentum as the preferred waste management option, a rethinking was required of the traditional mixed waste collection scheme. Separate collection schemes were introduced for all kinds of recyclables to minimize the contamination, material mixing and volume per type of waste stream. In combination with automated separation technologies these schemes improve the material recovery rates (i.e. the percentage of materials successfully separated) and lower the costs for the recycling plant. In turn, better separation improves the achievable recycling rates (i.e. the percentage of collected materials actually reused in new applications). Modern waste collection is no longer just a logistical operation as it should also motivate people to separate their waste in the easiest and best way possible. In turn, the main actors in the waste chain have the obligation to be transparent about the achieved recycling rates and inform people about their personal waste footprint. It is here that **synergy** can be created by combining efficient automated separation technology at the recycling plant with people's personal involvement in separate waste collection and applying a system of feedback to help people to reduce their waste footprint.

A separate collection scheme is aimed at isolating and safeguarding the valuable recyclables by using a system of separate bins or otherwise distinguishable collection means such as colored bags. This ensures they can be processed effectively in the recycling plant, provided people throw the right waste into the right bin. The waste pickup frequency will vary depending on the volume and nature of the recyclable. For example, smelly biogenic kitchen waste has to be picked up more often for hygienic reasons. Other recyclables have a quite low bulk density and appear in high volumes, which make their separate collection costly as more truck rides are required to collect the same total mass. Examples are packaging plastics and the more season-dependent garden waste.

For separate collection different options are available. People may separate and collect inside their home, or nearby in underground containers, or in a farther removed recycling street in case of more coarse types of waste. An example of a more recent approach of home-separation is the use of colored plastic bags. All these bags may be deposited in the same collection bin or underground container, but the recyclables remain isolated. Moreover, only one pickup moment is required for all the bags, which lowers the costs of collection. The collected bags are automatically separated at the recycling plant using sensor technologies. To limit the number of color bags needed in the home, different recyclables could be combined in one bag provided they can be efficiently separated at the plant using mechanical separators. For example, plastics, metals and drinking cardboards (called PMD fraction) form a large part of the daily household waste volume and are collected together in several EU countries. However, it is not yet clear if the recovery and quality of the plastics from PMD are comparable to what is possible using separate plastics collection.

As an extension to the colored bags system, it is technically possible to tag bags before they are distributed to the local population with an information carrier that can be read by automated sensor systems at the recycling plant. For example, a bar code may reveal the area and street of the pickup location. Other sensors can perform a quality inspection of bag contents and build a database with EOL information. This can be coupled to the area tag to reveal the efficiency of home separation and waste volume and consumption pattern per waste pickup location. This information can be used to steer the recycling chain, make local improvements to waste collection and inform people locally of their average waste foot print.

The option to separate only in bins outside the house can be a workable alternative in low-rise and open urban areas with sufficient outdoor spaces. But the success of any separate collection system depends on people's active involvement, and that depends on quite a few factors such as walking distance to the containers,[7] available space for temporary waste storage, effort and convenience, aesthetics, and waste pickup frequency in a trade-off with collection costs. Many different collection systems are employed and tried in the Netherlands. Some of these are combined with positive or negative financial incentives, such as differentiated waste taxes (called Diftar) and penalties for putting out waste at not approved locations or at the wrong time. There are a few distinct successes with separate collection and recycling in the Netherlands, for example packaging glass, paper and cardboard waste and metal scrap. However, there still is a major problem with packaging plastics as the costs of separate collection prove too high in combination with disappointing recycling rates. This indicates that, even many years after its first introduction, separate collection is still in a transition phase and warrants a higher level of academic involvement to understand the bottlenecks and to develop efficient solutions. Especially the connection between technology, social-cultural influences and human behavior is a crucial cross-road of academic studies that in collaboration can help the separate collection concept to become a successful system of the circular economy.[8]

Instead of asking people to sort their waste, one may also use automated technology to sort mixed waste at the recycling plant. This concept continues the mixed waste collection system, though shredding and high compaction in the waste truck can no longer be allowed for obvious reasons. Advantages of this system are the low cost of collection and it is convenient for big cities with old and crowded centers that cannot easily accommodate more waste bins and underground containers. The technology needed for mixed waste separation can be classical and readily available, which makes this system also less of a risk where it concerns return on investment. But because it also does not require any active collaboration from the people, it probably does not change people's perception of waste very much or their attitude towards reducing their waste footprint. The technical disadvantages are more cross-contamination and material mixing, which lowers the potential recycling rates in comparison to a well-implemented separate collection system. As such, mixed waste separation may be regarded as an evolutionary technological solution for the big cities and a temporary step towards advanced separate collection.

Lastly, manual separation cannot be ignored as a factor in modern waste separation. This human-based activity is still a necessary step in the separation of the largest solid waste streams, and not only in low-wage countries. Manual labor is, for example, deployed in the disassembly of electrical and electronic appliances which consist of a few large homogenous parts (e.g. casings) and smaller complex parts (e.g. wires, electronic circuit boards and power supplies, electric motors). Handpicking is a longstanding contribution of manual separation where one or more persons pick up and remove specified objects from a conveyor belt. The minimum object size a person can routinely pick up is about 25 mm, but bigger objects make for more efficient handpicking as they can be picked at a rate of up to 20–30 per minute. Manual labor in separation lines may seem to contradict the advancement of technological waste management, but a plausible motivation is the uncertainty about how long complex EOL will keep on evolving and diversifying. After all, electronics such as smartphones and smart household devices with sensors keep on changing and diversifying every year. As a consequence, recyclers postpone substantial investments in innovative technologies due to uncertainty if they will be able to cope with the continuous increase of EOL complexity. Instead, they rather rely on manual labor to break down complex EOL to waste streams that can be dealt with by proven technologies.

14.2 Theory of Solid Waste Separation

The main condition for separation of a group of targeted materials (objects) from a mixed waste stream is that the group must share at least one attribute that sets them apart from all the other waste objects. Based on this attribute, a separator could ideally isolate all the targeted materials. Unfortunately, separators can make two possible errors. A targeted

object that is not separated into the main product is called a **false negative**, and a non-targeted object that is unintentionally separated into the
main product is called a **false positive**. Though this terminology is mostly
used in connection to sensor-based sorting, the concepts can be consistently
extended to mechanical separation techniques. The cause of these errors is
either: the choice of attribute utilized for separation turns out to be not
that unique and is also shared by non-targeted objects (false positives); or
it is not shared by all the targeted objects (false negatives); or the separation process is not selective enough, which may lead to either error. We
discuss these a bit further.

The unique attribute and corresponding separator are chosen by the
plant operator on the basis of the value of materials and by sampling the
contents of the waste batch. However, for any waste batch there is a certain
probability that non-targeted materials share that attribute. In addition,
separation technology is effective only within certain physical ranges of
the attribute by which some targeted objects may be missed. For example
the eddy current separator (ECS). Its effectiveness in accelerating a non-
ferrous metal object through electromagnetically induced currents decreases
rapidly with size and electrical conductivity. On the other hand, it can also
not separate too large and heavy metal plates. In general, the probability
of an error may be reduced by utilizing more distinguishing attributes in
the separation. Unfortunately, mechanical separators are usually specialized
and aimed at one particular attribute. Therefore, if effective separation
requires more attributes, different separators need to be put in series and
that translates into higher processing costs. In contrast, sensor units can
be easily extended with other sensors and sorting can be done on the basis
of multiple distinguishing attributes in one processing step.

Feeding a continuous stream of solid objects into a separation process invites dynamic object-object and object-machine interactions. For
most part these result in the intended deterministic separation. However,
the dynamics of a large stream of solid particles allows for more compli-
cated dynamics that can negatively influence the result. We name here:
segregation (clustering of objects with similar attributes); **agglomeration** (different objects connecting/adhering and then behaving as one
composed object); incidental random occurrences (objects behaving in an
unpredictable fashion). If the latter random behavior results in a separation
error it is referred to as a **dropout**. Another undesired behavior of solid
objects is **entrainment**, where objects are locked in a stream and force-
carried into the wrong product. For example when a light object is enclosed

by heavy ones and forced to follow them into the heavy product. It is noted that entrainment can also have a positive connotation if the mechanism is utilized to extract or concentrate materials. Segregation can negatively affect the separation when it results in the formation of object clusters inside the machine that restrict the proper flow and intended motions of the targeted objects. Agglomeration negatively affects the separation when a targeted object connects to a non-targeted object. It is important to use the correct feeder that minimizes segregation and frees objects before the separation action takes place. This is usually a shaking feeder that evenly and homogenously distributes the materials into the machine. If agglomeration is a problem the responsible mechanism has to be determined. For example, moist materials may have to be dried first to free the individual particles. On average, the influences of segregation, agglomeration and entrainment on the separated products are fairly repeatable mechanisms for a specific combination of feed materials and separation process. As such, their average contribution to false positives and false negatives can be determined experimentally by processing a prepared test batch.

Dropouts are of a quite different nature as they occur incidentally and unpredictably. As such, their occurrence may be modelled using a fitting probability function, perhaps a Poisson distribution. An experimental test may also give insight into this type of error by examining the standard deviation of the detected false positives and negatives. Mechanical separators are always fed in bulk, which invites many possible interactions between the objects and the machine. The probability of dropouts may therefore be influenced by the feeding method and feed rate. But sensor sorters require a more controlled form of feeding, which limits the possible interactions and reduces the probability of dropouts.

14.2.1 *Attribute space and effective separation*

The performance of a solids separator can be mathematically described using separation system theory. This theory does not describe the physics, but rather the effective behavior of the machine for a specified waste batch and the resulting composition of the products. It can incorporate a continued separation line, binary and multiple separations in cases where a machine creates more than two products, and, though this seldom happens, a line where a part of the processed material is looped back to an earlier point in the separation line. We briefly introduce the basic concepts behind this theory.

First, all individual objects in the waste batch are **documented** according to a list of physical, geometrical and EOL-related attributes that may be relevant to their separation. Each physical or geometrical attribute is quantified by a parameter (e.g. length, density, RGB color values, magnetization). Among all the objects in the waste batch the parameter varies within a finite range of values. An EOL attribute (e.g. material type, color, description of shape) is specified by an identifier. The possibilities for the identifier must be arranged in a fixed order. For example, for a certain waste batch the attribute 'material type' may have three identifiers, ordered as 'plastic', 'metal' and 'wood'.

Next, each object is assigned to a **material class**, where each class forms a different marketable product. The exception is the residue that contains the objects that do not fit in any other material class. In practice, classes are chosen with a view towards the potential market price and the available separation techniques. A class may also be related to only EOL attributes. For example, the class 'plastic bottles' combines identifiers from the attributes 'material type' and 'description of shape'. The main issue for successful separation of all the objects in a material class is that they must share unique attributes that set them apart from all the other classes. Nevertheless, depending on the order of separation, different classes may have overlapping attributes without negative consequences for separation. For example, when separating the classes 'plastic' and 'plastic bottles', one should first separate the plastic bottles as they are more uniquely defined than 'plastic'.

Each listed attribute is assigned an axis in a multi-dimensional space, which we call the **attributes space**. To simplify the interpretation, the parameters are discretized using arbitrary axes (i.e. in the sense of non-linear axes), so as to span the multi-dimensional attribute space with simple shaped elements. This includes axes assigned to EOL attributes which are occupied by ordered identifiers.

To each element in attribute space the number of objects that comply with that specific set of attributes is assigned. Figure 14.1 shows a simplified example for different types of aluminum scraps. The distinguishing attributes are 'size', 'density' and 'conductivity', which span a 3D attribute space. For visualization we showed the projections onto two cross sections with 'size' as common axis.

An ideal separation technique is sensitive (as in effective) only to the unique attributes of the targeted material class. However, in reality a separation technique is also to some extent sensitive to other attributes. The

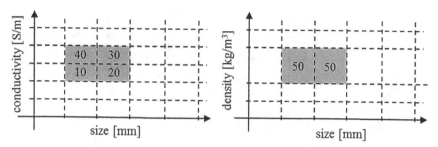

Fig. 14.1. Example of a 3D attribute space. Indicated are projections on two cross sections of this space, harboring 100 different aluminum scraps.

sensitivity spans a subspace in attributes space which we call the **separation subspace**. Note that, as a subspace, it conforms to the discretization of the attributes space. Each element in the separation subspace is assigned a sensitivity on the scale 0–1, where the complement to 1 defines the reduced success to separate an object with those specific attributes. For elements where the separation subspace coincides with the targeted unique attributes, the complement to 1 defines the chance of a false negative. For elements where the separation subspace does not coincide with targeted attributes, the sensitivity itself defines the chance of a false positive. As such, false negatives are due to a lack of sensitivity and false positives arise because the sensitivity extends to parts of attribute space that are occupied by non-targeted attributes from (most likely) non-targeted objects. It is noted that statistical spread in the sensitivity can readily be accounted for by introduction of probability distributions.

Figure 14.2 shows two examples of discretized 2D separation subspaces where the sensitivity is indicated in grey-scale. Figure 14.2(a) represents an eddy current separator (ECS), which is sensitive to non-ferrous metals. The sensitivity for metal scraps decreases with size and electrical conductivity, while too large and heavy scraps can also not be separated. This contributes to the false negatives. Because a high conductivity is mainly associated with metals, the chance of false positives in an ECS is mainly due to agglomerated objects that include a small piece of metal (e.g. a metal screw in a large piece of wood) and dropouts.

Figure 14.2(b) represents a float-sink tank where the float fraction is considered the main product. Materials that are generally separated well are much heavier or much lighter than the cut density and have a reasonable size. Small, heavy particles may still float due to frothing, adherent air bubbles, air inclusions, or agglomeration with light materials, in which case

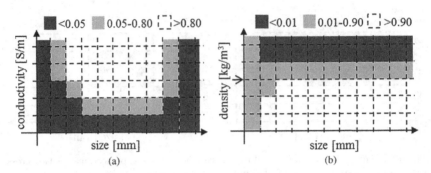

Fig. 14.2.   Examples of 2D separation subspaces. The sensitivity is shown in grey scale. (a) Eddy current separator. (b) Float-Sink tank, where the arrow indicates the cut density.

they contribute to the false positives. Materials with a density a bit higher than the cut-density sink slowly and have a tendency to be dragged by the flowing liquid into the product bin, in which case they also contribute to the false positives. Small particles lighter than the cut density may still sink due to agglomeration with heavy materials, which contribute to the false negatives.

Now that the basic concepts of separation system theory are explained we can give a precise definition of the effective separation technique. One selects the technique for which high sensitivity covers the part of attributes space occupied by the unique attributes of the targeted material class, while outside that part of space the sensitivity decreases rapidly to zero. If this type of separator is not available, an alternative is to first optimize the attributes space of the waste batch to fit the separation subspace of the available technique. This can be done by applying a **precursor process** (pre-process for short). There are two types of pre-processes: one that leaves the existing object attributes intact and one that changes them. The first is typically a separation process that removes the fraction of targeted objects from the batch whose attributes fall outside the separation subspace of the available technique. For example, a sieve can remove all the small objects from the batch in case the sensitivity of the available technique for the attribute 'size' is restricted to large objects. Moreover, sieving to size reduces the material volume that has to be processed by the available machine.

The second type of pre-process may be **comminution** (e.g. breaking, milling, rolling), **liberation** (e.g. shredding, hammering), or **surface**

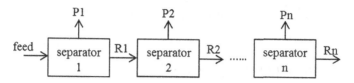

Fig. 14.3. A line of connected binary separators.

treatment (e.g. grinding, acid attack, spallation). Comminution aims to break objects down to prescribed size, liberation aims to break up heterogeneous objects into as big as possible homogenous objects, and surface treatment aims to modify surface properties like roughness, color or coating. These processes can change attributes as well as create new objects and new attributes, and therefore demand a new documentation and classification of the processed waste batch. A major research challenge is to detail the attributes space created by such a process on the basis of a transfer function and the known attributes space of the feed.

A series of different separators can turn a batch of mixed waste into a number of valuable products. Designing a separation line starts with knowledge of the material classes and their discerning attributes. From here on the challenge is to determine the products that can be concentrated successfully using the available techniques.

Figure 14.3 shows a generic line of binary separators that is continued on the non-concentrated fraction, which is here referred to as residue (labels R). The concentrated fraction is a product (labels P), while the other objects end up in a residue. It is noted that it is also possible to continue the line along a concentrated fraction if further separation is required. The separation line invites a bit of thinking about false positives and negatives. For example, if an object is erroneously assigned to the product of the first separator it is a false positive in that product. However, it is removed from the feed down the line and will never end up in the correct product, which we classify as a false negative for the correct product. Vice versa, if a targeted object is erroneously not separated in the first product, it can only cause a false positive down the line, except if it ends up in the final residue. What this example shows is that the definitions of false positive and negative are also consistent for separation lines, but that their cause may lie with another separator. Ergo, the ordering of the separators plays a role and it is efficient to set the most effective separator up front to reduce the chance of false positives down the line. In another strategy, it may also be efficient to put up front the separator that takes out a large material volume. This reduces the feed rate and capacity required of the machines down the line.

14.2.2 *Engineering considerations*

The throughput capacity of an industrial separator is limited in terms of volume and mass rate, which sets demands to how and how much material can be fed. Feeding can be done from a bunker or shaking feeder to bring objects in free fall into a machine or to deposit them on a conveyor belt and cover it with either a monolayer or a thick layer of materials. The feeding method serves to provide continuity to the separation process up to the maximum throughput capacity, and to render the separation effective within the required separation subspace. Machines come with variable settings that serve to facilitate limited variations to the separation subspace to gain a better match with the targeted part of attributes space. On the other hand, during prolonged operation the machine may be subject to wear, tear, and caking due to the continuous feeding of large objects and fine materials. These maintenance-dependent factors may change the separation subspace during operation and negatively affect the performance of the machine. The moisture content of waste can be a major point of concern. Water may adhere to objects, invade porous objects by the capillary effect, or loosely adhere as free water between objects by water bridges. The impacts of moisture are added mass and agglomeration of mostly small objects and fine materials (e.g. mineral grains), which adhere to moisture more effectively due to their relatively larger surface to volume ratio. In these ways, moisture can modify the attributes space spanned by the objects in the waste batch. On a similar note, adhered moisture on machine parts may affect the separation subspace of the machine due to stick or slip or due to caking of a multitude of fine materials.

Full separation of mixed waste involves several pre-processing and separation steps. To arrive at an all-encompassing theory for these connected processes the object attributes must be documented and the sensitivity of each machine must be determined. The practically infinite number of possible material/elemental combinations of complete and shredded EOL objects can only be addressed by the introduction of a system of materials classification, such as was introduced in the preceding section. To define classes one may take a pragmatic approach that is consistent with accepted recycling practices: *a material class forms a saleable SRM product that may be composed of either pure materials or of a specified mix of EOL parts and homogenous materials.* Within this definition, a material class may still require further solids separation and/or phase changing separation before it is pure enough to be acceptable for reuse. A few examples of mixed SRM are 'plastics', 'metal scrap', 'paper and cardboard', and 'printed circuit boards'. The criterion for selection of an object from a mixed waste batch in a material class is the question if the material variety brought in by that type of object and the combined mass of all the similar objects in the batch could negatively affect the quality or salability of the class. This implies that a small amount of 'material pollution' in the class is acceptable as long as one gains the targeted material, but one cannot accept a certain polluting object in the class if there are too many of those. If that is the case, they should be put in a class of their own. On a critical note, this selection criterion allows for dilution and associated loss of foreign elements and materials when their total mass contribution to the class is deemed negligible. However, it is consistent with the state of the art: it is unfortunately not economically feasible to separate with a resolution that recovers all minor materials in mixed solid waste.

14.3 Separation Techniques

The mechanical and sensor-based separation techniques are the enablers of modern solid waste separation. Both types are introduced and discussed in terms of potential and limitations. We also point out the design steps in which scientific research can join forces with industry to transform a promising principle for mechanical separation into a proven technology. Finally, we introduce a few successful principles that formed the basis for many contemporary mechanical separation technologies.

14.3.1 *Advanced data processing in sensor sorting*

The separation subspace of a sensor sorter is mainly determined by the sensor unit, consisting of the physical sensor and a stimulus. The stimulus excites the waste objects (e.g. optimized light source, x-rays, electrons, magnetic field) and the sensor detects their response. The response should be selectively sensitive for the unique attributes of the targeted objects. The raw sensor data is transferred to a real-time processing unit that contains software for data interpretation and decision making. The sensor system is complemented at the front end with an adequate system to present the objects to the sensor unit (e.g. a shaking feeder and conveyor belt) and at the back end with a mechanical actuator that removes the identified objects from the waste stream. The actuator should meet the resolution of the sensor unit and the speed of the data processing unit. The design of the software depends on how one can extract the information from the raw sensor data. for which there are roughly four strategies.

1. The data provide direct evidence of the unique attributes, e.g. if the detected signal exceeds an uncertainty level.
2. The data allow the unique attributes to be quantified. There are two different approaches. The first is calibration, where the detected signal can be compared to a documented reference, such as a lookup table or a formula. This has the advantage that it does not require knowledge of the physics behind the generated signals. The disadvantage is that the reference requires calibration samples to be prepared, analyzed and documented. The second technique is quantitative measurement, which requires some processing to translate raw data to quantitative attribute values. This has the advantage that one can utilize signal and selectivity enhancing measurement strategies such as compensating, modulating, and differential measurement techniques.
3. The data carry indirect quantitative information on the unique attributes due to their influence on a specified physical process, by which the attributes may be detected and quantified through their level of influence.
4. The data carry merely coherent information on the unique attributes due to their influence on several physical processes, but these influences are not well understood. Nonetheless, using statistical techniques the coherent influences of the attributes may be detected and classified with statistical certainty.

Option 1 forms the basis for a cost-efficient technique as it puts minimal demands on data quality and real-time hardware. In option 2 the calibration techniques share similar benefits, but have the major drawback of requiring extensive sampling and analyses of the targeted materials and attributes to adapt the lookup table every time these waste objects change significantly. The quantitative measurement techniques in option 2 set higher demands on data quality for a reliable translation from raw sensor data to a physical interpretation. Option 3 may also be called **model-based measurement**. There, a physical model is constructed that incorporates the influence of the targeted attributes to predict the collected sensor data with required accuracy.[9] By inverting the model equations the attributes can be quantified on the basis of the collected sensor data. In option 3, data quantity and quality play important roles. The main challenges are to suppress numerical sensitivity for small data variations (e.g. sensor noise) and to make sure the reconstructed attribute parameters form a unique solution. Both challenges are inherent to data inversion problems.[10] Option 4 allows for a range of statistical techniques to be applied. The mathematics behind the algorithms is generally aimed at isolating coherent features in the data and projecting them onto a virtual space in which objects with detected shared attributes are clustered.[11] Sorting of the clustered objects is effective only if these shared attributes are 'unique enough' so as to avoid different object clusters from overlapping.[12]

14.3.2 *Statistics of materials feeding*

Statistics are an important tool to help understand the feeding constraints of mechanical and sensor-based separators, because there is no mechanical system that is able to control the motions of each individual object when feeding them in large numbers. Here we focus on the main aspects of mono-layer feeding of a sensor, where the maximum feed rate depends on how densely the conveyor belt can be covered with objects without too many overlaps. A 2D sensor such as an infrared or visual light camera cannot always reliably distinguish overlapping or even touching objects, so some spacing is desired. A typical feed system involves a shaker that provides a uniform feeding along the width of the belt. To assure a uniform feeding in the length direction the belt must move faster than the objects falling from the shaker. The object distribution on the belt follows a probability distribution which we reconstruct using a Monte Carlo simulation. For simplicity, all objects are assumed flat and square with length d, and all objects fall perfectly aligned in a row along the width of the belt. The maximum feed rate for ideal 100% belt coverage for this case is $f_M = LV/E(d)^2$ objects/s, where L is the width of the belt and V the belt speed. The function E(d) is the statistical expected value of the length of the objects plus a required minimum spacing of 2 mm between any two objects. The feed rate is a fraction f of the maximum feed rate f_M. The parameters used in the three simulated cases in Fig. 14.4 are shown in Table 14.1. Case 1 uses only 5 cm objects and cases 2 and 3 use objects in the range 2.5–10 cm with an equal probability for each size. Figure 14.3(b) shows the function $A_O f f_M/(LvN_O)$, which is the percentage of the belt surface that could be perfectly covered with objects without any overlap. N_O is the total number of released objects and A_O is the total surface of the objects on the belt.

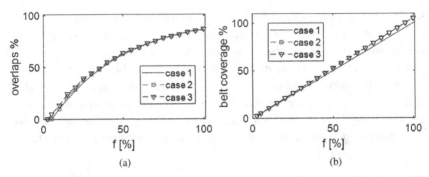

Fig. 14.4.   (a) Percentage of overlapping square objects as a function of fractional feed rate f: all 5 cm (case 1) or size range 2.5–10 cm (cases 2 and 3) (b) Average belt coverage.

Table 14.1:   Parameters in Three Simulations of 2D Monolayer Belt Feeding and Object Overlap.

Case	d [cm]	E(d) [cm]	belt width [m]	belt speed [cm/s]	f_M [objects/s]
1	5	5.2	1	10	37
2	2.5–10	6.45	1	20	48
3	2.5–10	6.45	2	10	48

A uniform random distribution determines where the next object is to be released along the width of the belt, while objects are released one by one with a constant time interval. This release strategy minimizes the probability of overlap, meaning that any other distribution or non-uniform time interval will on average result in more overlaps. The differences between mono and size-range feeding are surprisingly small in Fig. 14.4, but still noticeable at low feed rates. A similar conclusion applies to the two options for increasing the feed rate capacity, which are using a wider belt and a higher belt speed. If it is acceptable that at most 10% of the objects overlap, the feed rate should lie between 8–10% of the maximum in which case 8–10.5% of the belt surface will be covered with objects.

But not all overlaps result in a separation error. Overlaps between only targeted or only non-targeted objects may turn out ok. For mixed overlaps the sensor may generate false positives or false negatives, which affect either the purity of the main product or the recovery of the targeted objects. To better quantify the overlap error the simulation may be extended to keep track of the nature of each object and the overlaps. Combined with knowledge of the composition of the feed, the simulation can then predict the

Table 14.2: Examples of Mechanical Separation Techniques and Attributes.

Resistance techniques	Involved object attributes
friction	surface conditions; mass
drag	size; shape; Reynolds number
buoyancy	mass density; porosity
mesh or aperture	size; shape
impact	elasticity; mass; shape; damping
Stimulus techniques	**Involved object attributes**
fluid flow	mass density; drag
vibrations	mass density; elasticity; size
centrifugal force	mass
magnetic field	permeability; size
electric field	conductivity; permittivity; size
EM-induction	conductivity; magnetization; mass

percentage of false positives and negatives due to overlapping objects. Computer simulations can be extended even further to closely match the higher complexity of object dynamics during feeding. For example, it may account for varying object size and shape, randomness in position and alignment where objects can drop on the conveyor belt, and the probability that an object could bounce from the belt or from another object lying on the belt.

14.3.3 *Designing mechanical separation technology*

For mechanical separation, the targeted objects must be forced into a different trajectory than the rest of the feed on the basis of the discerning attributes. To this end, an attribute-specific force has to be applied that selectively obstructs (**resistance technique**) or accelerates (**stimulus technique**) the targeted objects. In both categories it may be a concentrated force or a small force exerted over a longer distance. A concentrated force tends to lead to a smaller machine that requires less space. Table 14.2 shows examples of the many possible techniques falling in the two categories and the associated main attributes. Different principles may be combined to form a hybrid technique, in which case more than one unique attribute may be utilized to improve the sensitivity of the separation.

Whichever approach is chosen, three phases may be discerned for the design of a mechanical separator. Phases 1 and 2 provide scientifically founded recommendations, leading to an assembly of the effective prototype in phase 3. One typically enters phase 1 on the basis of an inspired idea. Such a creative event can be the birth of novel technology, but it has no bearing on science other than that it sparks curiosity and stimulates the imagination.

Design phase 1: The physical principles are subjected to analytical assessment to establish the potential separation subspace and sensitivity. Directed physical experiments may be part of this phase to assess if all the main principles are carefully taken into account. Next, the mixed waste streams holding relevant material classes with unique attributes corresponding to the separation subspace may be identified, which requires the market knowledge of an industrial partner. Subsequently, the proposed technique can be judged on its potential for creating added value. This phase provides insights into the technical capabilities and the economic potential of the proposed technique.

Design phase 2: This phase involves more statistics to determine more reliably the optimum separation subspace and to improve the sensitivity. Included are the designs for a method for material feeding and a method for guiding the separated materials into the main product or residue. Attention has to be paid to measures that prevent blocking and minimize damage mechanisms such as fouling, wear and tear. First implementations of the different subsystems may be tested, separately or connected, for their performance.

Design phase 3: The subsystems are integrated into a functional prototype, which is performance tested using prepared batches and, finally, real mixed waste. Analysis of the test data may indicate shortcomings and bottlenecks. These require an in-depth investigation to understand the underlying physics in as much detail as is required for an efficient solution.

After the last design phase the prototype may be transferred from the research environment to the industrial partner, where engineering is usually required to meet special demands. For example, variable machine settings to improve operator control, automated sensor control and material flow monitoring capabilities, modifications to reduce or facilitate maintenance, and compliance with regulations for worker safety, noise and dust.

14.3.4 *Technology starts with a principle*

Seemingly simple physical principles can give rise to a diversity of powerful mechanical separation techniques by changing the conditions in some extreme way. Important examples are the **trajectory techniques.** There, objects are directed to follow a path during which they are subjected to a selective force that significantly modifies the path of the targeted objects to where they can be collected separately. For example, a path in air or water where objects are subjected to drag, or a path that leads objects along a rough surface or a surface equipped with attribute-selective obstacles (e.g. riffles). The mathematical analysis of shaped solid objects following a 3D path generally demands a computer to solve the equations. Fortunately, the physical principles are much easier to explain.

Ballistic separation: An object at some height above the floor is given an initial velocity under an upward angle after which it is left to gravity and drag to determine where the object will drop onto the floor. It is readily found that for equal-sized objects the heavy ones tend to travel farther from the launching point than the light ones. This forms a basis for separating light from heavy objects. Similarly, for objects with the same mass the ones with a small cross section in the direction of motion tend to travel farther than those with a large cross section. This forms a basis for separating objects according to shape (e.g. flat, round). A complication is that drag is influenced by both shape and size, which means both attributes play a role in ballistic separation. In a practical setting the choice of fluid (e.g. water or air) influences the amount of drag and therefore the space needed to house a ballistic separator.

Friction separation: An object is given an initial horizontal velocity at the top of an inclined surface, after which it is left to gravity and surface friction to determine where the object will drop into a collector. It is readily experienced that for equal-sized objects the ones with a low level of friction tend to travel farther from the launching point than the ones with a high level of friction. This forms a basis for separating objects on the basis of surface resistance. For example rubbers can be separated from smooth plastics, or round rolling objects from flat ones.

Sink-float separation: If an object is dropped into a water tank its density in relation to water determines if it eventually sinks or floats, as discovered by Archimedes. The light fraction can be collected by letting the water flow in a horizontal direction and spill over a lowered edge of the tank or by pulling a sieve along the water surface. This is the principle behind the sink-float techniques, for which different liquids are available. The throughput capacity depends partly on how fast the objects sink or move back to the liquid surface after being dropped into the tank, which is influenced by object size and drag.

The concept of sink-float may be extended by using a magnetizable liquid, called ferrous liquid, and applying a gradient magnetic field using either a permanent magnet or an electromagnet. In this configuration the magnetic field gradient produces a gradient in the effective density of the liquid. Objects with a targeted density will concentrate at a certain distance from the magnet surface where the effective density of the liquid matches the target density. Materials may be selectively extracted on the basis of density range by making the liquid flow into carefully positioned open slots or by pulling sieves at selected depths through the liquid. This technique is called **magnetic gradient density separation**, and allows the production of multiple products with different density ranges in one processing step.

Terminal velocity separation: If we take a water tank and generate a vertical flow from bottom to top the conditions are set for terminal velocity separation. If we drop an object in the tank it is subjected to gravity, buoyancy and the enlarged drag exerted by the flowing water. An object that sinks will eventually reach a maximum velocity called terminal velocity, which is determined in a balance between gravity, buoyancy and drag. This may be compared to a falling raindrop that also reaches a terminal velocity. The terminal velocity depends on the Reynolds number, because water flows differently around an object depending on its shape and velocity. If the density of the object is too small it will get flushed from the top of the tank into the light product. If the density is high enough it will sink to the bottom of the tank where it is collected into the heavy product. By varying the water flow one can set the optimum cut-point between the heavy and light materials. This principle is the basis for techniques like elutriation, where heavy objects are separated from light ones, and where objects with large cross sections are separated from those with small cross sections. Though both shape and density play a role, the advantage is that water can be used as a cheap and available medium to separate much denser objects such as metals.

In a variation on this principle one may replace gravity with a centrifugal force by feeding a drum that is rotated around the vertical central axis. The chosen acceleration and terminal velocity may be smaller or larger than the acceleration of gravity. If it is smaller, the flow speed needed for separation must also be smaller, thereby saving some energy on pumping. If the centrifugal acceleration is larger than the acceleration of gravity, the flow needed for separation must be faster, which allows for a higher throughput capacity.

Other forced flow techniques: The fact that a flowing fluid can lift or move objects is utilized in a wide variety of techniques. Examples are: air suction (concentrates light and airy materials); air knives (concentrates light and air-resistant objects); fluidized beds (separates effectively as in sink-float, using air or a liquid and a homogenous bed of selected density and size particles); cyclones (separates small, light and airy particles from heavy ones), spirals (using water: separates heavy particles from light ones; in air: separates particles that roll from those with high surface friction), and the water table (separates heavy particles from light ones and/or small particles from large ones using a flow of water over an inclined surface; the surface may also be equipped with riffles as a selective obstruction to small, high density particles that may be separated into an additional product).

References

1. Eurostat. (n.d.). Eurostat waste databases. Retrieved from http://ec.europa .eu/eurostat/web/waste
2. Meadows, D.H., Meadows, D.L., Randers, J., and Behrens, W.W. III. (1972). *The limits to growth: A report for The Club of Rome's project on the predicament of mankind.* Washington D.C., USA: Universe Books.
3. European Commission. (2008). The Waste Framework Directive 2008/98/EC. Retrieved from http://ec.europa.eu/environment/waste/framework/.
4. Van Der Valk, H.J.L., Braam, B.C., and Dalmijn, W.L. (1986). Eddy-current separation by permanent magnets part I: Theory. *Resources and Conservation,* 12(3–4), 233–252.
5. Scott, D.M. (1995). A two-colour near-infrared sensor for sorting recycled plastic waste. *Measurement Science and Technology,* 6(2), 156–159.
6. Ellen MacArthur Foundation. (2013). Towards a circular economy. Cowes, UK: Ellen MacArthur Foundation.
7. Rousta, K., Bolton, K., Lundin, M., and Dahlén, L. (2015). Quantitative assessment of distance to collection point and improved sorting information on source separation of household waste. *Waste Management,* 40, 22–30.
8. Rousta, K., Ordoñez, I., Bolton, K., and Dahlén, L. (2017). Support for designing waste sorting systems: A mini review. *Waste Management and Research,* 35(11), 1099–1111.
9. Rahman, A. and Bakker, M.C.M. (2012). Hybrid sensor for metal grade measurement of a falling stream of solid waste particles. *Waste Management,* 32, 1316–1323.
10. Isakov, V. (1993). Uniqueness and stability in multi-dimensional inverse problems. *Inverse Problems,* 9(6), 579–621.
11. Jain, A.K., Murty, M.N., and Flynn, P.J. (1999). Data clustering: A review. *ACM Computing Surveys,* 31(3), 264–323.
12. Xia H. and Bakker, M.C.M. (2014). Reliable classification of moving waste materials with LIBS in concrete recycling. *Talanta,* 120, 239–247.

Chapter 15

Primary Production and Recycling
of Critical Metals

Yongxiang Yang
Department of Materials Science and Engineering,
Delft University of Technology,
Mekelweg 2, 2628 CD Delft, The Netherlands

Metals are important engineering materials, and are produced from non-renewable natural resources. This chapter starts with the criticality and scarcity issue of metals and metal resources. Then various available extraction and refining technologies and processes are introduced for primary production and recycling of metals. The principles of pyrometallurgy, hydrometallurgy and electro-metallurgy (electrowinning and electro-refining) are described for primary production of metals. The applications of metallurgical technologies to metals recovery from secondary resources (metal scrap and waste residues), i.e. recycling of metals, are discussed. As the last part, the extraction and recycling of cobalt, one of the most critical metals, is given as an example to illustrate how different metallurgical technologies are used and combined to extract a metal from different types of raw materials.

15.1 Introduction

The definition of critical metals or materials is a dynamic process. The list of critical materials depends on the region of end use or consumption and will change with time. Globally there are four countries and regions where critical materials have been defined in recent years: EU, USA, Japan, and China. Table 15.1 lists the critical materials (including metals) defined by different countries and regions. There are big overlaps and also many differences.

As the definitions of critical metals and materials have been discussed in the previous chapters of this monograph, this chapter focuses on the extraction and recycling technologies for "critical metals". Because of the dynamic nature of critical materials selection, this chapter will try to provide a

Table 15.1: List of Critical Metals and Materials Defined by EU, U.S.A., Japan, China.

Country or region	Critical materials list	Remarks
European Union (2010)	**REEs, PGM, Ge, Ga, In, Ta, Nb, Co, W, Be, Sb, Mg**, graphite, fluorspar	EU 2010[1]
(2014)	**REEs** (HREEs, LREEs), **PGM, Ge, Ga, In, Nb, Co, W, Be, Sb, Mg, Li, Cr, Si**, magnesite, borates, coking coal, natural graphite, fluorspar, phosphate rock	EU 2014 update[2]
(2017)	**REEs** (LREEs, HREEs), **Sc, PGMs, Ge, Ga, In, Bi, Be, Sb, Ta, Nb, Hf, Co, W, V, Mg, Si**, He, baryte, borates, natural graphite, natural rubber, fluorspar, phosphate rock	EU 2017 update[3]
United States (DOE) (2010–2011)	Critical: **REEs** (**Dy, Eu, Nd, Tb, Y**) Near Critical: **Li, Te, In, Ce, La**	US DOE 2011[4]
China (2014)	17 **REEs**, Rare metals: 4 light metals (**Li, Rb, Cs, Be**), 6 **PGMs**, 15 heavy and refractory metals (**W, Mo, Cr, Sn, Sb, Co, Ni, V, Ti, Zr, Hf, Nb, Ta, Sr, Be**), 8 scare metals (**Ga, In, Ge, Se, Te, Re, Tl, Cd**)	Recommendations only. No official list.[5]
Japan (2008–2009)	31 types of metals (47 elements in total): 17 **REEs**, 30 rare metals (**Li, Rb, Cs, Be, Sr, Ba, Ti, V, Zr, Hf, Nb, Ta, Cr, Mo, W, Re, Mn, Co, Ni, Pt, Pd, B, Ga, In, Tl, Ge, Sb, Bi, Se, Te**)	General coverage of all rare and REEs[6]
UNEP (2009)	In, Ge, Ga, Ta, PGM, Sb, Co, Li, REEs	UNEP report, 2009.[7]

broader coverage of the metals which have been selected as "critical" or could become "critical" in the future. The description of their primary production and recycling will be based on the similar properties shared by each group of metals.

Although different countries or regions have their own critical materials agenda from a geopolitical perspective, materials "criticality" or "scarcity" is also a global issue. It should be addressed globally, and a more healthy and sustainable solution is required. The ideal solution would be joint efforts which are binding on all the parties in the production and consumption world. Therefore the following actions are strongly recommended:

(1) *Curbed/reduced consumption (austerity)*: However, this is one of the most difficult actions. People face all kind of attractive new products in modern society, and it is very challenging or difficult to reduce the living standard once people have enjoyed a better life. The austerity measures imposed during or after the financial crisis in 2008–2009 encountered huge problems, e.g. in southern Europe. *There is quite often a conflict of interest between material/metal producers and societal needs due to financial benefits/profits.*

(2) *Increasing the "recycling" rate*: From both the production waste/scrap and from end-of-life (EOL) products. This is the most relevant option, and has been taking place at different levels. However, there is still a large potential to improve. It is both a short-term and a long-term solution. Furthermore, recycling should be promoted all the time for both resource saving, under the conditions that it is technologically and economically feasible, and has less environmental footprint. Recycling is part of the circular economy, and takes place at various steps in the whole life cycle of materials, from material production and product manufacturing, to the end-of-life phase.

(3) *Development of "substitutional" materials*: Using a more abundant and less critical material/metal to replace a more critical material or element. In fact, material substitution has been happening all the time in human civilization. However, this normally takes a very long time, and sometimes not all materials can be replaced with another material with the same functionality and performance.

(4) *Increase resource efficiency of materials*: This should be practiced by both materials producers and product manufacturers.

In practice, a harmonic development among the three pillars of materials/metals supply is highly recommended: primary production, recycling, and substitution, in order to balance the increased demand and consumption with limited and reduced resource availability. Quite often, the primary production of critical metals involves the use of secondary raw materials from the recycled source, and thus this chapter will address both the primary production and recycled production for critical metals.

15.2 Metallurgical Processes for Metals Extraction & Refining

Metals are primarily produced from concentrates, after mining and mineral processing, through various metallurgical processes. Primary metallurgical production is the starting point of the whole life cycle of metals, and together with recycling, it closes the metals cycle from both ends, as is illustrated in Fig. 15.1.

During primary production of metals, also called *"Extractive Metallurgy"*, minerals or compounds of the targeted metals are converted to pure metals or alloys through various chemical reactions, using many different steps. Different technologies are used: *pyrometallurgy, hydrometallurgy*, quite often in combination with *electrowinning* or *electrorefining*. These technologies are also used to separate and refine metals from secondary resources such as scrap or production wastes of various forms.

At present, primary production and secondary production (recycling) have many overlapping issues. Many secondary raw materials or solid residues are processed as part of raw materials in primary smelters. For example, copper scrap of different quality is refined in a copper smelter at the converting or fire refining stages. At a primary zinc smelter, secondary zinc oxide from electric arc furnace (EAF) flue dust or zinc-bearing residues from other industrial processes can be treated together with primary zinc sulfide concentrates during or after roasting. However, for some metals such as aluminum, the metallurgical recycling takes place exclusively at secondary smelters. It does not make sense to melt Al scrap (pure or contaminated) in the primary Al smelter which relies exclusively on molten salt

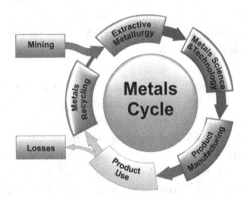

Fig. 15.1. Primary production (extractive metallurgy) and recycling of metals close the metals cycle.

electrolysis technology. Steel scrap is refined exclusively in primary steel-making, either in the integrated BOF (basic oxygen furnace) or EAF pro-cesses. The latter utilizes 100% steel scrap as the raw material. When the raw material is from secondary sources, such as scrap or production waste, the separation and refining of metals is called "*secondary metallurgy*" or "*recycling metallurgy*". However, in steelmaking, "*secondary metallurgy*" means the refining of steel after BOF or EAF crude steel production.

15.2.1 *General introduction of metallurgical processes*

Metal extraction takes place either at near room temperature or at very high temperatures. We can divide these metallurgical operations into two different categories: (1) *hydrometallurgy* (low or near room temperature in aqueous media), and (2) *pyrometallurgy* (at high temperatures of a few hundred degrees or well above 1000°C), depending on the temperatures and state of the materials involved (feed and products).

In addition, there is the 3rd type of operation, *electrometallurgy*, (we often call it *electrolysis*), in which the metallurgical reactions (deposition of metals) take place under the application of an electric potential difference. *Electrolysis* can take place in aqueous solutions at lower temperatures, but it can also take place in molten salts at higher temperatures, which is nor-mally regarded as part of *pyrometallurgy*. There are two types of electrolysis: (1) "*electrowinning*" for metal deposition from dissociated species (com-pounds), (2) "*electrorefining*" for converting impure metals to pure metals (this can be applied to aqueous solutions and also to molten salt systems).

Pyrometallurgy is the science and technology of extracting metals from minerals or refining metals at high temperatures. It is a general term for all metallurgical operations at high temperature above a few hundred degrees Celsius. Pyrometallurgy consists mainly of the following 3 steps or unit operations:

(1) Feed preparation
(2) Smelting
(3) Fire-refining or pyro-refining

The products of pyrometallurgical operations include *metals* (crude or refined) and *metal alloys* as the main products, various *metal compounds* as intermediate products (e.g. copper or nickel matte for converting, $TiCl_4$ for sponge titanium production, or $MgCl_2$ for molten salt electrolysis of Mg), and *slags* and *off-gases* as wastes or by-products. Off-gases consist of high temperature gases, smoke and flue-dusts. Off-gases normally go through cleaning processes to recover thermal energy via heat exchangers or waste-heat boilers (as steam, or electricity), de-dusting through waste-heat boilers, cyclones, electrostatic precipitators, and other gas purification units such as acid or basic scrubbers (e.g. Venturi scrubbers). For sulfide smelting–extraction of metals from sulfidic ores–the off-gas is basically SO_2 arising from oxidation of metal sulfides, and is used to produce sulfuric acid (H_2SO_4). Therefore, many pyrometallurgical smelters also include a chemical plant. A good example is ironmaking and steelmaking, and all iron and steel of more than 1.6 billion tons in the world are produced by pyrometallurgical technologies.

Hydrometallurgy is the science and technology for metal extraction through dissolution of metal ions from the minerals, crude metal or metal scrap and precipitation of the metals from the purified and concentrated solutions. Hydrometallurgical processes are conducted mostly in aqueous solutions at lower temperatures below the boiling point of water. Some operations take place at higher temperature in pressurized vessels (autoclaves) or using organic solvents (e.g. solvent extraction).

Hydrometallurgical processes normally consists of the following unit operations:

(1) Leaching
(2) Liquid–solid separation
(3) Solution purification
(4) Metal precipitation

The hydrometallurgical unit operations also generate some solid residues and waste water or effluent which need proper processing and final disposal. Some of the solid residuals contain valuable metals and compounds, and are often used for further processing to recover the remaining valuable products. Waste water could be easily purified and used internally in the process. Nowadays, hydrometallurgical processes are often operated as a closed system, and if electrowinning is used for metal precipitation, leaching agents are also regenerated during electrolysis. A good example is the hydrometallurgical production of zinc, which accounts for 80% of total zinc production (over 13.5 million tons per year).

Electrometallurgy or electrolysis is quite often the last step in hydrometallurgical or pyrometallurgical processes. For example, zinc electrowinning is the last part of the hydrometallurgical production of zinc; copper electrorefining is the last step of the pyrometallurgical production of copper. Molten salt electrolysis is conducted at high temperatures and can also be viewed as a pyrometallurgical operation; quite often the feed materials (oxides or chlorides) require purification from ore minerals with hydrometallurgical and/or pyrometallurgical processing. For molten salt electrolysis of aluminum, the raw materials of pure aluminum oxide (Al_2O_3) are produced through refining bauxite ore using hydrometallurgical processes (Bayer process). Due to the limited scope, electrometallurgy will not be discussed separately in this chapter.

In industrial practice, different types of metallurgical technologies are used in combination to produce one or more metals from primary ore concentrates or scrap metals and waste residues. Figure 15.2 shows a typical flowsheet from Umicore's multi-metal production (a) and rechargeable battery recycling (b) in Hoboken (Belgium), processing both primary concentrates and secondary raw materials (electronic waste, and spent rechargeable batteries).[8]

15.2.2 *Pyrometallurgy*

In general pyrometallurgical processes include 3 types of unit operations: (1) feed preparation such as roasting, sintering, (2) smelting such as sulfide smelting of copper concentrates and reduction smelting of iron ore, and (3) fire-refining e.g. copper fire-refining, and steelmaking.

Feed preparation: The following operations are used to upgrade and transform the nature or characteristics of the feed materials so as to make the metals extraction easier and more effective. They normally include *feed drying, feed mixing, granulating and pelletizing, calcinations, roasting and sintering*. The first 3 operations are more physical processes, and last 3 operations are chemical processes.

For certain smelting operations such as flash smelting of copper, the water content in the concentrates (originally $\sim$8%) must be very low ($<$1%, normally 0.1–0.2%) before feeding to the flash furnace. The drying of the concentrates is normally arranged in rotary and flash dryers by burning natural gases or using waste off-gases. Pelletizing and sintering are common pre-steps for reduction smelting. Examples include ironmaking and lead-producing blast furnaces. Grate sintering is the dominant technique. For the bath-type of sulfide smelting, the furnace can take the feed with 6–8% H_2O content, and normally pelletizing is needed. Roasting is the oxidation

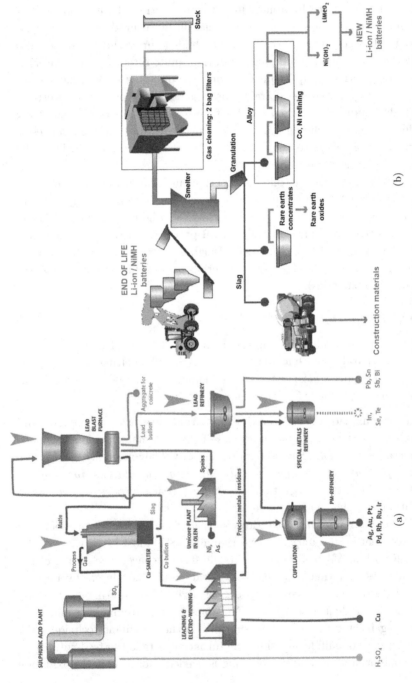

Fig. 15.2. (a) Flowsheet of Umicore's multi-metal extraction, refining and recycling plant, 8, with permission from Elsevier. (b) Flowsheet of Umicore's battery recycling plant, 8, with permission from Elsevier.

of metals sulfides to oxides/sulfates, and is needed for hydrometallurgical production of metals from their sulfide concentrates such as zinc, or for reduction of metal oxides from their sulfide minerals (e.g. pyrometallurgical production of zinc). Roasting can take place in fluidized-bed roasters or multi-hearth furnaces, and normally the process is autogenous (energy self-sufficient).

Smelting: In the smelting process, minerals (usually metal sulfides and oxides) are heated beyond their melting point, and are chemically processed to purify them while in the liquid phase. Smelting operations in pyrometallurgical processes include the oxidation and reduction processes in a molten state, to further concentrate or upgrade the metal content or extract as crude metals. The smelting processes include matte smelting of sulfide concentrates, reduction smelting of oxidic ores or oxide materials converted from sulfide ores. Applications of smelting technologies are numerous. A few examples are given below in Table 15.2.

The metals produced through smelting operations normally contain various impurities, and the purity is normally around 99%, and the products are called crude metals (each metal may have its own name: pig iron or hot metal, blister copper, lead bullion etc.). They need to be refined by either pyro-refining (fire-refining) or electrorefining (at low temperatures), or by both such as for copper.

Table 15.2: Example of Smelting Technologies and their Applications.

Process examples for smelting	Process characteristics
Matte smelting of copper and nickel	Transforming sulfide concentrates to copper matte or nickel matte or Cu-Ni matte, and molten slag by using oxygen. The matte can be converted to crude metal such as copper
Lead or lead-zinc blast furnace smelting	Producing lead or lead and zinc metals, and molten slag after roasting from their sulfide ores, using coke as reductant
Reductive smelting of tin concentrates	Producing crude tin metal and slag from the oxide concentrates using coke as reductant
Reductive smelting of laterite concentrates	Production of ferronickel using carbonaceous reductant
Blast furnace ironmaking	Reduction of iron oxides to crude iron — called "hot metal" using coke and pulverized coal as reductants
Ferroalloy production by submerged arc furnaces (SAF)	Reduction of iron and another metal from their oxide ores using coke

Fire-refining, also called pyro-refining: The fire-refining operation is carried out at high temperatures to remove impurities according to the property differences among the metallic elements. Examples are the fire-refining of crude lead to remove Cu, As, Sb, Sn, Au, Ag and Bi; the fire-refining of crude tin to remove Cu, Fe and Pb; the fire-refining of copper to remove O and S, As, Sb and Pb; the ladle-refining of crude steel; the vacuum degassing of liquid steel to remove hydrogen and nitrogen; the vacuum refining of niobium and tantalum; the re-melting and fire-refining of secondary metal scraps such as aluminum and copper. Steelmaking is generally classified as a refining operation. After fire-refining, the metal purity can generally reach from 99% up to 99.999% to meet the market requirements. Sometimes, electrorefining is needed to obtain the required purity. Copper is a typical example where high purity of at least 99.99% is needed, which cannot be provided by fire-refining operation and electrorefining must be used.

Metallurgical furnaces: Pyrometallurgical operations require high temperatures and are carried out at various reactors heated with various fuels or other energy sources. These types of reactors are normally lined with refractory materials and are called metallurgical furnaces, to keep the high temperature environment and to prevent significant heat losses. Table 15.3 below lists the main types of metallurgical furnaces and their typical applications.

Table 15.3:   Main Furnace Types and their Typical Applications.

Furnace type	Example of applications
Fluidised-bed roaster	roasting of zinc concentrates
Rotary kiln	roasting and drying of concentrates, zinc fuming
Multi-hearth furnace:	roasting of molybdenum concentrates
Reverberatory furnace:	copper smelting, tin smelting
Blast furnace	ironmaking, lead & zinc smelting
Electric arc furnace (EAF)	steelmaking, stainless steelmaking
Submerged arc furnace (SAF)	copper and nickel smelting
Flash smelting furnace	copper and nickel smelting
Bath smelting furnaces	Noranda and Mitsubishi for copper smelting, QSL, Kivcet etc. for lead smelting, TSL (Ausmelt or Isasmelt) for copper, lead and tin smelting
P-S converter	copper and nickel converting
BOS converter (BOF or LD)	steelmaking (also called LD converter)
Fuming furnace	zinc fuming

15.2.3 *Hydrometallurgy*

Hydrometallurgical processes consist of 4 types of unit operations: (1) leaching of minerals and ores or scrap metals, (2) liquid–solid separation, (3) solution purification, and (4) metal precipitation from solutions.

Leaching: Leaching is a dissolution process of minerals in an ore or scrap metals through use of a leaching agent such as an acid or alkaline solution. After leaching, the remaining non-dissolvable gangue materials will be separated from the metal containing ionic solutions. In another word, leaching is also a separation process for the targeted metals from the impurities and the gangue materials. Thus, leaching selectivity is very important for the targeted metals over the gangue minerals or impurities (unwanted species).

Leaching can be operated with various leaching agents: mineral acid (such as H_2SO_4 and HCl), alkaline (such as $NaOH$ and NH_4OH), or salt solutions (such as NH_4Cl and $NaCl$). The leaching system can be arranged at atmospheric pressure (below the boiling point of water), or in pressurized vessels (autoclaves) to accelerate the leaching process at higher temperatures above the boiling point of water. Sometimes bacteria could also be used to help dissolve certain metals from ores or scrap (bio-leaching).

Liquid–solid separation: During the leaching process, not all minerals will be dissolved into solution, and in fact it is only the targeted metals that are preferred to dissolve. The rest of the minerals will remain as solids, which need to be separated from the leach solution. Liquid–solid separation can be performed with thickeners, vacuum filters, and press filters. The liquid–solid separation process requires that the solid residues have a large particle size and a good crystallization structure.

Solution purification: The objective of the solution purification is to remove the co-dissolved impurities from the leach solution which will otherwise pollute the main metal product or make the precipitation of the targeted metals difficult or less efficient. In addition, the impurities co-dissolved in the solution may be of great market value, and the solution purification will also separate the impurities into a valuable secondary resource for other metals production. The typical solution purification technologies include: crystallization, distillation, precipitation such as hydrolysis, cementation, solvent extraction, ion-exchange, etc.

Metals precipitation: The last step in the hydrometallurgical process is to reduce the metallic ions in the purified solution to elemental form. To do this, one can use "cementation", in which a more electro-negative metal is used or sacrificed as the reducing agent to precipitate the targeted metal from the solution. For large scale production, electrowinning is used where a DC current is passed through the purified solution and the targeted metal will be precipitated on the cathode if the electrode potential is properly controlled. However, only relatively electro-positive metals could be electrodeposited in aqueous solutions. For more electro-negative metals such as aluminum and magnesium, a non-aqueous molten salt electrolyte has to be used. Furthermore, a gaseous reductant can be used to reduce the metal in the solution. For instance hydrogen reduction in aqueous solutions is used to produce metallic powders of copper, nickel and cobalt.

15.2.4 *Waste treatment and utilization of secondary resources*

During pyro- and hydrometallurgical operations, there are also various by-products and wastes generated together with the metal production. These by-products and wastes have to be effectively recovered and processed to increase the value of production and to control the environmental pollution.

Off-gas treatment: The off-gases handling is an important part of the pyrometallurgical operation and involves the removal and recycling of the flue dust, the recovery of the thermal energy, and the removal of environmentally hazardous compounds such as SO_x, NO_x etc. From the flue gases other chemical products can be produced e.g. production of H_2SO_4 or liquid SO_2 from sulfide smelting processes. Energy recovery as high pressure steam and thus electricity is not uncommon in pyrometallurgical smelters. Many sulfide smelters are also small power plants, and they often can partly provide the power required to operate the whole smelter. They produce also steam, which can be used for various heating purposes.

The main equipment used in an off-gas treatment system include: cyclones, bag filters, electrostatic precipitators (ESPs) for dust removal, and the total dust removal efficiency can reach $95\% \sim 98\%$. The energy recovery is through various heat exchangers, e.g. hot stove for blast furnace ironmaking, and waste-heat boilers (WHBs) for sulfide smelting. In addition, wet scrubbers (e.g. Venturi scrubbers) are used to remove gaseous compounds such as SOx and fine dust. It is worth noting that the off-gas treatment system often is larger than the smelting furnace, and the total investment of the off-gas handling system accounts for 25% to 50% of the total plant investment.

Treatment of metallurgical slags: Slag formation plays a critical role in pyrometallurgical metal production. Metallurgists often say that if you can make a good slag, then you will automatically make a good metal. Slags are more often complex silicates (similar to glass in structure) formed from various metal oxides of the gangue materials in the ores together with the added fluxing agents such as silica (SiO_2) and lime (CaO). Ironmaking slags are basically $CaO-SiO_2-Al_2O_3-MgO$, and slags from non-ferrous smelting are generally $SiO_2-CaO-FeO-Al_2O_3$. Slags in steelmaking systems contain mainly $CaO-SiO_2-FeO$. The relative amounts or concentrations of different components depend on the process selection and operating conditions.

The amount of generated slag can be very high in comparison with the produced metal, depending on the main metal grade in the ores. Production of 1 ton of pig iron generates 300–600 kg of slag, and production of 1 ton of steel generates 120–150 kg of slag. However, production of 1 ton of copper will generate 1 to 3 tons of slag; for lead smelting this figure is 1 to 1. On one hand, the slag production rate depends on the metal grade in the concentrates. On the other hand, it also depends on how much fluxing agent such as silica or quartz, lime, dolomite, and fluorspar (CaF_2) is added.

Metallurgical slags have found various applications, but the main application is their use as construction materials. Sometimes, the smelting slag contains relatively high valued metals. In this case, the metal needs to be recovered before the final disposal of the slag, and thus there are slag leaning operations. For instance, copper smelting slags containing higher than 0.5% Cu must be cleaned, and this is carried out in electric furnaces such as in copper flash smelting. Another example is the lead smelting slag, and it normally contains 10% or more zinc which is recovered through slag fuming. In some special cases, the slag is the intermediate product but not a waste, e.g. the production of titanium slag through reduction smelting of titanium-bearing iron ore. The titanium slag is further used to extract titanium dioxide or titanium metal.

Waste water treatment: In hydrometallurgical processes, water is an important diluent for the leaching agents: acid, alkaline, or salt. Water is also used to wash the leach residues and capture the solid particles during gas–solid separation in off-gas treatment. Water is also used in pyrometallurgical processes for cooling the furnace systems in the forms of water jackets or simply water spray. The waste water may contain heavy metals, arsenic or mercury. Various waste-water treatment technologies are available, and have been successfully applied in waste water purification from metallurgical smelters.

Recovery of minor but valued elements: It is very common during the extraction of non-ferrous metals, particularly for heavy metals (such as copper, lead and nickel), that various precious and rare metals are recovered. Those precious metals come from the concentrates together with the bulk metal, and get enriched during smelting and refining. They are then further recovered after refining operations. These metals include silver (Ag), gold (Au), platinum group metals (PGM), and the scarce metals of selenium (Se), tellurium (Te), indium (In) and germanium (Ge). Many sulfide smelters also treat low-grade gold and silver ores together with bulk metal concentrates. For example, approximately 60% of the silver in the world is produced from lead smelters. In the non-ferrous smelters, the recovery of the precious metals can reach 95% to 99%.

Figure 15.3 shows a modern copper smelter using flash smelting and converting technologies.[9] As can be seen, flue dust, slags, and hot off-gases are generated, but most of them are internally circulated first and exit the process at a single point. The off-gases from smelting and converting are firstly cleaned at waste heat boilers (WHBs) to remove coarse dust and recover waste heat as high-pressure steam for power generation. The

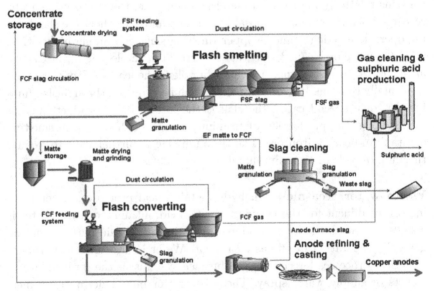

Fig. 15.3.   Illustration of a modern copper smelter with Outokumpu/Outotec Flash smelting and Flash converting technologies,[9] with permission from TMS.

cleaned SO_2-bearing off-gas is further cooled and cleaned in electrostatic precipitators (ESPs) and after that the cleaned off-gas is sent for H_2SO_4 acid production before discharge to the ambient air. Converter slag with high copper content is circulated back to the smelting furnace for copper recovery, and the smelter slag is cleaned in a slag cleaning furnace before exiting the process.

15.2.5   *Current status and future perspective*

Pyrometallurgy versus hydrometallurgy: Worldwide, iron and steel are produced exclusively via the pyrometallurgical route, and almost all lead (Pb), tin (Sn), antimony (Sb), titanium (Ti), and the majority of nickel (Ni) are also produced by pyrometallurgical processes. In addition, 85% copper and 20% of zinc are produced by way of pyrometallurgy. On the other hand, various hydrometallurgical processes involve also pyrometallurgical unit operations, such as the roasting of sulfidic zinc and cobalt concentrates for the hydrometallurgical recovery of zinc and cobalt. In the same way, the pyrometallurgical production of various metals such as copper, nickel and lead involves electrorefining as hydrometallurgical or electrometallurgical operation.

Hydrometallurgy is the main technology to produce zinc metal (80%) through leaching and electrowinning process. However, sulfide roasting as unit operation of pyrometallurgy is almost always used to prepare the raw materials (calcine) for the leaching operation, although direct pressure leaching is also available for sulfide zinc concentrates. Hydrometallurgical processes are used to produce gold and silver through leaching, cementation or electrowinning. Hydrometallurgical processes are particularly suitable for recovery of metals from low grade ores and solid wastes and residues. The recovery of copper from low grade copper concentrates or oxidic copper ores occurs mainly through hydrometallurgical processes (15% total copper production). Recently, the full hydrometallurgical production of copper has been practiced by direct pressure leaching of normal sulfide copper concentrates in the Phelps Dodge copper smelter (Bagdad, USA),[10] now part of Freeport-McMoRan.

Main characteristics of pyrometallurgical operations: the following lists some main features of pyrometallurgical processes, both advantages and disadvantages.

- Fast reaction rate at high temperatures, and thus with high productivity and low investment and operating cost.

- Possible utilization of the chemical energy contained in the sulfide concentrates, and thus with relatively low energy consumption.
- Good ability to capture precious metals by matte and heavy metals, and thus with high recovery of valued metals coming from the concentrates.
- Slags as the by-product with relatively stable compositions and properties, and thus with less environmental impact.
- Large volume of high temperature off-gases, requiring expensive treatment before final disposal to the air.
- Relatively poor working environment due to the presence of thermal radiation and emissions, which need to be improved.

Main characteristics of hydrometallurgical operations:

- Efficient technology for the treatment of low grade and complex ores.
- Better utilization of all valuable contents of the ore due to high selectivity over different metal species in the raw materials.
- Better environmental conditions (absence of waste off-gases).
- Lower productivity and larger facilities and higher investment cost in land use.
- Generation of larger amount of waste water and solid residues.
- Higher energy consumption in particular for sulfide concentrates, compared to pyrometallurgical processes.

Perspectives and future developments: The following lists a number of points which show the further needs in process improvement and development.

   For pyrometallurgical processes:

- Process intensification through more use of oxygen and oxygen-enriched air: this can significantly reduce the off-gas volume, and reduce the off-gas processing cost.
- Process intensification through more use of fluidization and injection techniques.
- Expanding use of suspension smelting and bath smelting for sulfide smelting processes, for more efficient utilization of the energy content in the sulfide ores.
- Development of more efficient oxygen production techniques and new refractories to serve the needs of the process intensification.
- Development of more advanced reactors and furnaces, and improvement of process automation and process control, as well as working environment.

For hydrometallurgical processes:

- Increasing the metal recovery and selective extraction, for lower grade and multi-metal complex concentrates.
- Intensification of the leaching rate, in particular more use of pressure leaching as an efficient leaching technology.
- Efficient use of metal separation technologies
- Development of more efficient separation technologies for multi-species and low concentration solutions.
- Innovation in new technologies for more effective extraction of minor metals and scarce metals from EOL products.

The development of new energy-efficient technologies for the electrowinning of metals is needed. New anode materials and new electrolyte systems for low cell voltage deserve more investigation.

15.3 Metallurgical Processes for Metals Recycling

The former discussion in section 18.2 focuses mainly on the introduction of metallurgical processes for the extraction of metals from primary raw materials, i.e. ores or concentrates. For metals recycling, the raw materials are different in nature in comparison with the primary ores. Generally speaking, there are two types of secondary resources for metals recycling: (1) *metal scrap*, and (2) *wastes of residues and sludge or solutions*. To convert metallic scrap and waste residues into pure metals or alloys, different processing routes are taken. Comparing metallurgical recycling to primary extraction and refining technologies, there are a lot of similarities and many extraction and refining technologies can be used more or less directly apart from some fine tuning. There are also distinctive differences where different technologies are used particularly for scrap metals. Figure 15.4 illustrates different processing routes to transform metal scrap or the waste of residues back into pure metals or alloys.

15.3.1 *Metallurgical recycling of metal scrap*

Metal scrap is already in metallic form, and it can take relatively pure form arising from product manufacturing (new and production scrap), or it can be in a very contaminated and complex form from EOL products (old scrap).

New scrap generated from product manufacturing is often specifically collected and re-melted in the plant or sold to the re-melters for production

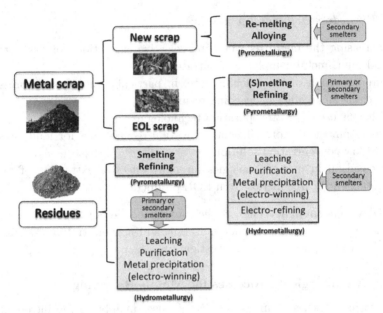

Fig. 15.4. Overview of processing routes for metals recovery from secondary resources.

of the same type and quality of metals or alloys. This is a common practice for aluminum and copper. Re-melting is a simple physical process of heating – melting – casting operation. What's important is to keep the metals from oxidation loss during re-melting. The main operating cost is the energy which can be electrical or fossil fuels (oil or gas). Since no refining is required, the consumption and operational cost is very low. It has to be kept in mind that recycling of new scrap does not generate new metals, and it is one of the internal material loops or circles, as considered from a circular economy perspective. However, for steelmaking, new and old scrap are both recycled through the primary production of steel via either the integrated steelmaking route (BOF) or mini mills of the EAF route. Heat treatment is always required after steel re-melting or refining, and casting.

Production scrap from within the smelters are normally rejects or off-specs. The quality is not high enough for remelting to produce the end quality of metals. In this case, the scrap is refined at different stages and comes out together with the primary production process. Internal scrap from smelters is normally treated together with old scrap together at a proper refining stage. This is a typical operational strategy for copper smelters. The copper anode rejects due to physical defects or incorrect mass (2–3%), and the undissolved remaining anodes from electrorefining (12–20%) are normally remelted in the anode refining furnace and cast again into fresh anodes for electro-refining.

Scrap from EOL product recycling (old scrap) is contaminated at different levels. It needs refining using different refining technologies: pyrometallurgical, hydrometallurgical, and electrometallurgical methods. During pyrometallurgical refining, the targeted metal for recycling will normally not be oxidized, and only impurity metals or nonmetallic contaminations are removed through e.g. oxidation or salt treatment. For aluminum, the recycling of old scrap is completely organized in separate aluminum recycling plants (aluminum refineries). Normally a refining is conducted at ca. 800°C in a rotary furnace or box furnace by use of molten salt of $NaCl + KCl$ in the presence of a small amount of fluoride such as in cryolite (Na_3AlF_6).

Old scrap of copper is partially recycled and refined within primary copper smelters. Depending on the contamination level of the scrap, it could be smelted during the matte smelting and converting stage such as for electronic scrap. Then the scrap goes through the smelting and two stage refining steps (fire-refining and electro-refining) for a final quality of at least 99.99%. Old copper scrap is also refined to new copper of virgin quality in secondary copper smelters. In the secondary copper smelter, the contaminated old copper scrap first goes through smelting to produce "black copper" (Cu contaminated with iron). Then the black copper (70–90% Cu) is converted to blister copper (96–98% Cu) in a Peirce–Smith (P-S) converter or a Top-Blown Rotary Converter (TBRC) through oxidation and slagging of iron and other impurities. The blister copper is further fire-refined and electro-refined to produce the final quality of electrolytic copper of 99.99%, quite similar to the quality achieved by the operations of the primary smelter. Copper scrap from cables, wires, automotive parts or motors, and electronic waste, are typical EOL scrap and are recycled and refined either in primary smelters or secondary smelters. Good examples in Europe are Umicore, New Boliden, and Aurubis, which are large scale metal producers of copper and precious metals from both primary and secondary copper resources.

If metal scrap is very complex and heavily contaminated, hydrometallurgical recycling or electro-chemical processing could be used. However, during hydrometallurgical recycling, the metals are chemically or electrochemically dissolved or oxidized into solutions (acid or alkaline), and the purification and separation of metal species in the solutions are required.

The purified metal-containing solutions (containing metal cations) need to be processed though electrowinning or other precipitation reduction operations such as cementation or hydrogen reduction to produce pure metals. Hydrometallurgically recycling can be flexible and robust, however, it involves re-oxidation and reduction of the targeted metals and from energy consumption perspectives it is not always favorable. If the scrap metal is not heavily contaminated, electrorefining could be used. For example, high quality copper could be electro-refined without converting and fire-refining. Furthermore, anodic dissolution can be used to selectively dissolve the targeted metals (less noble compared to the impurities or precious metals) in the scrap. However, on the cathode no metal is precipitated except for H_2 evolution for more electro-negative metals. Following the solution purification the targeted metals can be produced through electrowinning or cementation, or a metal compound is produced for further reduction. Printed circuit boards (PCBs) from WEEE or e-waste can be treated hydrometallurgically through acid (or alkaline) leaching, in this case an oxidizing agent may be needed e.g. oxygen for acidic dissolution of copper, depending on their relative redox potential compared to H_2. Leached copper using e.g. H_2SO_4 will be purified to remove other impurities e.g. through solvent extraction, and finally copper will be reduced to pure metallic metal through electro-winning, similar to the hydrometallurgical production of copper from low grade primary copper ores.

15.3.2 *Metal recovery from wastes of residues*

Another type of secondary resource for metal recovery comes from waste residues, sludge and solutions containing metallic elements in the oxidation state in the form of various compounds. They include the following categories:

(1) Tailings from mining and mineral processing operations.
(2) Metallurgical slags and residues, as well as flue dust.
(3) Solid residues from waste processing and energy production such as bottom ashes and fly ashes from municipal solid waste (MSW) incineration and pyro-power plants.
(4) Waste sludge and residues from chemical and other process industries.

Metal-containing solid waste and residues as well as sludge and solutions are normally smelted together with ore and concentrates in primary smelters for recovery of valuable metals. However, separate processing in secondary smelters is also possible. This will depend on which metals are to be recovered.

Spent lead-acid batteries are smelted mostly in secondary lead smelters e.g. by using lead blast furnaces. Zinc-bearing neutral leach residues and electric arc furnace (EAF) steelmaking dust are normally processed through Waelz kilns or bath fuming to produce higher grade ZnO raw materials, which are then fed into primary zinc smelters for Zn recovery. The modern hydrometallurgical zinc smelters can treat up to 30% secondary zinc raw materials in the form of ZnO.

Furthermore, primary metals smelters recycle their flue-dust and slags within their plant to different processing stages. In primary copper smelters, converter slag contains often high copper (2–15%), and it is normally recycled back to the matte smelting stage. Anode furnace slag will return to the converter. The only slag which is discarded is the matte smelting slag, containing less than 1% copper (normal range: 0.3–0.5% Cu). In comparing the copper grade in the copper ore, which is at 0.5% level nowadays, to the discarded copper slag, it becomes clear that the copper grade of discarded copper slag is comparable to the copper ore. Therefore, the discarded copper slag could become a future copper resource, which is at present limited by the processing cost and economics. The flue dust in smelters contain normally high toxic compounds as volatiles such as arsenic in copper smelters. Internal treatment without removing toxic compounds will cause accumulation of the toxic elements. They are often cleaned externally and returned to primary smelters or treated in secondary smelters for metal recovery. When the economics of doing this is not feasible, they have to be landfilled with fees.

Ironmaking and steelmaking flue dust are internally recycled to certain extent to recover ferrous materials, fluxing agents and/or carbon. However, the internal recycling is limited by the accumulated zinc content of up to 4–5% which causes operational problems in the blast furnace ironmaking system. Therefore, these types of dusts may have to be landfilled at a certain cost. Different technologies have been developed to remove zinc in the dust, however, they are also limited by the economic feasibility of the operation. For instance, rotary hearth technology can be used to remove zinc and the iron oxide is also reduced to crude iron from the steelmaking dust. Nevertheless it has not yet gained wide application due to some remaining technology issues and relatively high investment and operational costs.

15.4 Production of Critical Metals: Example of Cobalt

The critical metals (in a broader perspective) referred to in this chapter cover mainly those from the EU definition plus selections from the US, Japan and China. However, some other metals which fall into the same type or group may also become critical such as Mo, Zr, and Re together with W, V, Hf, Nb and Ta as the refractory metals group, and Se, Te together with Sb, Bi, In, Ge, and Ga as the scarce metal group. REEs both heavy and light are among the most critical metals, as are introduced in a separate chapter of this book, and the same accounts for PGMs. Chromium and silicon are two important metals in the critical metal family, and both of them are mainly produced and used in steelmaking as alloying elements in the form of ferroalloys (FeCr, FeSi). But they are also produced as pure metals of chromium and silicon for special applications such as solar grade and electronic grade silicon for photovoltaic and electronic applications.

A description of the production technologies and processes for all these critical or near critical metals is not possible in a chapter within this book. Instead, cobalt as an example from the critical metal list is used to illustrate briefly how various metallurgical processes and technologies are applied in the production and recycling of typical metals, in particular critical metals.

15.4.1 *Primary production of cobalt*

Cobalt is a critical metal used in rechargeable battery electrodes, superalloys for gas turbine engines, cemented carbides and diamond tools, and more. Cobalt is mostly produced as a by-product or co-product with nickel (50%) and copper (44%).[3] Only less than 6% of Co total primary production is mined as a main product (in Morocco). The Democratic Republic of Congo has the largest reserves of cobalt in the world accounting for 48.5% (3.4 million tons out of 7 million tons), and the second largest reserve comes from Australia with 15% of the world total, according to USGS.[11] World mine production reached a record level of 126,000 tons in 2015. Congo (Kinshasa) remained the leading producer of mined cobalt, supplying 50% of world mine production, followed by China (6%), Canada (5.5%), Russia (5%), and Australia (5%).[11] In 2015, world production of refined cobalt reached a record level of 97,400 tons according to USGS,[12] contributed mainly by China (50%), Finland (10%), Belgium (6.5%), Australia (5%) and Japan (4%).

Cobalt is mainly extracted from nickel-cobalt and copper-cobalt concentrates and occasionally directly from the ore itself, by hydrometallurgical, pyrometallurgical, and electrometallurgical processes.[13] Although most methods of extraction are based on hydrometallurgy, cobalt concentrates, mattes, and alloys have been reduced to metal by pyrometallurgical methods. The hydrometallurgical processes involve (1) the leaching of concentrates to generate a cobalt-containing solution, (2) the separation of cobalt from the other metal ions in solution, and (3) the reduction of cobalt ions to cobalt metal. Electrolysis is used in the electrowinning of the metal from leach solutions and in refining the cobalt that has been extracted by hydrometallurgical or pyrometallurgical methods.

Figure 15.5 illustrates the nickel and cobalt extraction from Ni-Co sulfide concentrates at Harjavalta smelters (Norilsk and Boliden), Finland, using well known Outokumpu/Outotec DON process.[14,15]

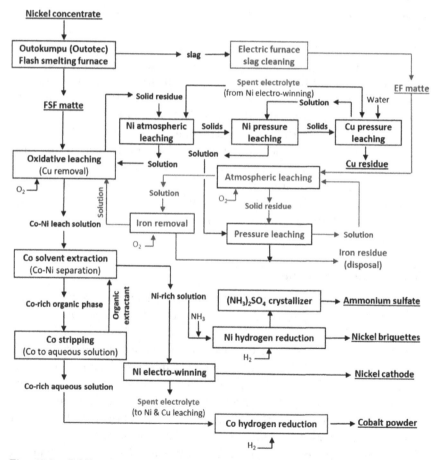

Fig. 15.5. DON-process flowsheet for nickel and cobalt production at Harjavalta smelter, Finland,[14] adapted from Grundwell *et al.*

Pyrometallurgical matte smelting is the first step to produce Co-bearing Ni-Cu matte using both flash smelting (having low Fe content) and electric smelting (high Fe content) at Boliden smelter, and both mattes produced at Boliden smelter are treated at Norilsk smelter in the neighborhood by the atmospheric and pressure leaching of nickel and copper. Flash smelting furnace (FSF) matte is dissolved in 4 steps: copper removal with atmospheric Ni-Co leaching, atmospheric and pressure leaching of remaining nickel, and copper pressure leaching. Cobalt is produced by hydrogen reduction as cobalt powder from all leach solutions after the separation of nickel with solvent extraction. After the solvent extraction of cobalt, nickel is produced as a cathode by electrowinning and briquette by hydrogen reduction. The electric furnace (EF) matte is leached in a combined atmospheric-autoclave leach step in the acidic solution coming from the nickel pressure leach step. The leach solution is returned to the atmospheric FSF matte leach step after separation of the iron residue in the form of goethite (FeOOH). The filtered residue containing copper sulfides and PGMs coming from the nickel pressure leach step is fed to the adjacent copper smelter of Boliden for further processing. Alternatively the copper leach residue can be pressure leached followed by PGM separation and copper electrowinning.[15]

Figure 15.6 shows the cobalt extraction from Cu-Co sulfide concentrates.[14] The extraction and separation of cobalt involve sulfate roasting, atmospheric H_2SO_4 acid leaching, solvent extraction to separate copper from cobalt, the precipitation of cobalt hydroxide and the re-dissolution of the cobalt hydroxide followed by cobalt electro-winning. Copper extracted to the organic solvent is stripped to aqueous solution for copper production by electro-winning.

15.4.2 *Cobalt recycling*

Cobalt scrap generated from product manufacturing is well recycled. Major EOL cobalt scrap is also recyclable and recycled. Cobalt used in the main applications of super-alloys, hard metals, batteries and even spent catalysts (80–85% of total cobalt) can be collected and either reused or recycled. However, dissipative use of cobalt as pigment in glass, ceramics and paints are not recycled. There are no reliable global statistics on the recycling rates of cobalt. However, if using the U.S. as a reference or benchmark, the end-of-life recycling rate for the U.S. is about 68%, and the recycled content or recycling input rate (including both new and old scrap) is 32% for the total cobalt production.[3] This implies that about 1/3 of cobalt is supplied from the recycled source.

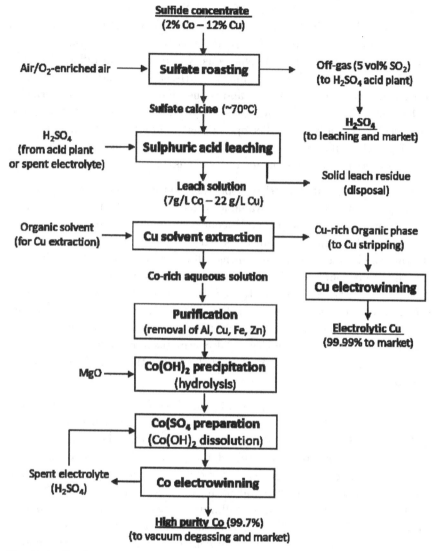

Fig. 15.6. Production of high-purity cobalt from cobalt-copper sulfide concentrate,[14] adapted from Grundwell *et al.*

Cobalt from alloy scrap: Old scrap cobalt from super-alloys and other alloys can be recycled both in primary smelters and in secondary smelters.[16] In the primary nickel sulfide smelter (e.g. Xstrata's Falconbridge in Sudbury), various types of cobalt-bearing scrap are added to different stages of matte smelting or converting. Co and Ni from the scrap alloys will be reported to low or high grade Ni matte, and cobalt will be separated and recovered from nickel refineries. In the secondary smelter, various types of cobalt-based alloy and miscellaneous Ni/Co reverts in different shapes (solid scrap, spills, grindings, turnings, swarfs, sludges, oxides in the form of dust and pellets) can be melted and refined into new alloys. Furthermore, cobalt scrap from Ni-Co alloys is also recycled by using hydrometallurgical techniques, typically at smaller operations, mainly for recovering and separating nickel and cobalt, and/or recovering other valuable metals which would otherwise be lost in a smelter (such as Ta, W, Re).

Cobalt from spent batteries: depending on the type of cobalt-containing battery, various options are available to recycle the cobalt within primary sulfide smelters[16] like Xstrata's Falconbridge smelter, or in dedicated secondary smelters such as Umicore's battery recycling plant in Hoboken. During the high temperature melting, all the base metals are produced as an alloy that could be refined and separated into individual constituents. Lithium or REEs are reported to slag and are not recovered at the present due to economic reasons, however, technologies are available for recovering Li and the REEs when economics are favorable in the future. Furthermore, hydrometallurgical technologies are also available to recover Co and Ni from spent Li-ion battery and Ni-M Hydride batteries, through various leaching and precipitation processes.

Cobalt from spent catalysts: Similar to cobalt-containing alloys and spent batteries, this type of secondary cobalt can be recovered either in primary sulfide nickel smelters[16] like Xstrata's Falconbridge, or in dedicated secondary smelters like Umicore's Hoboken plant. Spent cobalt catalysts can also be regenerated in-situ or externally by burning off the carbon and sulfur deposited on the catalyst during operation and gaining 70–80% of its original capacity. Hydrometallurgical processes are also available for recovery of cobalt and other metals in the spent catalysts through leaching and precipitation as high grade sulfide cobalt concentrates for further cobalt extraction in sulfide smelters.

Cobalt from metallurgical wastes and residues: This is another low-grade source of secondary cobalt, and can be used for cobalt extraction. This type of cobalt bearing waste includes[16]:

(1) flotation tailings from the mineral processing of cobalt-bearing ores
(2) slags generated during the smelting of Cu-Co and Ni-Co ores

(3) nickel refinery sludge residues

(4) zinc smelter waste streams

Both pyrometallurgical smelting and hydrometallurgical leaching-based technologies are used to recover cobalt and other valuable metals from this type of waste. For detailed technology information, the reader can refer to the review paper by Ferron,[16] or the monograph by Rao for a broader coverage of metals recovery from metallurgical wastes and scrap.[17]

15.5 Concluding Remarks

Metals are produced both from primary resources after mining and from various secondary resources (urban mining). A large number of technologies are available for the extraction and refining of metals from both types of metal resources (concentrates, scrap and residues): pyrometallurgy, hydrometallurgy, and electrolysis (electrowinning and electro-refining). These metallurgical technologies are quite often used in combination to produce refined metals by using their best merits.

For metals production from recycled sources, a distinction is made between metal scrap (in metallic form) and process residues (metal in different forms of compounds). Metal scrap is better directly re-melted to new metal if it is relatively pure and clean (production or manufacturing scrap), or melted and refined to pure metal or alloys without unnecessary oxidation and reduction. For very complex scrap, hydrometallurgical extraction and separation are also used to produce pure metals. EOL metal scrap can be refined both in primary smelters (such as steel, copper, nickel and cobalt) at different process steps, but can also be refined in standalone secondary smelters (such as copper and aluminum). For process residues, both pyrometallurgical and hydrometallurgical technologies are used, very similar to those used in primary production. Quite often the residues are processed in primary smelters together with concentrates.

Critical metals are normally produced as by-products or co-products of the bulk or big metals. Their use is generally in trace amounts and at low concentrations. Therefore, EOL recycling of critical metals is much more challenging than recycling the big metals. More efficient and cost effective extraction and refining technologies are needed in the future.

The extraction and recycling of cobalt, one of the most critical metals, is discussed as an example to illustrate how different metallurgical technologies are used together to extract a metal from many different types of raw materials.

References

1. European Commission (2010). *Critical raw materials for the EU*. Brussels, Belgium: European Commission.
2. European Commission (2014). *Report on critical raw materials for the EU*. Brussels, Belgium: European Commission.
3. European Commission (2017). *Study on the review of the list of Critical Raw Materials*. Brussels, Belgium: European Commission.
4. US DOE (2011). *Critical Materials Strategy*. Washington, D.C., USA: US DOE.
5. He, X.J. and Zhang, F.L. (2014). Considerations and suggestions on early planning of strategic emerging minerals development (in Chinese). *Natural Resources Economics of China*, 2014(4), 4–8.
6. Kawamoto, H. (2008). Japan's Policies to be adopted on rare metal resources. *Science and Technology Trends Quarterly Review*, 27, 57–76.
7. UNEP (2009). *Critical metals for future sustainable technologies and their recycling potential*. Nairobi, Kenya: UNEP.
8. Rombach, E. and Friedrich, B. (2014). Recycling of rare metals. In E. Worrell and M.A. Reuter (Eds.), *Handbook of recycling — state-of-the-art for practitioners, analysts, and scientists*, Amsterdam, the Netherlands: Elsevier.
9. George, D.B., Nexhip, C., George-Kennedy, D., Foster, R., and Walton, R. (2006). Copper matte granulation at the Kennecott Utah copper smelter. In C. Harris, H. Hanein, and T. Warner (Eds.), *Granulation of molten materials*. Pittsburgh, Pennsylvania, USA: The Minerals, Metals, and Materials Society.
10. Dreisinger, D. (2006). Copper leaching from primary sulfides: Options for biological and chemical extraction of copper. *Hydrometallurgy*, 83(1), 10–20.
11. USGS. (2017). *Mineral Commodity Summaries: Cobalt*. Retrieved from https://minerals.usgs.gov/minerals/pubs/commodity/cobalt/mcs-2017-cobal.pdf. 52–53.
12. Shedd, K.B. (2017). Cobalt. In K.B. Shedd (Ed.), *Minerals yearbook 2015*. Reston, Virginia, USA: United States Geological Survey.
13. Donaldson, J.D. and Gaedcke, H., (2007). Cobalt. In F. Habashi (Ed.), *Handbook of extractive metallurgy, volume II*. Weinheim, Germany: Wiley-VCH.
14. Crundwell, F., Michael, M., Ramachandran, V., Robinson, T., and Davenport, W.G. (2007). *Extractive metallurgy of nickel, cobalt and platinum group metals*. Amsterdam, the Netherlands: Elsevier.
15. Svens, K. (2013). By-product metals from hydrometallurgical processes — an overview. Paper presented at *III International Conference: By-product Metals in Non-ferrous Metals Industry*. Wrocław, Poland: Institute of Non-Ferrous Metals.
16. Ferron, C.J. (2013). The recycling of cobalt from alloy scrap, spent batteries or catalysts and metallurgical residues — an overview. In T. Battle *et al.* (Eds.), *Ni-Co 2013*. Pittsburgh, Pennsylvania, USA: The Minerals, Metals, and Materials Society.
17. Rao, S.R. (2006). *Resource recovery and recycling from metallurgical wastes*. Amsterdam, the Netherlands: Elsevier.

Chapter 16

Recovery of Rare Earths from Bauxite Residue (Red Mud)

Chenna Rao Borra*,§,¶, Bart Blanpain†, Yiannis Pontikes†, Koen Binnemans‡,
and Tom Van Gerven§

*Department of Material Science and Engineering, TU Delft,
Delft, The Netherlands

†Department of Materials Engineering, KU Leuven, Leuven, Belgium

‡Department of Chemistry, KU Leuven, Leuven, Belgium

§Department of Chemical Engineering, KU Leuven, Leuven, Belgium

¶c.r.borra@tudelft.nl

The management of bauxite residue (BR) is a major issue for the aluminum industry because of its high alkalinity and the large volumes generated. Therefore, the recovery of rare earth elements (REEs) with or without other metals from BR and utilization of the generated residue can contribute to a solution on the management problem of BR and it can be one of the options to meet the demand of REEs. In view of the above, the selective recovery of REEs over major elements such as iron by direct acid leaching was studied initially. From the leaching results, either the recovery of REEs was low or the dissolution of iron was high. To address that, iron was removed from BR by smelting. The slag generated after smelting was leached with mineral acids. The selectivity of REEs over iron was greatly improved. However, the high level of alumina presence in BR required a large amount of fluxes thereby increasing the energy consumption in smelting. Hence, the removal (and recovery) of alumina from BR by sodium carbonate roasting was carried out. The sample, after alumina removal, was smelted and the REEs were successfully recovered from slag by leaching with mineral acids. An alternative process, called sulfation–roasting–leaching, was also developed by which the REEs can be selectively leached. The scandium recovery, however, was low. Preliminary energy and economic analysis showed that alkali roasting–smelting–leaching and sulfation–roasting–leaching were the most promising processes for the treatment of BR.

16.1 Introduction

Bauxite residue (BR, also known as "red mud" in the slurry state) is a waste product generated during the Bayer process of alumina production from bauxite.[1-4] About 140 million tonnes of BR are generated annually and almost 4×10^9 tonnes are already stockpiled. Long-term storage of such waste not only occupies valuable land resources but it is also potentially harmful to the environment, which in turn incurs major liabilities and costs to the alumina industry. The utilization of BR in other applications could be a sustainable solution for the BR problem. However, only 2–3% of this material is currently being used in cement and ceramic applications. The high residual sodium content and the presence of other alkaline solids in BR restricts its use in other applications.

The major elements that are present in BR are iron, aluminum, silicon, titanium and occasionally calcium. It also contains some valuable but minor elements such as gallium and rare-earth elements (REEs). Therefore, BR can be viewed as a potential polymetallic secondary raw material. Despite this potential, metal extraction has not been practiced so far, due to the fact that the concentration of many elements is too low to make recovery economically feasible. If it is not financially viable to recover all the metals, it might still be reasonable to recover at least the valuable metals, after which the remaining solid residue could be used for other applications like building materials or cementitious binders.

REEs are critical metals with high supply risks and their demand is increasing annually. The main applications of REEs are in green technologies[5]: the production of permanent magnets, lamp phosphors, rechargeable NiMH batteries, catalysts, alloys and other applications.[6] Scandium is the most valuable element among the REEs in BR (>95% of the economic value of REEs in BR).[2] The price of scandium oxide (99.95%) was 4200 US$/kg.[7] Secondary sources like BR can potentially be considered as one of the options to meet the REEs demand. REEs present in the bauxite ore end up in BR during the Bayer process.[2] BR generated from karst bauxite ores are more rich in REEs compared to that generated from lateritic ores.[8] Karst bauxite ores are mainly found in Europe, Jamaica, Russia and China. Only 13% of total bauxite reserves are rich in REEs. The recovery of scandium together with other REEs and perhaps additional metals combined with the utilization of the remaining residue can partly solve both the supply problem of REEs and the storage problem of BR.

There is literature available on the recovery of REEs from BR via direct hydrometallurgical processes, while very few studies involve combined pyro- and hydrometallurgical methods to recover REEs together with other metals.[3] In the direct leaching studies, the majority of the literature focuses on scandium and very few on the other REEs.[9] Furthermore, the dissolution of major elements was not studied in detail during leaching. There exist few studies on the removal of iron via a pyrometallurgical process, followed by a hydrometallurgical process to recover the REEs. However, large volumes of flux were used in these studies and there are no detailed data available on leaching of the REEs. Conceptual flow sheets for the recovery of iron, aluminum, titanium and REEs from BR have been reported in the literature.[3] However, most of these studies have not yet been tested experimentally. There exists no complete study available on the combined recovery of iron, aluminum, titanium and REEs yet. Furthermore, most of the flow sheets contain a magnetic separation step after alumina removal, which does not allow complete iron recovery.

Greek BR was used in the present study as it is rich in REEs. In this work, we present different routes with a focus on recovering REEs to develop an economically viable, near-to-zero waste process. The different routes explained in this chapter to recover REEs from BR are shown in the form of a combined flow sheet (Fig. 16.1). In the first route, direct acid leaching of BR was studied to evaluate the yields and selectivity for the recovery of REEs. In the second route, iron was removed by smelting, followed by REEs and titanium leaching from the slag. In the third route, alumina in the

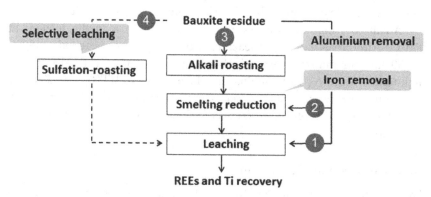

Fig. 16.1.   Conceptual flow sheet for the processing of BR, with an indication of the four alternative processes.

BR was removed by alkali roasting prior to smelting and subsequently the residue from the alumina removal process was smelted without any additional flux. The slag generated after smelting was leached for recovery of REEs and titanium. A fourth route (called sulfation–roasting–leaching) was developed to selectively leach the REEs from BR. This process can selectively leach the REEs with very low amounts of major elements co-dissolved. The different routes will be described in the following sections, but more details can be found in the individual journal papers from which this chapter was derived.[9–12]

16.2 Materials and Methods

The BR studied in this work was provided by Aluminum of Greece, Greece. Analytical grade chemicals were used in the present study. Chemical analysis of major elements in BR was performed using wavelength dispersive X-ray fluorescence spectroscopy (WDXRF, Panalytical PW2400). Chemical analysis of minor elements was performed after complete dissolution of BR by alkali fusion and acid digestion in a 1:1 (v/v) HCl solution, followed by Inductively Coupled Plasma Mass Spectrometry (ICP-MS, Thermo Electron X Series) analysis.

The room temperature leaching was carried out in sealed polyethylene bottles by constant agitation using a laboratory shaker (Gerhardt Laboshake) at 160 rpm and 25°C. High-temperature leaching experiments were carried out in a 500 mL glass reactor fitted with a reflux condenser and placed on a temperature-controlled ceramic hot plate with a magnetic stirring system. The leach solution sample was filtered using a syringe filter (pore size of 0.45 μm) and diluted with deionized water for ICP analysis.

Alkali roasting experiments were carried out in a muffle furnace at 950°C. After roasting, water leaching experiments were carried at 80°C for 60 min for alumina removal. Smelting studies were carried out in a high-temperature vertical alumina tube furnace (Gero HTRV 100–250/18). Sulfation–roasting experiments were carried out in a muffle furnace. Full experimental details of the processes developed in this study can be found elsewhere.[9–12]

16.3 Results and Discussion

The chemical analysis of the BR sample used in this study is shown in Tables 16.1 and 16.2. BR contains high concentrations of iron oxide and alumina (Table 16.1). The total REE concentration in the BR is about

Table 16.1:   Major Chemical Components in
the Bauxite Residue Sample (Excluding LOI).[9]

	Concentration (wt%)
Fe_2O_3	44.6
Al_2O_3	23.6
CaO	11.2
SiO_2	10.2
TiO_2	5.7
Na_2O	2.5

Table 16.2:   REEs Composition
of the Bauxite Residue Sample.[9]

	Concentration (ppm)
Sc	121
Y	76
La	114
Ce	368
Pr	28
Nd	99
Sm	21
Eu	5
Gd	22
Tb	3
Dy	17
Ho	4
Er	13
Tm	2
Yb	14
Lu	2

0.1% (Table 16.2). XRD analysis showed that it contains different phases
like hematite, goethite, gibbsite, diaspore, calcite and cancrinite.[9]

Route 1: Direct acid leaching[9]

Selective recovery of REEs compared to iron was studied in the case of direct
acid leaching with mineral acids (HCl, HNO_3, and H_2SO_4). It was found
that acid leaching at low acid concentrations (<1 N) yields low recovery of
REEs (scandium and LREEs <50%, HREEs <70%). The recoveries were
similar for all the mineral acids at low acid concentrations (<1 N). The
yields could be improved (HREEs ~80%, scandium and LREEs > 70%) by
increasing the acid concentrations (6 N), especially for HCl (Fig. 16.2), but
the dissolution of iron (~60%) also became high. It was also found that

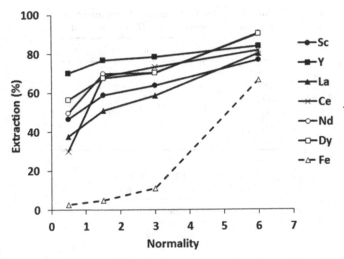

Fig. 16.2.   Effect of HCl concentration on leaching of REEs from BR (T: 25°C, L/S: 50, t: 24 h).[9]

scandium is associated with the iron oxide phases in the residue, which makes it difficult to dissolve unless the iron is dissolved. Moreover, a large part of the major elements (Fe, Al, Ca, Si, Ti and Na) was also dissolving during direct leaching, which can generate large amounts of effluents. The large amount of iron going into solution during leaching poses problems in the downstream processes because of the similar behavior of iron and scandium in solvent extraction or ion exchange processes. Therefore, the removal of iron from BR by smelting reduction was studied prior to acid leaching.

Route 2: Smelting-leaching[10]

Smelting experiments of BR were carried out with the addition of wollastonite ($CaSiO_3$) as flux and graphite as reductant. Addition of wollastonite decreased the slag melting temperature and the viscosity, enhancing slag-metal separation. Graphite contents were higher than the optimum level reduced part of the silica and titania, which hindered the slag-metal separation. Iron was separated from the slag in the form of a nugget (Fig. 16.3). The optimum conditions were 1500°C, 20 wt% of wollastonite and 5 wt% of graphite. More than 95% of the iron was separated from the slag. The slag obtained after iron removal was treated with different mineral acids to extract REEs. The recovery yields for REEs were low (<70%) when the

Fig. 16.3. Smelted sample (20% wollastonite, 5% graphite, 1500°C), with the iron nugget separated from the slag.[10]

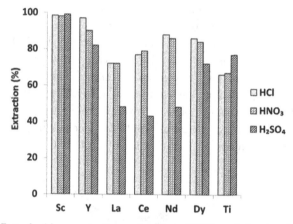

Fig. 16.4. Effect of acid concentration on leaching of REEs and titanium from slag (acid conc.: 3N, T: 90°C, t: 1 h, L/S: 50).[10]

leaching was done at room temperature. Hence, high-temperature leaching (90°C) experiments were also carried out. The selectivity of REEs over iron during slag leaching was improved compared to in direct leaching as >95% of iron was removed during smelting. All of the scandium, most of the other REEs and about 70% of the titanium could be leached using different mineral acids at 3 N acid concentration, a L/S ratio of 50:1, a temperature of 90°C and a reaction time of 1 h (Fig. 16.4). The REEs recovery was lower in H_2SO_4 solution and the effect is higher with increasing ionic radius, which is due to the formation of a solid product layer ($CaSO_4$).

Direct smelting of BR rich in alumina requires a large amount of fluxes and large energy consumption during smelting, and subsequently, large acid consumption during slag leaching. As a result, the cost of the overall process increases and make the sustainability of the process questionable. Therefore, the removal (and recovery) of alumina from BR by Na_2CO_3 roasting prior to smelting was carried out.

Route 3: Alkali roasting-smelting-leaching[11]

BR was roasted with sodium carbonate at 950°C with a BR to alkali ratio of 1:0.5. The roasted mass was leached with water at 60°C to dissolve aluminum in the solution. The residue after alumina removal was smelted at 1500°C without any added flux and it was possible to obtain a clear slag-metal separation (Fig. 16.5). The slag after grinding was leached with different mineral acids at 90°C. However, the recoveries of the REEs, with the exception of scandium, were drastically lowered (<50%) in the alumina-poor slags, compared to the direct smelting slags. This is due to the formation of a perovskite ($CaTiO_3$) phase, which strongly binds REEs (except scandium). Perovskite is a stable phase, does not dissolve in acids under normal conditions and ends up in leach residue. Therefore, the slag was cooled by water quenching. Leaching after quenching can dissolve REEs successfully from alumina-poor slags, even at room temperature. Most of REEs and about 90% of titanium could be leached from the quenched slag using mineral acids at a L/S ratio of 50:1, a temperature of 25°C

Fig. 16.5.   Smelted sample (stoichiometric carbon, 1500°C). Reproduced from Borra et al..[11]

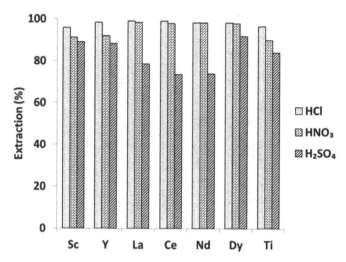

Fig. 16.6. Effect of type of acid on the leaching of REEs and titanium from a quenched slag. (acid conc.: 1 N, T: 25°C, t: 24 h, L/S: 50). Reproduced from Borra *et al.*[11]

and a reaction time of 24 h (Fig. 16.6). The alkali-roasting temperature can be decreased to <500°C by using sodium hydroxide instead of sodium carbonate.

Route 4: Sulfation-roasting-leaching[12]

An alternative process, called sulfation–roasting–leaching, was also developed to selectively leach the REEs from BR, while leaving iron, titanium, aluminum and calcium undissolved in the residue. In this process, BR was mixed with water and concentrated H_2SO_4, followed by sulfation, roasting and finally water leaching of the roasted product in water. It was found that most of the oxides were converted to their respective sulfates during the sulfation process. In the subsequent roasting stage, unstable sulfates (mainly $Fe_2(SO_4)_3$) decomposed to their respective oxides. Rare-earth sulfates, on the other hand, were stable during roasting and dissolved in water leaving the iron oxides in the residue.

Figure 16.7 shows the effect of roasting temperature on the extraction of REEs and major elements. The extraction yields of all REEs slightly decrease with increasing the roasting temperature. The dissolution of aluminum, iron, and titanium also decreases with increasing roasting temperature. However, the dissolution of sodium and calcium did not change as their sulfates decompose only at very high temperatures. About 60% of

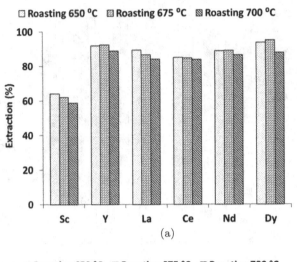

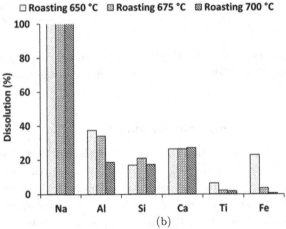

Fig. 16.7.   Effect of roasting temperature on leaching of (a) REEs (b) major elements (acid to BR ratio: 1:1, roasting duration: 1 h, non-agitated leaching duration: 7 days, L/S ratio: 50). Reproduced from Borra et al..[12]

scandium and more than 90% of the other REEs could be dissolved at optimum conditions (roasting temperature: 700°C, roasting duration: 1 h, acid to BR ratio: 1, non-agitated leaching duration: 7 days, L/S ratio: 5), while only a very small amount of iron (<1% of total iron) was solubilized. The residue after leaching was found to be rich in Fe-, Al-, Si- phases and $CaSO_4 \cdot 0.5H_2O$ and it is suggested that this residue can be used for instance in iron-rich cementitious binders.

Recovery of REEs and valuable metals from leach solutions

The recovery of REEs and other metals from the leach solutions can be performed by solvent extraction, ion exchange, precipitation, neutralization, hydrolysis etc. In solvent extraction and ionic exchange processes, conventional or new reagents can be investigated for selective scandium and other REEs recovery.[13,14] Titanium can be precipitated from the leach solutions as $TiOSO_4$ at a temperature of ca. 140°C. Then TiO_2 can be recovered from $TiOSO_4$ by hydrolysis.[15]

Comparative analysis of the different processes[16]

A preliminary cost analysis of the developed processes is shown in Fig. 16.8 with two options, i.e. with current scandium price (4200 US\$/kg)[7] and at 50% of the current price. This analysis includes only the costs of consumed reagents and energy and values of the generated products and does not include the capital and operating costs (CAPEX and OPEX) and metal recovery cost from solutions. This cost analysis shows that alkali roasting–smelting–quenching–leaching is the most promising process for treatment of BR and it is justified to further study this process at a pilot-scale to develop detailed techno-economics. MYTILINEOS S.A. (formerly Aluminum of Greece) and its project partners are going to study a process similar to alkali roasting–smelting–leaching process in a pilot scale project (RemovAL). The sulfation–roasting–leaching process could also be an economic process if the

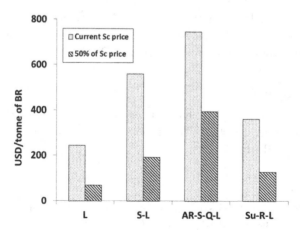

Fig. 16.8. Operational margins for different processing routes (L: leaching, S: smelting, AR: alkali roasting, q: quenching, Su: sulfation, R: roasting).[16]

scandium recovery was higher. In a recent study, Narayanan *et al.* were able to increase the recovery of scandium to about 90%.[17]

16.4 Conclusions

It is our hope that the study of the proposed processes (which are relevant for BR generated from karst bauxite ores) can contribute to the development of a near-zero-waste flow sheet for the processing of BR, including the recovery of major (Al, Fe and Ti) metals, the utilization of the residue in building applications and simultaneously the production of valuable REEs, thus partially decreasing the supply risk of these elements.

Several processing routes have been developed to recover REEs and valorize BR: 1) direct leaching; 2) smelting-leaching; 3) alkali roasting-smelting-leaching; 4) sulfation-roasting-leaching.

REEs recovery was low during direct acid leaching at low acid concentrations (1 N). High acid concentration can improve the recovery, especially for HCl, but the dissolution of iron ($\sim$60%) was also high. Direct smelting can allow recovery of more than 95% of the iron from BR and slag leaching can recover most of the REEs and titanium. Roasting in the presence of Na_2CO_3 at 950°C can remove $>$ 75% of alumina. Smelting of the alumina removed sample, without any added flux, can remove $>$ 95% of iron from slag. Most of REEs and about 90% of titanium could be leached from the quenched slag using mineral acids. In sulfation–roasting–water leaching about 60% of scandium and more than 90% of the other REEs can be dissolved at optimum conditions. Alkali roasting–smelting–leaching and sulfation–roasting–leaching processes should be studied at a pilot scale for detailed techno-economic analysis. We expect that these two processes can solve the BR problem and produce valuable REEs.

Acknowledgements

This work was supported by KU Leuven (DBOF grant to CRB, IOF-KP RARE3 and GOA/13/008). The authors thank Aluminum of Greece for providing the BR sample.

References

1. Evans, K. (2016). The history, challenges, and new developments in the management and use of bauxite residue. *Journal of Sustainable Metallurgy*, 2(4), 316–331.

2. Binnemans, K., Jones, P.T., Blanpain, B., Van Gerven, T., and Pontikes, Y. (2015). Towards zero-waste valorisation of rare-earth-containing industrial process residues: a critical review. *Journal of Cleaner Production,* 99, 17–38.

3. Borra, C.R., Blanpain, B., Pontikes, Y., Binnemans, K., and van Gerven, T. (2016). Recovery of rare earths and other valuable metals from bauxite residue (red mud): A review. *Journal of Sustainable Metallurgy,* 2(4), 365–386.

4. Borra, C.R. (2016). *Recovery of rare earths from bauxite residue (red mud)* (Doctoral thesis). KU Leuven, Belgium. Retrieved from https://lirias.kuleuv en.be/handle/123456789/537871.

5. European Commission. (2014). *Report on critical raw materials for the EU.* Brussels, Belgium: European Commission.

6. Binnemans, K., Jones, P.T., Blanpain, B., *et al.* (2013). Recycling of rare earths: A critical review. *Journal of Cleaner Production,* 51, 1–22.

7. Rare earth metals. (n.d.). Retrieved December 22, 2017, from http://minera lprices.com/.

8. Mordberg, L.E. (1993). Patterns of distribution and behaviour of trace elements in bauxites. *Chemical Geology,* 107(3–4), 241–244.

9. Borra, C.R., Pontikes, Y., Binnemans, K., and van Gerven, T. (2015). Leaching of rare earths from bauxite residue (red mud). *Minerals Engineering,* 76, 20–27.

10. Borra, C.R., Blanpain, B., Pontikes, Y., Binnemans, K., and van Gerven, T. (2016). Smelting of bauxite residue (red mud) in view of iron and selective rare earths recovery. *Journal of Sustainable Metallurgy,* 2(1), 28–37.

11. Borra, C.R., Blanpain, B., Pontikes, Y., Binnemans, K., and van Gerven, T. (2017). Recovery of rare earths and major metals from bauxite residue (red mud) by alkali roasting, smelting, and leaching. *Journal of Sustainable Metallurgy,* 3(2), 393–404.

12. Borra, C.R., Mermans, J., Blanpain, B., Pontikes, Y., Binnemans, K., and van Gerven, T. (2016). Selective recovery of rare earths from bauxite residue by combination of sulfation, roasting and leaching. *Minerals Engineering,* 92, 151–159.

13. Roosen, J., van Roosendael, S., Borra, C.R., van Gerven, T., Mullens, S., and Binnemans, K. (2016) Recovery of scandium from leachates of Greek bauxite residue by adsorption on functionalized chitosan-silica hybrid materials. *Green Chemistry,* 18(7), 2005–2013.

14. Onghena, B., Borra, C.R., van Gerven, T., and Binnemans, K. (2017). Recovery of scandium from sulfation-roasted leachates of bauxite residue by solvent extraction with the ionic liquid betainium bis(trifluoromethylsulfonyl)imide. *Separation and Purification Technology,* 176, 208–219.

15. Petersen, A., Shirts, M., and Allen, J. (1992). *Production of titanium dioxide pigment from perovskite concentrates, acid sulfation method.* Washington D.C., USA: US DOI.

16. Borra, C.R., Blanpain, B., Pontikes, Y., Binnemans, K., and van Gerven, T. (2016). Comparative analysis of processes for recovery of rare earths from

bauxite residue. *The Journal of the Minerals, Metals, and Materials Socety,* 68(11), 2958–2962.

17. Narayanan, R.P., Kazantzis, N.K., and Emmert, M.H. (2017). Selective process steps for the recovery of scandium from Jamaican bauxite residue (red mud). *ACS Sustainable Chemistry and Engineering,* 6(1), 1478–1488.

Author Biographies

S. Erik Offerman (editor and Chapters 1 and 10) has been working as an associate professor at the department of Materials Science and Engineering of Delft University of Technology (TU Delft) in the Netherlands since 2008. He likes to unravel the underlying mechanisms of kinetic processes that govern the evolution of the microstructure of metals — in particular, steel — during thermal and thermomechanical processing. In recent years he has applied this knowledge to the sustainable development of metals and the processing of metals by substituting critical materials, which are used as alloying elements in major metals, by alternative materials that are less/not critical and that give the major metal similar properties. Erik received the Sawamura Award of the Iron and Steel Institute of Japan in 2015 for research on the effect of alloying additions on the evolution of the microstructure and hardness of martensitic steel during tempering. He obtained PhD (2003) and MSc (1999) degrees in Materials Science and Engineering from TU Delft. Erik has been a member of the daily board of the Leiden-Delft-Erasmus Centre for Sustainability since 2014. He serves as an editorial board member for Scientific Reports. Erik served as the chairman of the Western European steering committee of the European Institute of Innovation and Technology (EIT) on Raw Materials (2015–2017). He has been the chairman of the Transdisciplinary Science and Technology Strategic Event 'Materials in a resource-constrained world', which was supported by COST (Cooperation in Science and Technology) in 2013.

Michel Rademaker (Chapter 2) is the Deputy Director of HCSS. He has a degree in Transport and Logistics, which he obtained at the University of Tilburg. He has fifteen years of hands-on experience as an officer in The Royal Netherlands Army, where he held various military operational and staff posts and also served a term in the former Yugoslavia. After leaving the armed forces, Mr. Rademaker went on to work at the Netherlands Organization for Applied Scientific Research (TNO) as a project and program manager and senior policy advisor for ten years. As NATO RTO project leader, he and his team developed new serious gaming assessment methods for defense. He has conducted several assessments of security technologies, including disruptive forms, and worked on numerous strategic security topics. He has written several book chapters, and numerous research reports and articles. He served as secretary of the pilot joint venture TNO Clingendael Centre for Strategic Studies for three years and is one of the co-founders of its successor, HCSS.

At HCSS he is responsible for business development. He is particularly interested in setting up multistakeholder projects on new and upcoming security and politico-economic themes. For example, he and his team supported the development of the Dutch national security strategy and crisis management exercises.

Mr. Rademaker was also an initiator of The Hague Security Delta, the biggest security cluster in Europe, and continues to lead projects for this cluster. His fields of expertise include security strategy, policy, concepts and doctrines, technology surveys and assessments, and serious gaming techniques. One of his fields of expertise includes the geopolitical and economic security implications of raw materials and energy supply, and the development of policies. He frequently acts as a moderator in workshops, as a trainer, as a guest lecturer at various universities and a speaker at conferences, and as a commentator in news media.

Sijbren de Jong (Chapter 3) was — at the time of writing — a Strategic Analyst at the Hague Centre for Strategic Studies, a lecturer in Geo-Economics at Leiden University and a writer for EUobserver specializing in Europe's relations with Russia and the former Soviet bloc. Born in 1982, Sijbren studied Geography and Conflict Studies at the Universities of Groningen, the Netherlands and Leuven, Belgium. He holds a PhD in EU external energy security relations from the University of Leuven. From 2007 until 2008 Sijbren worked as Head of Outreach for the Atlantic Initiative, a security and defense think tank based in Berlin. Between 2008 and 2012 he was staffed as a Research Fellow with the Leuven Centre for Global Governance Studies in Belgium working on EU external relations with Russia and the CIS countries. From 2012 until late 2017 he worked as an analyst for the Hague Centre for Strategic Studies and as a lecturer at Leiden University.

Erik Kelder (Chapter 4) Returning from a Danish battery company, Danionics A/S, Erik started on a patented Li-ion battery system using a novel densification technique (Magnetic Pulse Compaction). His research focused on Li-ion battery materials, components, and cells, including those for down-hole application. Research for other supporting industries was conducted such as stimulating introduction of separator materials (DSM Solutech) and development of critical technologies for high-temperature Li-ion batteries (Shell IEP). Various EU projects were carried out, e.g. a 42-volt board-net Li-ion polymer battery coordinated by Erik (SLION, FP5), creating an aging model for Li-ion batteries using electrochemical impedance spectroscopy (LIBERAL, FP5), setting up a durable collaboration on Li-ion batteries (ALISTORE, FP6), deposition of thin layers for high-aspect ratio micro-batteries (E-Stars, FP7), and fabrication of novel Li-ion batteries for EVs (EuroLiion, FP7) also coordinated by Erik. From a national perspective, several projects started within a project called ADEM on advanced materials for energy systems of which Erik is the battery theme leader. Erik was also involved in an initiative of the Dutch Ministry of

Foreign Affairs to set up a project concerning a Lithium Industrialization Project for Bolivia's state mine company Comibol. Since two years ago, Erik has been a member of the program committee for the DUBBLE beam line at the ESRF in Grenoble, France. Today, his research interests are lithium and magnesium batteries, advanced polymer electrolytes for both batteries and fuel cells, and advanced diagnostic tools such as synchrotron analysis and scanning electrochemical microscopy in an AFM set-up.

David Peck (Chapters 5 and 9) is Senior Research Fellow at TU Delft. He researches and teaches in the field of circular built environment and critical materials. David works in the faculty of Architecture and the Built Environment, Department of Architectural Engineering and Technology. His research objective is the development of a knowledge framework for a circular materials economy that enables the circular design of future cities and buildings. He is a member of the faculty based Circular Built Environment team. David advises the European Commission on Critical Materials and Circular Economy issues via the EIP Raw Materials and also supports the Netherlands Government in this work.

David is also a visiting Professor with Coventry University and an adjunct Professor at MIP Politecnico di Milano, Graduate School of Business, both roles on circular cities and critical materials. David is the TU Delft lead for the pioneer university status with the Ellen MacArthur Foundation for a circular economy. He is the TU Delft lead manager for a Horizons 2020 project, ProSUM — Prospecting Secondary raw materials (Critical Materials) in the Urban Mine and mining waste and ERN — European Remanufacturing Network. David is the TU Delft representative for the EU KIC EIT Raw Materials (sustainable exploration, extraction, processing, recycling and substitution). He leads a number of projects in this 2 billion Euro programme that has a focus on circular and critical materials. Specific projects cover Remanufacturing, International Round Table on Critical Materials, Materials History, Conflict Materials, Circular On-line Learning, and Circular Start-up companies.

Thomas Graedel (Chapter 6) is Clifton R. Musser Professor Emeritus of Industrial Ecology in the School of Forestry and Environmental Studies, Yale University. His research is centered on developing and enhancing industrial ecology, the organizing framework for the quantification and transformation of the material resource aspects of the Anthropocene. His textbook, *Industrial Ecology and Sustainable Engineering*, coauthored with B. R. Allenby, was the first book in the field and is now in its third edition. His current interests include studies of the flows of materials within the industrial ecosystem, and of evaluating the criticality of materials. Graedel's books (17) and papers (~380) have been cited by colleagues more than 24,000 times, putting his citation record in the upper 1/4 of 1% of all active scientists. He was elected to the U.S. National Academy of Engineering in 2002 for "outstanding contributions to the theory and practice of industrial ecology", and was a founding member of the UNEP International Resource Panel.

Barbara Reck (Chapter 6) is a Senior Research Scientist at the School of Forestry and Environmental Studies at Yale University. Her research centers on mapping the use of materials as a means to identify opportunities for improving the material and energy efficiency across the material cycle. She has studied metal cycles extensively, including detailed case studies on nickel and stainless steel. These studies formed the basis for in-depth studies on metal recycling and its energy implications, metal criticality, and scenarios on the future supply and demand of the major metals — topics that inform both environmental and resource policy. She regularly serves on advisory boards of international research projects and is an active member of the International Society for Industrial Ecology. Dr. Reck serves as node lead of Systems Analysis and Integration at the newly established REMADE Institute, whose mission it is to reduce the embodied energy and carbon emissions associated with industrial-scale materials production and processing.

James Goddin (Chapter 7) is Market Development Manager at Granta Design. He is a materials scientist and engineer with over 13 years' experience of creating and managing significant, multi-disciplinary, multi-national, strategic research and development projects. He received his PhD from the University of Bath in 2005 in Materials Science and Engineering. He has a long-standing passion for innovation, and is responsible for the early stage-gate processes for innovative and emerging technology areas of value to Granta's business which currently include Additive Manufacturing, Integrated Computational Materials Engineering (ICME), Collaborative Materials Information Management, Eco-design, critical/conflict minerals and the Circular Economy. He has led activities to develop risk metrics for engineering materials in collaboration with many of the world's leading advanced engineering companies to enable the rapid screening of materials in use and to inform decision making and strategy in response to supply constraints. Dr. Goddin represents Granta on various industry and standards groups including the European Innovation Partnership on Raw Materials, the European Rare Earths Competency Network, the EU/US/Japan Trilateral Cooperation on Critical Materials, the ADS Design for Environment working group, and the Circular Economy 100. He is a registered Chartered Engineer, a Chartered Environmentalist, as well as a Chartered Member of Engineering New Zealand. He is also a Fellow of the Institute of Materials, Minerals, and Mining (IOM3) where he is actively involved in membership activities, professional development, and various specialist interest groups.

Ton A.G.T.M. Bastein (Chapter 8) has a PhD in chemistry (Heterogeneous Catalysis, Leiden, 1988). He started as a research manager and product developer of detergents at Unilever. He has worked at TNO (The Netherlands Organization of Applied Scientific Research) since 1993 in many areas as research manager (ranging from cleaning technology to nanotechnology, semiconductor equipment development and defense technology). Since 2010 he has focused on research in the area of raw materials availability, resource efficiency, and circular economy, developing, initiating, and conducting research activities. He founded the Platform on Materials Scarcity in order to ensure the topic of raw materials availability received a higher awareness in the Dutch academic and policy landscape. Amongst others he wrote reports for the Dutch government on "Opportunities for a circular economy in The Netherlands" (2012/13) and "Critical Materials for the Dutch Economy (2015). He is also involved in recent analyses concerning the potential of the circular economy on a municipality scale (Amsterdam, Nijmegen, Rotterdam) and a regional scale. He was involved in several European R&D projects (POLFREE, RECREATE, CRM_InnoNet) and service contracts (Ramintech) for the EC. Together with colleague Elmer Rietveld, he was involved in the European Commission's recent 2nd revision of the Critical Raw Materials list. The basis for these activities is the drive to quantify the impacts of sustainability, circular economy, and raw materials issues on job creation and environmental issues. Bastein therefore strongly believes in working with multidisciplinary teams (business administration, macroeconomics, environmental economics, environmental assessments, and technology experts in relevant fields), both within TNO (the broad Netherlands Applied Research Organization) and outside TNO (government, academic institutions, employers organizations, individual companies, and consultants).

 **Elmer Rietveld (Chapter 8)** is a researcher at the Netherlands Organization for Applied Scientific Research. He started his career in spatial and transport modelling, civil engineering, and project appraisal. His work has had a strong focus on resource efficiency research since 2008. At TNO, he participated in advanced research and modeling in the fields of critical raw materials, the circular economy, product service systems, regional spatial economics, material flow analysis and cost-benefit analyses. He was a member of the core team executing the 2nd revision of the critical raw material list of the EU in 2017. Most of his work is based on official statistics (Eurostat, OECD, UN, ESPON), CGE, and EE-IO models. A key contribution of Elmer was to introduce SCBA techniques in an early stage to resource efficiency studies such as "Opportunities for a circular economy in The Netherlands (Ministry for the Environment,

2012/13) Another was to improve data analysis research by linking existing public data in an innovative way, for instance in the Materials in the Dutch economy (2014 and 2015), a study into criticality of the Dutch economy in relation to materials supply. The application of shadow prices in macro-economic models based on LCA techniques was a third innovative contribution in his research field. He was the main author of the report on product durability for the European Parliament IMCO committee published June 2016.

Conny Bakker (Chapter 9) is professor of Design Methodology for Sustainability and Circular Economy at TU Delft, faculty of Industrial Design Engineering. Her research field is the development of design methods for products that have multiple lifecycles. She explores strategies such as product lifetime extension, reuse, remanufacturing and recycling, and the business models that enable these strategies. A second research interest is the field of user centred sustainable design, which focuses on exploring the relationships between consumer behaviour, sustainability, and design. Conny co-leads the European H2020 project ReCiPSS, that helps designers and manufacturers understand how collection, remanufacturing and reuse of products can lead to more profitable, resource-efficient and resilient business practices. She coordinates and teaches several courses in Sustainable Design and Design for the Circular Economy, including a popular MOOC (Massive Open Online Course) which has attracted over 20,000 learners so far.

Ruud Balkenende (Chapter 9) is professor of Circular Product Design. In 2015 he started at the TU Delft after 25 years of experience at Philips. He has published about 40 papers and holds 30 patents. He has been active in several EU-funded projects and was the coordinator of GreenElec, a project that successfully developed and implemented design for recycling. His research, projects and teaching concentrate on the connection between product design and the circular economy with a focus on the engineering aspects of implementation. His research interest is in improving the resource efficiency of products through design for recycling and design for the circular economy.

 **Marcel den Hollander (Chapter 9)** is a Dutch industrial designer and design researcher. He has been working for over twenty years in industrial design for internationally renowned design studios and clients. The vast spectrum of design projects he has been involved in range from fast moving consumer goods, such as (food) packaging, to durable consumer goods, such as electronics, (office) furniture, and professional products, such as retail interiors and exteriors, museum showcases, means of transport, industrial equipment, and logistic systems. His long-standing interest in sustainable product design, combined with the changing landscape of business and industrial design, has led him into the field of design research in order to explore the options for making design for sustainability an integral part of commercial industrial design practice and design education.

Marcel was a researcher in the Design for Sustainability Department at the Faculty of Industrial Design Engineering of his alma mater, the Delft University of Technology, receiving his PhD in Industrial Design Engineering in 2018 from that same university. His research specialization is in strategic product design and design methodology for circular business models.

He is co-author of the book *Products That Last — Product design for circular business models*, that is currently being used at the Faculty of Industrial Design Engineering at the Delft University of Technology in their Towards Circular Product Design course and that was adopted by the Dutch Ministry of Infrastructure and Environment to promote the circular economy concept and circular product design principles to small and medium sized enterprises (SMEs) in the Netherlands.

To disseminate the results of his research, Marcel has been giving guest lectures at design and business schools in the Netherlands, presented his research at a number of international conferences and consulted on product design for circular business models.

He currently works at the Amsterdam University of Applied Sciences on the development of the Amsterdam Design Center, a multidisciplinary initiative for design education and design research. The center aims to bring together insights and methodologies from Industrial Design, Computer Sciences, and the Humanities to stimulate researchers and practitioners in these domains to look beyond their traditional toolsets and disciplinary boundaries to find new ways to address contemporary societal challenges in an increasingly volatile, unpredictable, complex, and ambiguous world.

Zaloa Arechabaleta (Chapter 10) is currently a Senior Researcher in the Department of Industry and Transport at Tecnalia. She received her PhD in 2013 in Materials Science from the University of Navarra (Tecnun). In 2014, she joined the Department of Materials Science and Engineering at Delft University of Technology, where she worked as a postdoctoral researcher for three years. During this period of time she worked on two main topics: the understanding of the anelastic behavior of Advanced High Strength Steels (AHSS) and the interaction of the austenite to ferrite phase transformation with the nano-precipitation of vanadium carbonitrides in a new type of AHSS. In July 2017, Zaloa joined Tecnalia as a senior researcher. Currently, she carries out research on topics related to automotive manufacturing, including forming and multi-material joining, among others.

Amal Kasry (Chapter 11) received her PhD (2006) in Materials Science at the Max Planck Institute for Polymer Research (MPIP) in Mainz, and the Johannes Gutenberg University, Mainz, Germany. Her major work was in the field of optical biosensors based on surface plasmon fluorescence spectroscopy (SPFS), where she's developed a biosensor which is of 22 times higher sensitivity than the conventional SPFS sensor. After a one year postdoc at the MPIP, she was a postdoc fellow at the Center for Cell Analysis and Modeling (CCAM) at UConn health center, Connecticut, USA, where she worked on live cell imaging by dark field light scattering.

She was then appointed as a research associate at the Department of Bioscience in Cardiff University, UK. During that time she led the activities of designing biochips based on DNA Nanotethers to study protein-protein interactions on the surface in collaboration with GE Health Care. In 2009, she was appointed as a senior research scientist in the Egypt Nanotechnology Center (EGNC) in a collaboration project between the Egyptian government and IBM Research; she performed her research at IBM T. J. Watson Research Center in Yorktown Heights, NY, USA. During that time, besides performing her research related to carbon nanomaterials, specifically graphene, she was involved in the strategic conceptual design of the EGNC in Cairo. After a short time as a senior research scientist in the R&D division of the Nitto Denko Asia Technical Center (NAT) in Singapore, she joined the Biosensor Technologies Department at the Austrian institute of Technology (AIT) in Vienna as a scientist to work on Organic Field Effect Transistors biosensors besides graphene related research. Currently she is a member of the Basic Science Department at the faculty of Engineering, the British University in Egypt, and was, for three months, a visiting professor at the Melbourne Center for Nanofabrication in Melbourne, Australia.

She is now leading a research group in the field of nanomaterials, where graphene related research is the major scope. This includes studying graphene optical and electrical properties for several applications. Biosensing is one of them.

Amal Kasry is a corresponding author and co-author of several peer reviewed articles and holds several patents and patent applications in the fields of photonics, optical biosensors, protein-protein interactions, fluorescence spectroscopy and carbon nanomaterials, She is also an author of one book and a co-author of seven book chapters. She was an invited speaker at and organizer of several international meetings and institutes. She is also a reviewer for AIP, ACS and Wiley, and DAAD. She received a fellowship for her PhD from the Max Planck Society and has several awards and recognitions inside and outside Egypt.

Ahmed Maarouf (Chapter 11) received his PhD in theoretical condensed matter physics at the University of Pennsylvania, Philadelphia, in 2002. His work was on modeling the electronic properties of various carbon nanotube structures. He then joined Hilbert Technology Inc. in Newtown, Pennsylvania, working on the mathematical modeling of databases using vector space models. He then worked as an assistant professor of physics at the Department of

Physics at Cairo University, Cairo, Egypt, from 2003 till 2006, when he joined Hilbert Technology Inc. again until 2008. He then worked at IBM Watson Research Center from 2009 to 2013, where his research was on the computational modeling of graphene-based structures. In 2014, he worked as an associate professor at Zewail City of Science and Technology, Cairo, Egypt, where in 2016 he received the Professor of Excellence Award. In 2017, he joined the Institute for Research and Medical Consultations at Imam Abdulrahman Bin Faisal University, Dammam, KSA. His research experience is in the field of analytical and computational modeling of materials properties at the atomistic and mesoscopic scales, with many published research papers and granted international patents. His recent research has focused on using first principles calculations for studying the properties of some novel materials to utilize them for various industrial and technological applications. These materials include graphene/carbon nanotubes hybrids, graphene nanomeshes, two-dimensional boron nitride, and molybdenum disulfide. Other current research interests include the physics of low dimensional systems and bioinformatics.

Mike Buxton (Chapter 12) obtained his PhD in geology. He spent 17 years in the South African mining industry involved in exploration geology, mine geology, business improvement, project due diligence and technical innovation. He worked on multiple commodities (including gold, platinum, diamonds, copper, zinc, iron ore and coal) in multiple countries in southern, central, eastern and western Africa.

He joined Delft University of Technology in 2011 as head of the Resource Engineering Section. He has initiated, coordinated and participated in several multi-partner, multinational EU funded projects and multiple PhD research projects. He is chairman of the Federation of European Minerals Programs and is coordinator for the European Mining Course MSc program in collaboration with partners from RWTH-Aachen and Aalto Unversity.

His primary interest is the use of sensing technologies for material characterization in the minerals industry, specifically to distinguish ore from waste, for mine operational process control and for mine waste modelling.

He is a Fellow of the Geological Society of South Africa.

Jack Voncken (Chapter 12) passed his 'Doctoraal Examen' (equivalent with obtaining an MSc) in Geology with (Igneous) Petrology as a major and Exploration Geochemistry and Economic Geology as minors at Utrecht University, the Netherlands, in 1984. After his graduation he embarked on PhD research, also at Utrecht University, on the synthesis, characterization and crystallography of two rare ammonium-aluminum-silicates (buddingtonite, or $NH_4AlSi_3O_8$, and tobelite, or $NH_4Al_2Si_3AlO_{10}(OH)_2$ and their Rb- and Cs-analogues.

He obtained his PhD in 1990 at Utrecht University with a thesis entitled 'Silicates with Incorporation of NH_4^+, Rb^+ or Cs^+'. Between 1988 and 1990, he was employed at Twente University of Technology, Enschede, the Netherlands, where at the Faculty of Chemical Technology, in the section Inorganic Materials Science, he developed a method to measure drying stresses in thin drying gel coatings.

In 1990 he entered Delft University of Technology. He started at the Faculty of Mining and Petroleum Engineering, in the Raw Materials Technology (later renamed Resource Engineering) section, as an assistant professor in Mineralogical Aspects of Raw Materials. From 2006 until 2011, after the cancellation of the Resource Engineering section, he worked in the section Applied Geology. After the re-establishment of the section Resource Engineering in 2011, he returned to that group, again as an assistant professor.

Jack Voncken specializes in Economic Geology, in Ore Minerals and Industrial Minerals, and in analytical techniques and characterization techniques for minerals and rocks. In the course of his interest in Ores and Industrial Minerals, he has focused since 2009 on Rare Earth Elements.

He is a member of the Society of Economic Geologists, and the International Mineralogical Association — Commission on Ore Mineralogy (IMA-COM).

Jan-Henk Welink (Chapter 13) is a project manager at the TU Delft on European projects concerning the circular economy and critical raw materials. Born November 26th 1967, he graduated from the University of Twente in waste incineration. He worked for several organizations including semi-governmental ones on project development, business development, and feasibility studies on the conversion of waste and biomass into energy. He worked for the energy agency of the Dutch Ministry of Economic Affairs on stimulation programs for further development of renewable energy industry and waste management, and advised on projects and policy. Since 2011 he works on sustainable resource management, and set up a knowledge platform on this topic, where he organizes seminars, master classes, workshops, and courses. He co-wrote two (Dutch) books on energy form biomass and waste and on recycling.

Maarten C.M. Bakker (Chapter 14) is a professor in the research group Resources and Recycling of the Delft University of Technology (TUD). He received an MSc in Electrical Engineering in 1994 and a Ph.D. in Ultrasound Inspection of Steel Structures in 2000, both with distinction, and has been working in Delft University ever since. In 2002 he extended his research to Non-Destructive Measurement Techniques for aircraft structures, and high-temperature testing of metallic components for space re-entry vehicles in the department of Aerospace Engineering. In 2007 he switched to the current group to work on solid waste recycling where he specializes in the fields of sensor technologies and materials separation. Fundamental research involves high power laser–matter interaction (LIBS), shape and material identification using electromagnetic fields, and the behavior of mixtures of solid particles. Dr. Bakker holds three patents on sensor-based technologies and advises industry and municipalities on issues ranging from technological innovation to sustainability. As a teacher he runs a variety of B.Sc. and M.Sc. courses on topics such as recycling, technologies, waste separation physics, and circularity.

Yongxiang Yang (Chapter 15) is associate professor and group leader for Metals Production, Refining and Recycling (MPRR) at the Department of Materials Science and Engineering, Delft University of Technology (TU Delft). He got his BSc. and MSc. in extractive metallurgy at Northeastern University, China (1982, 1988), and obtained his Licentiate and Doctor of Technology in process metallurgy and materials processing at Helsinki University of Technology (HUT), Finland (1992, 1996). He has about 30 years of experience in metallurgical engineering, research, and education in China, Finland and the Netherlands. His recent activities are focused more on the fundamental studies in blast furnace and new ironmaking technologies, and technological innovations in the recycling of critical metals from e-waste and other secondary resources, as well as process modelling through computational fluid dynamics (CFD) simulation. Dr. Yang has been participating in a variety of EU FP7 and H2020 consortia for REE recycling and critical metals recovery. He is responsible for the education and lectures in extractive metallurgy and metal recycling at TU Delft (the only program in the Netherlands) since 2005. Dr. Yang is actively involved in the EIT (KIC) Raw Materials programs and research projects. He has been a scientific committee member of PROMETIA — Mineral Processing and Extractive Metallurgy for Mining and Recycling Innovation Association — since 2014. Dr. Yang has published more than 190 papers in refereed international journals, international conferences, and co-authored book chapters. He has been an adjunct professor at the Norwegian University of Science and Technology (NTNU) for hydrometallurgy and metals recycling since 2017.

Koen Binnemans (Chapter 16) is an inorganic chemist and has more than 25 years of expertise in the field of rare earths. He is head of the Laboratory of Inorganic Chemistry at KU Leuven (Belgium). His main research lines are: (1) critical metals, with a focus on rare earths, (2) ionic liquids, and (3) solvometallurgy. Koen Binnemans has published more than 400 papers in international journals. His work has been cited 17,000 times (h-index = 63). In 2016, he was awarded an ERC Advanced Grant

(SOLCRIMET) on the use of solvometallurgy for the recovery of critical metals. He is a key player in SIM^2 KU Leuven, a leading, interdisciplinary research cluster at KU Leuven uniting the research groups working on Sustainable Inorganic Materials Management.

Bart Blanpain (Chapter 16) is a professor of Materials Engineering at the KU Leuven (Belgium) where he is coordinator of the "High Temperature Processing and Industrial Ecology" research group and the head of the "Sustainable Metals Processing and Recycling" division. He obtained a MSc in Metallurgical Engineering (KU Leuven) and a MSc and PhD in Materials Science and Engineering (Cornell University). His research interests are high temperature metals production and refining, metallurgical slag engineering and zero waste metal recovery schemes. He acts as a director for the NSF I/UCRC Center for Resource Recovery and Recycling and is an editor-in-chief for the *Journal of Sustainable Metallurgy*. He also coordinates the EIT label multi-university master program in sustainable materials and is the chairman of Leuven Materials Research Center as well as co-founder and chairman of the board of the KU Leuven spin-off company InsPyro.

Chenna Rao Borra (Chapter 16) is currently a postdoctoral researcher at the Department of Materials Science and Engineering, TU Delft. His research interests include the recovery of metals from minerals, waste materials and end of life products (recycling). Borra received a diploma in metallurgical engineering from Govt. Polytechnic, Vijayawada, India followed by a Bachelor's degree in metallurgical engineering from the Indian Institute of Metals and a Master's degree in extractive metallurgy from IIT (BHU) Varanasi, India. After his Master's, he worked as a project scientist/engineer for about three years at NFTDC, Hyderabad, India. Then he moved to R&D, Tata Steel (India) and worked as a researcher for five years. In 2013, he moved to Belgium to pursue his doctoral degree from KU Leuven and obtained his PhD in June 2016.

Yiannis Pontikes (Chapter 16) is a BOF-ZAP associate professor at the Department of Materials Engineering, KU Leuven, Belgium. Prof. Pontikes is leading the Secondary Resources for Engineered Material (SREMat) research group that consists of approximately 15 postgraduate researchers. SREMat is developing "sustainable processes for sustainable materials" and has built an expertise on the valorization of residues towards cement and inorganic polymer (geopolymer) formulations, from the level of binder synthesis all the way to full-scale prototypes. The scientific output of SREMat exceeds 10 peer-reviewed journal papers a year. Professor Pontikes has been a co-author in approximately 70 peer-reviewed journal papers, is currently work package leader or project coordinator in approximately 10 national and international projects and is participating in different fora and networks (such as SIM2 at KU Leuven and CR3 in USA). In 2015, Professor Pontikes was one of the founders of the *Journal of Sustainable Metallurgy*, published by Springer, where he serves as the managing editor.

Tom Van Gerven (Chapter 16) is Professor in Process Intensification at KU Leuven and Head of the Department of Process Engineering for Sustainable Systems. His expertise is situated in the use of alternative energy forms (ultrasound, light) for chemical processes, in particular crystallization, extraction, and leaching. He is General Coordinator of the MSCA-ITN project "Continuous Sonication and Microwave Reactors (COSMIC)" and of the EIT-KIC Raw Materials Network-of-Infrastructure "Intensified Flow Separator Infrastructure and Expertise Network (INSPIRE)". He is or was also involved in several other European projects, e.g. the EU-FP7-NMP "Alternative Energy Forms for Green Chemistry (ALTER-EGO)", the MC-ITN "European Rare Earth (Magnet) Recycling Network (EREAN)" and two running MSCA-ITNs (REDMUD and SOCRATES). He is Chairman of the EFCE Working Party on Process Intensification since 2013 and Vice-President of the European Society of Sonochemistry since 2018. He also serves as member of the Expert Panel on Chemical Engineering and Material Science of the Flemish Science Fund (FWO).

Index

Printed in the United States
By Bookmasters